MODERN ORGANIC ELEMENTAL ANALYSIS

MODERN ORGANIC ELEMENTAL ANALYSIS

T. S. Ma

Professor of Chemistry
City University of New York
New York City, New York

Robert C. Rittner

Senior Research Associate
Analytical Department
Olin Corporation
New Haven, Connecticut

Marcel Dekker, Inc.   New York and Basel

Library of Congress Cataloging in Publication Data

Ma, Tsu Sheng, [Date]
    Modern organic elemental analysis.

    Includes bibliographical references and indexes.
    1. Chemistry, Analytic--Quantitative. 2. Chemistry,
Organic. I. Rittner, Robert C., [Date] joint author.
II. Title.
QD271.7.M29   547'.34   79-11573
ISBN 0-8247-6786-1

Marcel Dekker, Inc.
270 Madison Avenue, New York, New York 10016

Current printing (last digit):
10 9 8 7 6 5 4 3 2 1

Printed in the United States of America

Preface

The purpose of this book is to organize in a single volume the methods of organic elemental analysis applicable to different sample sizes and the techniques for the determination of all of the elements in organic materials. It is intended for various groups of users including (a) chemists in small laboratories who occasionally perform organic analysis; (b) practicing analysts in large establishments who may encounter difficult samples; (c) research analysts who wish to have a comprehensive review of the existing methods; (d) graduate students in analytical and organic chemistry who should have some knowledge of the evolution and current status of organic analysis; (e) teachers of advanced quantitative techniques who need a textbook containing a large variety of experiments; and (f) organic chemists who do not analyze their samples but wish to know how the analysis can be done.

At the present time, it can be said that there is a method for the determination of any element which may be present in an organic substance. The book is written to assist the reader in finding the method and in choosing the best method if more than one is available. The individual elements are systematically treated in Chaps. 2-9. Chapter 10 deals with the simultaneous determination of two or more elements using one sample. The basic principles for each method of analysis are fully discussed, as well as the applications and limitations. Most of the procedures described are at the milligram level, which is the current practice. However, some macro procedures are also mentioned, such as Kjeldahl determination of nitrogen. Chapter 11 presents methods using less than milligram amounts of sample; such methods are recommended when a larger quantity of the working material cannot be obtained. Finally, Chap. 12 deals with trace analysis of specific elements which, understandably, involves the use of large samples.

Experimental details are given separately in the last part of each chapter. This arrangement is convenient for the practicing chemist who wishes to try out a procedure and for the student who uses this book as a laboratory manual.

In the selection of laboratory experiments, I have tried to include both classical and new methods. Procedures using elaborate instrumentation and physicochemical measurement devices are presented. On the other hand, determinations that can be accomplished by means of simple equipment and techniques are also described; these methods can be employed with advantage in small laboratories. Most of the experiments have been thoroughly tested by my students or are being used in the laboratory of Dr. Robert C. Rittner. Dr. Rittner contributed a great deal in the experimental sections of this book. He took my course in quantitative organic analysis at New York University 25 years ago and has since been in charge of several analytical laboratories in industry.

It should be noted that this book is not limited to the experimental methods of one school. A number of detailed procedures are based on information that I obtained from other analytical laboratories; these procedures were confirmed by the original workers named in the respective experimental sections.

My involvement in organic analysis began in 1934 when I entered the University of Chicago to study organic synthesis under Professor Julius Stieglitz. In our first meeting, the professor told me about the Pregl combustion trains that Oskar Wintersteiner had recently introduced at the laboratory and advised me to learn the microchemical techniques which should be taught in the Orient some day. Consequently, by the time I received by Ph.D I was already well-versed in organic analysis. During World War II, I was supervisor of the microchemical laboratory at the University of Chicago, where I saw thousands of samples processed and many new methods developed.

After the war, I had the opportunity to lecture in Europe and Asia and made four extended lecture tours in the last fifteen years. On those tours, I visited numerous laboratories, observed the operations of many practicing and research analysts, and discussed the merits and drawbacks of various methods. The consensus of our discussions was that (1) the analyst can no longer restrict himself to procedures for only one range of sample sizes because he now deals with pure organic compounds as well as complex mixtures containing minute amounts of elements of interest; (2) the types of samples received for analysis vary from laboratory to laboratory; and (3) the facilities and maintenance services available differ greatly from one location to another. These discussions and conclusions stimulated the idea of a book to meet the needs implied. At the invitation of my revered friend Dr. Maurits Dekker, this book was begun. Dr. Dekker deserves the credit if this book is a success, while I shall bear responsibility if it does not fulfill our expectations.

I wish to express my heart-felt appreciation to my colleagues in organic analysis who have generously shared with me their experience and expertise. I thank the many journals, publishers, and instrument manufacturers for their kind permission to reprint the illustrations in this book; the source of each illustration is indicated under the figure caption.

I am deeply indebted to my wife Gioh-Fang whose encouragement and patience have been indispensable for the progress of my work. Our children Chopo and Mei-Mei helped in typing my manuscripts and reading proofs.

T. S. Ma

Contents

List of Tables

MODERN ORGANIC ELEMENTAL ANALYSIS

Introduction

1.1 OBJECTIVES OF ORGANIC ELEMENTAL ANALYSIS

Organic elemental analysis is concerned with the determination of the elements present in organic materials. Generally speaking, the objectives of such quantitative measurements can be classified into three categories.

(1) The analysis may be carried out on a single compound (known as the pure sample) for the purpose of determining its elemental composition. If the sample is a synthetic product formed in a known reaction, the analytical data can be checked against the values calculated from the expected molecular formula. Usually the determination of one or two elements in the compound suffices. On the other hand, if the sample is a new substance (e.g., obtained from nature, or an unexpected synthetic product), determination of all elements present is recommended so that the sum total of elemental composition adds up to 100%.

(2) The analysis may be performed on a mixture for the purpose of finding out the content of a particular element, or the amount of certain compounds, in the analytical sample. For example, fertilizers are analyzed for nitrogen by the Kjeldahl method; sulfur determinations are carried out in the petroleum industry in crude oils and gasolines; nonprotein nitrogen (NPN) analysis is made on certain biological fluids. Similarly, the amount of fluoroacetic acid in plant leaves is usually measured by the fluorine content of the material, while the extent of chlorobiphenyl residue left in the field which has been sprayed by these insecticides is indicated by organic chlorine analysis.

(3) The analysis may be undertaken for the purpose of assessing the purity of the organic substances in question, such as pharmaceutical preparations, fine chemicals, and high-quality solvents. In this case, the element to be analyzed is usually the one which is present in the contaminant and not in the pure product. It should be noted that product purity is

frequently determined by functional group analysis [1, 2] or by chromatography [3]. There are occasions, however, when quantitative elemental analysis provides a simple method to attain this objective. For instance, the determination of sulfur serves to indicate the amount of thiophene in benzene.

1.2 HISTORICAL BACKGROUND

Quantitative analysis of the elements in organic substances was initiated by Lavoisier [4] two centuries ago. After learning the experimental technique from Priestly [4] in 1774 to prepare oxygen, Lavoisier burned alcohol and other combustible organic compounds in oxygen; from the weight of water and carbon dioxide produced, he calculated their compositions. Understandably, Lavoisier's results were imperfect, partly because of impurities in the substances investigated, partly because of defects in his methods, and partly in consequence of his incorrect values for the compositions of water and carbon dioxide. About five decades later, Liebig [5] developed the method for the determination of carbon and hydrogen (1831) by combustion in the presence of copper oxide, while Dumas [5] analyzed nitrogen (1830) in organic compounds by measurement of the nitrogen gas produced. It is of interest to note, however, that much emphasis had been given to the quantitative aspects of organic analysis prior to the publication of the procedures of Liebig and Dumas. Thus, in the "Handbuch der Allgemeinen und technischen Chemie" published in 1829, discussion of each organic compound was accompanied by its elemental percentage composition as reported by the investigators. For example, the data for urea were as follows [6]:

Investigators	%C	%H	%N
Fourcroi and Vauquelin	14.7	13.3	32.5
Prout	19.975	6.650	46.650
Berard	18.9	9.7	45.2
Ure	18.57	5.93	43.68
Prevost and Dumas	18.23	9.89	42.23

When compared with the currently accepted values [7] (CH_4N_2O: C, 20.00%, H, 6.71%, N, 46.65%), we can appreciate the remarkably accurate results obtained 150 years ago by Prout who analyzed a sample of urea which was dried over concentrated sulfuric acid in a vacuum desiccator.

The methods for determining sulfur and halogens were developed by Carius [8] (1860) who discovered that heating organic compounds with nitric acid in a sealed tube quantitatively converted these elements into ionic forms (see Chaps. 5 and 6). For many decades, the Liebig, Dumas, and Carius procedures were standard techniques practiced by organic chemists. It should be mentioned that the determination of oxygen was not performed, although oxygenated compounds were the first crop of organic substances isolated by Scheele [5,9] in 1770.

A great advance in organic elemental analysis took place at the turn of this century when Pregl [10] worked out micromethods using milligram quantities of samples. Much of Pregl's contribution is discussed in the succeeding chapters. It is interesting to note, however, that the general acceptance of the Pregl micromethods has deprived most students of organic chemistry of the opportunity to acquire the experimental skill of quantitative analysis.

1.3 RECENT DEVELOPMENTS

A. Modification and Expansion of Pregl Procedures

The Pregl school for organic elemental analysis was characterized by detailed directions and precise laboratory operations. The apparatus and reagents to be employed, as well as the experimental procedures, were strictly prescribed and followed. Such precisely defined instructions were essential in order to attain the same precision and accuracy using a 3- to 5-mg sample as in elemental analysis using a 200- to 500-mg sample. Consequently, the Pregl school tended to discourage variations and innovations. Modifications of the Pregl procedures began during World War II because of various reasons: (1) the unavailability of the Pregl apparatus from Europe led to the production of microbalances and microapparatus in America; (2) the increase in the number of samples to be analyzed fostered the development of rapid methods (e.g., the micro Kjeldahl procedure developed by Ma and Zuazaga [11]); (3) the dearth of chemists having delicate manipulative skill necessitated the introduction of less demanding experimental operations. Thus, workers in several countries including Czechoslovakia [12], Germany [12a], Japan [13,14], UK [15,15a], USA [16], and USSR [17,18] have published their respective modifications of the Pregl micromethods.

The field of organic elemental analysis has expanded in recent years. First, the methods for the determination of oxygen and fluorine, respectively, have been perfected; thus it is now possible to determine any and every element that may be present in an organic sample. Secondly, the rapid increase of interest in the study of organometallic compounds has created new problems since these compounds are not always amenable to the existing methods of analysis [19]. Thirdly, public attention on pollution control and environmental protection has demanded reliable methods for trace analysis in complex systems [20]. Finally, it may be mentioned that some macromethods using decigram or larger samples are still recognized as official procedures, e.g., the Kjeldahl procedure in US Pharmacopeia [20a].

B. Development of Rapid and Automated Procedures

Rapid procedures of organic elemental analysis are usually achieved by eliminating certain operations and/or devising techniques for fast decomposition of the sample and measurement of the reaction products. For example, in the micro Kjeldahl procedure (see Sec. 3.3), distillation of ammonia by means of a rapid stream of water vapor and absorption of ammonia in 2% boric acid instead of standardized 0.01 $\underline{N}$ hydrochloric acid shortens the experiment considerably [11]. Decomposition of halogenated compounds in the closed flask (see Sec. 5.3) is practically instantaneous, compared to the slow operation of oxidation by means of nitric acid or in a current of oxygen [21]. Simultaneous determination of more than one element using a single weighed sample (see Chap. 10) frequently eliminates some duplicate operations and hence speeds up the entire analytical process.

Automated procedures were designed to relieve the manipulative strain of the analyst [22]. They are also labor-saving and faster than manual operations. The commercial automated machines for C,H,N determinations are described in Chap. 2. Recently Ubik and co-workers have proposed methods for automatic determination of oxygen [23] and for simultaneous determination of oxygen and nitrogen [24]. The scheme of the apparatus for nitrogen analysis [25] is shown in Fig. 1.1. The procedure involves pyrolytic decomposition of the organic compound in a stream of hydrogen at 1000-1100°C; the hydropyrolytic products are then passed through layers of Mn(II) oxide and nickel at 950°C. Nitrogen gas is liberated on the nickel. After removal of interfering substances (CO, H_2O, hydrogen halide), the amount of nitrogen formed is measured by thermal conductivity. Thus the basic Dumas principle to convert organic nitrogen to elementary nitrogen is retained. Ebel [26] has published a comprehensive survey on automated C,H,N determinations. Salzer [27] has reviewed the physicochemical measuring methods which are amenable to automation [28]. Scheidl and Toome [29] have described an automated potentiostatic procedure for the determination of chlorine, bromine, or iodine. A combustion furnace, a buret and a titrator are modified and connected to an apparatus. A solid-state programmer using monostable and multivibrator circuits to generate various time delays completes a halogen determination in 4 min. Fraisse and Richard [29a] have fabricated an automated introduction device that can be used for various determinations.

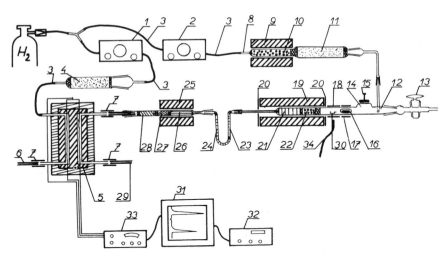

FIG. 1.1. Scheme of the apparatus for automated nitrogen determinations. 1,2, Manostats; 3, rubber tubings; 4, device with the molecular sieve 5A; 5, catharometer; 6, glass capillary; 7, rubber tubings; 8, cleaning tube; 9, cleaning furnace, 500°C; 10, activated copper; 11, molecular sieve 5A; 12, reaction tube; 13, valve; 14, chamber; 15, magnet; 16, insert; 17, safety furnace; 18, decomposition furnace; 19, reaction furnace, 950°C; 20, asbestos; 21, nickel; 22, manganese(II) oxide; 23, Anhydrone; 24, soda asbestos; 25, furnace, 120°C; 26, triiodic acid; 27, silver sponge; 28, soda asbestos; 29, glass capillary; 30, platinum boat; 31, recorder; 32, integrator; 33, Wheatstone bridge; 34, inlet for cooling air. (From Ref. 25, courtesy of Microchem. J.)

Combination of the computer with automated analytical procedures can further speed up organic elemental analysis. For instance, Maciak and co-workers [30] have described a computerized on-line system for C, H, N and O determinations. Four elemental analyzers and two microbalances are interfaced to the computer via scanner and DVM. Calculations of percentages found and theoretical are collected on two teletypes. Two technicians operate the two stations and the daily output of analyses is approximately 75 C, H, N and 25 O determinations.

The senior author recently visited several laboratories where organic elemental analyses are computerized. According to the microanalysts in charge [30a-d], the results have been very satisfactory and there is much saving in time and manpower. This will be the trend of centralized analytical service operations where thousands of samples are processed every month and quick answers are demanded. Merz [30e] has reported on the use of data processing in rapid methods for determining C, H, N and O.

C. New Approach to the Determination of Elemental Composition

Since it is now possible to determine all elements present in an organic compound, the elemental composition (i.e., the atomic ratios of the various elements) of the compound can be obtained by calculating the molar ratios of the corresponding analytical products (e.g., CO_2, H_2O, N_2) from the respective elements. For example, Haberli [31] and Stoffel and Grade [31a] have determined the empirical formulas of compounds using the C,H,N analyzer without weighing the samples. Liebman and co-workers [32,33] have described on-line elemental analysis of gas-chromatographic effluents whereby C, H, N, and O are determined and empirical formulas calculated.

A number of physicochemical methods [34] can be employed to determine the atomic ratio of a substance without performing the conventional combustion analysis. Griepink and Dijkstra [35] have applied the information-theory concepts to the consideration of the amount of information obtainable from some analytical results. In particular, the number of formulas corresponding to a measured molecular weight known to a specified degree of precision are considered for hydrocarbons and for compounds containing C, H, and N.

Very precise molecular weights can be measured with the aid of a high-resolution mass spectrometer. Hence it is possible to obtain atomic ratios from mass spectra [36-39]. Walisch et al. [39a] have used mass spectrometry to study combustion processes and Hammar and Hessling [40] have described a procedure for elemental analysis of compounds in submicrogram quantities without prior isolation. Van Leuven [41] has proposed organic multielement analysis with a mass spectrometer as detector. Vanderborgh and Ristau [42] have obtained the carbon/hydrogen ratio of organic compounds by laser pyrolysis. Merritt [43] has applied the laser pyrolysis technique to determine the C,H,N ratios of the surface of organic materials (e.g., plastics and fibers).

Another tool for estimating the carbon/hydrogen ratio is nuclear magnetic resonance spectroscopy [44-47]. Radioactivity has been utilized by Ricci [48] to determine nitrogen/carbon ratios in biological substances. Asai and Ishii [49] have determined carbon/oxygen ratios by reacting the sample with fluorine and measuring the quantities of CF_4 and oxygen formed. Recently Morrison and Nadkarni [49a] have proposed neutron activation for multielement determinations and Hutson et al. [49b] have advocated muon X-rays for the same purpose, excepting the element hydrogen.

D. Collaborative Studies on Micromethods

Concerted efforts have been, and are being, made to evaluate the various methods for organic elemental analysis with a view to improving the procedures. National and international collaborative studies continue to be undertaken. Thus, the following reports have appeared during the past few years: A comprehensive report on errors in organic elemental analysis has been published by the International Union of Pure and Applied Chemistry [50]. Sources of errors are tabulated. Based on the suggestions of many practicing analysts, means of minimizing the errors are discussed for the determination of C, H, N, O, halogens, S, P, B, Ti, Al, Fe, and Zr respectively. Meyer and Geissler [51] have made comparative studies on ammonia-distillation sets for Kjeldahl nitrogen determinations. Comparison between laboratories of the Kjeldahl method for nitrogen in milk has been made [51a]. The Association of Official Analytical Chemists and the Office International du Cacao et du Chocolat have jointly worked out the final Kjeldahl procedure for the determination of total nitrogen in cacao and chocolate products [52]. Lalancette and co-workers have reported on collaborative studies on bromine, chlorine [53], and iodine [54] by closed-flask combustion. Denney and Smith [55] have compared the potentiometric and mercurimetric methods for determining bromine. Scroggins [56] has evaluated four modes of finish for sulfur after closed-flask decomposition. Collaborative studies of phosphorus in organic compounds [57] and in fertilizers [58] have been reported. Several studies on arsenic determination have been completed [58a-c]. The Society for Analytical Chemistry has published the specifications for the composition and analysis of 36 reference substances and 28 reagents [59]. Martin and Thompson [60] have investigated the volatility of organic microanalytical standards; for benzoic acid, trifluoroacetanilide, and α, α, α-trifluoro-m-toluic acid, the rates of loss of weight for 5-mg portions at 24°C were 15, 18, and 14 μg per day, respectively.

1.4 GENERAL REMARKS ON ORGANIC MICROANALYSIS

A. Microchemistry is an Expanding Discipline

As pointed out by Ma and Guttelson [61], microchemistry is concerned with the principles and methods of chemical experimentation using the minimum quantity of material to get the maximum amount of chemical information. Organic microanalysis, therefore, is an extending and not a confined field [62]. As new methods, techniques, and apparatus become available, the weight of analytical sample required for the determination can be reduced. It is worthy of note, however, that organic elemental analysis was performed at the decigram region for about 150 years. Since the Pregl micromethods prevailed, the sample size has been maintained at the milligram level for over 50 years. While determination of the elements in organic substances below the milligram region is feasible (see Chap. 11), it is not commonly practiced.   The main difficulties are in processing the sample in such small quantities without contamination and in keeping the organic compounds without alteration prior to elemental analysis [3].

B.   Use of Outside Analytical Service

There are a number of commercial microanalytical laboratories which perform organic elemental analysis at reasonable fees. It is convenient to use their service when the need for analysis is infrequent and when the required apparatus is expensive and sophisticated. On the other hand, it is prudent to be aware of the method used by the outside service as well as the limitations of the method. For instance, Hodgson and co-workers [63] have reported that commercial analyses of the lanthanide complexes with cyclo-octatetracene-dianion were unsatisfactory. Crease and Legzdins [64] have shown that some organolanthanide adducts are extremely sensitive to air and moisture, even decomposing over a period of several days in an atmosphere of prepurified nitrogen. Schneider and Nold [65] have discussed the determination of carbon, oxygen, and nitrogen in organohafnium compounds. Colombo and Vivian [66] have found that carbonitrides, carbides, and refractory nitrides present difficulties in an automated C, H, N analyzer.

Before the change of the chemistry curriculum in the universities in recent years, chemists were generally taught some phase of organic elemental and functional group analysis. Presently, students of organic chemistry usually learn the quantitative analytical methods only when their research projects require these techniques. In consequence, the research organic chemist tends to rely on the analyst in his institution or the commercial laboratory to perform analysis of his samples. While this is a matter of expediency, it should be realized that the analyst is not familiar with the nature of the research samples, and that the sample may be decomposed or contaminated en route to analysis.

C.   Suggestions to Those Who Perform Elemental Analysis

When the need for routine determination of certain elements or materials arises, it is advisable to assemble the particular apparatus and carry out the analysis in one's own laboratory. For example, Kjeldahl nitrogen determinations may be carried out with the Labconco 6-flask microdigestion and microdistillation apparatus [67], or the Kjel-Foss Macro Automatic Machine [68]. Patterson [69] has described an automated Pregl-Dumas technique for the determination of carbon, hydrogen, and nitrogen in atmospheric aerosols, which has been applied to the analysis of these elements in size-fractionated atmospheric particulate matter [70].

It is also advisable for the research chemist to perform his own analysis when a series of compounds are synthesized which can be determined with simple equipment and techniques. For instance, chlorine- or sulfur-containing compounds can be analyzed by means of the closed-flask technique and titrimetry.

In a research institution which produces numerous and varied organic compounds, the establishment of a microanalytical laboratory is recommended. Understandably, the organization of the laboratory depends on the needs of the research staff and the projects being conducted. As a rule, it is advisable to utilize apparatus which is versatile, such as the combustion assembly of Herrman-Moritz [71, 72] which can be used for the analysis of C, H, N, S, F, P, Cl, Br, O, or I, and the combustion furnace of Heraeus [73] suitable for the determination of C, H, N, or O. The recent report by Marzado and co-workers [74] exemplifies the work of a small laboratory where C, H, N, O, halogens, S, P, B, Hg, and other metals are determined. The reader is referred to a review on automated machines written by Fish [75] in 1969 for comparison with present-day equipment in order to appreciate the rapid changes in the commercial instruments.

1.5 EXPERIMENTAL: PREPARATION OF THE SAMPLE FOR ORGANIC ELEMENTAL ANALYSIS

Proper preparation of the sample is an essential step in the experimental procedure for organic elemental analysis. Neglect of this matter often leads to disappointment and argument between the analyst and the research chemist who submitted the sample. For instance, a new C, H, O compound might have been obtained by the proposed synthetic route, but the analysis, even though the C-H analytical procedure employed is checked against benzoic acid from the National Bureau of Standards, may not give the expected C and H values due to the fact that the substance is slightly hygroscopic.

Preparation of the sample is dependent on the nature of the organic substance as well as the objective of the analytical experiment. The sample may be (1) a pure compound, i.e., single entity, or (2) a mixture containing more than one organic compound, or consisting of organic and inorganic materials. It should be recognized also that a pure compound does not preclude its possible combination with the solvent of crystallization.

Prior to analysis, a pure solid compound is usually dried in a drying apparatus. The apparatus shown in Fig. 1.2 is recommended because it permits the temperature in the drying compartment to be automatically controlled between 25 and 195°C to within ±1°C. By drying the organic substance at different temperatures and taking the respective samples for analysis, the results will reveal the presence of surface moisture or solvent of crystallization in the original sample. The apparatus is operated

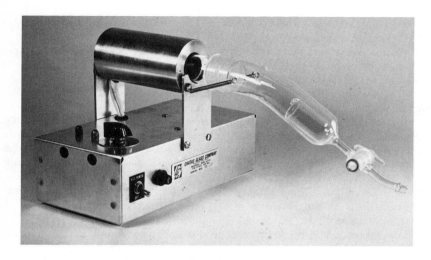

FIG. 1.2. Automatic temperature-controlled drying apparatus. (Courtesy Chatas Glass, Inc.)

in the following manner [76]: The material to be dried is placed in a container and pushed inside the drying chamber heated by the electric oven. The drying agent (e.g., silica gel, phosphorus pentoxide, paraffin) is placed in a test tube and put in the vessel outside the oven. After the specified temperature is set, the stopcock is opened to connect the apparatus to the vacuum pump. After evacuation, the stopcock is turned to close the system, and the organic substance is kept at the specified temperature for 1 h. Upon cooling to room temperature, dry air is introduced through the stopcock and the material is taken out for analysis. If drying in a special atmosphere (e.g., nitrogen) is desired, an inlet tube is added to the outer vessel so that the particular gas can be conducted through the system.

When the organic substance submitted for elemental analysis consists of a mixture, it should not be dried, as the volatile matter in the original material will be lost. The chief problem in handling mixtures is to obtain a representative sample [3]. Since only milligram quantities of sample are taken in the analytical procedure, the original material usually requires pulverization and thorough mixing. In contrast, a sample of pure compound should not be mixed; furthermore, apparent extraneous matter (e.g., filter paper) should be discarded while the sample is being taken.

1.6 EXPERIMENTAL: TECHNIQUES OF WEIGHING

A. The Analytical Balances

Various designs of balances have been introduced in recent years for analytical purposes. Understandably, the choice of the balance for organic
elemental analysis is dependent on the experimental method employed, the
condition of the laboratory, and funds available. A balance manufacturer
often markets several types of analytical balances which differ in certain
characteristics. These balances are classified as "analytical" (commonly
known as "quantitative"), "semimicro," and "micro." In the opinion of the
senior author [62], they are more appropriately designated by the names
"decigram-balance," "centigram-balance," and "milligram-balance,"
respectively, in order to indicate the level (weight of sample or product)
at which accurate and precise weighing is required.

   Most analytical balances currently available on the market are single-
pan balances. They are simple in operation and give results more rapidly
than the two-pan (equal-arm) balances. Two examples of single-pan microbalances are shown in Fig. 1.3A, B. It should be recognized that weighing
in the single-pan microbalance is always carried out at its maximum load
(i.e., low sensitivity). The zero point of the single-pan microbalance is

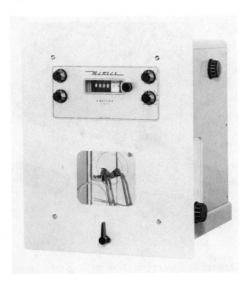

   FIG. 1.3A. Mettler microbalance with pan extractor. (Courtesy
Mettler Instrument Corp.)

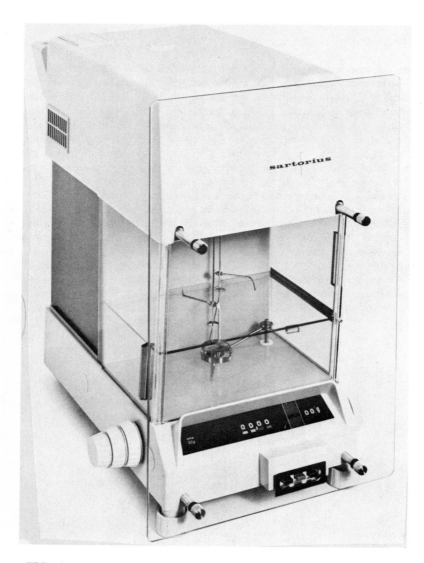

FIG. 1.3B. Sartorius microbalance with digital readout. (Courtesy Brinkmann Instruments.)

very temperature-sensitive, hence requires frequent checking. For these reasons, the conventional two-pan microbalances are still favored by some microanalysts. The reader is referred to Ref. 1 for information on the use and care of these balances. Figure 1.3C shows a bench and cabinet for the microbalance and some tools used in microanalysis. Figure 1.3D illustrates a two-pan microbalance currently produced in Japan; it has a wide front window which can be opened, thus providing more room for maneuvering large vessels.

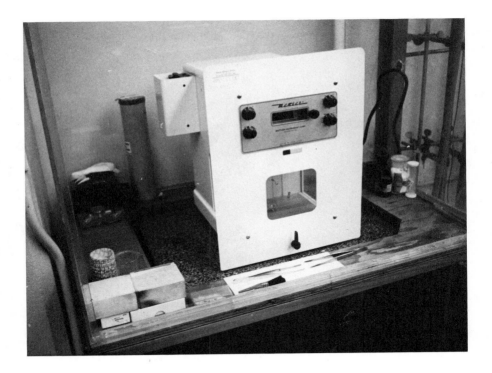

FIG. 1.3C Cabinet and bench for microbalance.

FIG. 1.3D. Two-pan microbalance. (Courtesy Chyo Balance Corp.)

Another group of new balances are known as electronic balances in which the weight of the object is indicated by electrical signals, instead of by optical means. Figure 1.4A, B shows two types of electronic balance recommended for use in organic elemental analysis. Generally speaking, these balances are best suited for measuring weight differences below the milligram level. It should be noted that the pans of the electronic microbalances are not constructed to carry large vessels such as absorption tubes. A distinct advantage of the electronic balances is that they are not sensitive to vibrations, and hence do not require special benches.

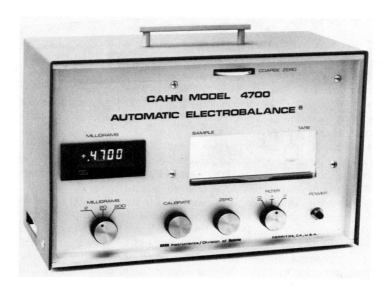

FIG. 1.4A. Cahn automatic electrobalance. (Courtesy Cahn Instruments.)

FIG. 1.4B. Sartorius electronic microbalance. (Courtesy Brinkmann Instruments.)

B. Weighing of Solid Samples

One way to weigh a required quantity of solid sample for elemental analysis
is to use the vessel in which chemical reaction will be subsequently carried
out. The vessel may be a microboat or crucible, made of platinum (see
Fig. 1.5), porcelain, or quartz. In use, the empty clean vessel is first
accurately weighed. A portion of the sample is then added. When the
amount is close to the desired quantity, the weight of vessel and contents
is accurately determined.

    After the solid sample has been weighed in the microboat or crucible,
the vessel and contents are stored in the microdesiccator. Three types of
desiccators are shown in Fig. 1.6. Models (a) and (b) are commercially
available. Model (c) can be conveniently made from an aluminum or copper
cylinder of 50-mm diameter; it is machined to give a slightly concave
surface depressed about 3 mm to accommodate the glass cover which is cut
from a 100-ml beaker to a 40-mm length and polished. The outside edge of
the metal block is grooved so that it can be held by the fingers without
slipping. In general, no desiccants are placed in microdesiccators, and
the ground-glass cover is not lubricated. If the organic material is sensitive to oxygen, weighing should be carried out in a controlled atmosphere.

FIG. 1.5. Platinum microboats and microcrucibles. (Courtesy A. H.
Thomas Co.)

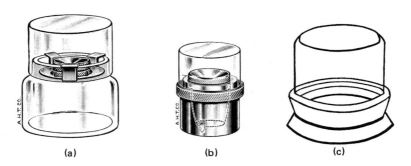

(a)   (b)   (c)

FIG. 1.6. Microdesiccators. (a, b courtesy A. H. Thomas Co.)

For example, Nuti [77] has employed a "dry box" filled with pure carbon dioxide for weighing samples for micro Dumas nitrogen analysis. Cousin and Muller [77a] have described a device for weighing solids or liquids under argon.

Another technique to measure solid samples utilizes the weighing tubes shown in Fig. 1.7. These are made by drawing out 10-mm glass tubings. The narrow end is closed and sealed to a 2-mm glass rod. It is important that the closed end is round (not pointed), and wide enough for the microspatula to reach the bottom. Weighing is carried out as follows: (1) Hold the weighing tube by its knob end; (2) introduce the sample to near the bottom by means of a microspatula; (3) when the amount is right, determine the weight of the tube and contents accurately; (4) transfer the sample into the reaction vessel (e.g., micro Kjeldahl flask, see Sec. 3.3, or micro-bomb, Sec. 5.7); (5) reweigh the tube and calculate the sample weight by difference.

C.   Weighing of Liquid Samples

The recommended technique to weigh an organic liquid and subsequently deliver it into the combustion stream is to use the two-chamber weighing capillary [78]. Figure 1.8 illustrates the steps in making the two-chamber capillary: (a) Draw 10-mm glass tubing to form capillary tubes of about 2-mm bore; (b) heat the capillary to produce a fine constriction; (c) seal one section; (d) draw out to form the air chamber, 15 mm long, and handle, 20 mm long; (e) soften the other side; and (f) draw out to form the liquid chamber, 25 mm long, and a fine tip, 20 mm long.

The procedure to fill the weighing capillary and place it in the combustion tube is as follows: (1) Accurately weigh the empty capillary. (2) Holding the handle, warm the liquid chamber over a small flame and then immerse the tip into the liquid sample, as shown in Fig. 1.9a. The sample enters the capillary due to the reduction of the gas volume inside the liquid chamber by cooling. (3) When the sample reaches the wide part for a length

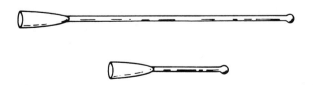

FIG. 1.7. Weighing tubes. (Courtesy J. Wiley & Sons, Inc.)

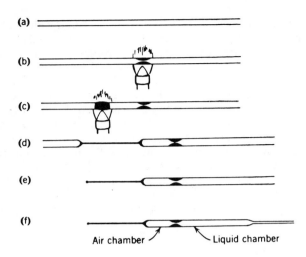

FIG. 1.8. Preparing the two-chamber weighing capillary. (From
Ref. 1, courtesy of J. Wiley & Sons, Inc.)

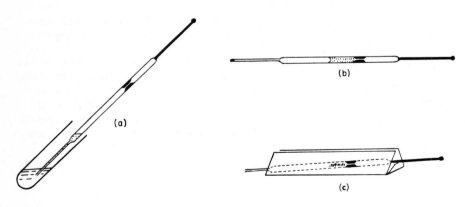

FIG. 1.9. Using the weighing capillary. (a), Filling the capillary.
(b), Sealed capillary before weighing. (c) Open capillary placed in pla-
tinum foil. (From Ref. 62, STANDARD METHODS OF CHEMICAL ANA-
LYSIS, Vol. 2A, 6th Ed., edited by Frank J. Welcher,   1963 by Litton
Educational Publishing, Inc., reprinted by permission of Van Nostrand
Reinhold Company.)

of about 2 mm, withdraw the capillary and take its approximate weight.
(4) If the amount of sample that has entered is suitable, hold the handle of
the capillary by one hand, and gently tap this hand against the other hand,
whereupon the liquid glides down along the liquid chamber and stops at the
fine constriction. (Fig. 1.9b). (If the amount of sample that has entered
is too much, expel a portion of it onto a filter paper by warming the lower
empty section of the liquid chamber. Contrastingly, if the amount of
sample is not enough, expel all liquid by warming, and reimmerse the
capillary tip into the liquid to draw in more sample.) (5) Now seal the tip
of the capillary in the flame and accurately weigh the capillary and contents.
(6) When the sample is ready for introduction to the combustion tube (e.g.,
for carbon and hydrogen determination by the manually controlled procedure, Sec. 2.6, or for sulfur determination by catalytic oxidation, Sec.
6.6), place a platinum foil (bent along its length to form a triangular shape
and having one end slightly higher than the other) at the mouth of the combustion tube, cut open the end of the capillary, insert the capillary into
the platinum triangle (see Fig. 1.9c) and then push the latter inside the
combustion tube. (7) In order to expel the sample from the weighing capillary for combustion, place the burner under the air chamber of the
capillary; expansion of the air in the air chamber forces the liquid to move
forward and deposit on the walls of the combustion tube. The stream of
oxygen then carries the sample vapor to the oxidizing area of the combustion tube.

The technique described above permits easy control of the sample size
to be weighed and also of the movement of the sample during combustion.
It is considerably better than the single-chamber techniques [79] in which
the sample should occupy a quarter of the space available and should be
situated centrally in the chamber.

If the organic liquid boils near the room temperature, fill the capillary
by immersing its tip into the liquid and then touching the liquid chamber
surface with a piece of Dry Ice. Contraction of the gas volume in the liquid
chamber causes the sample to enter the capillary. Schwarzkopf and
Schwarzkopf [80] have described a technique for weighing liquids which boil
below 20°C under atmospheric pressure by condensing the sample into a
capillary kept in a liquid nitrogen bath. The apparatus (Fig. 1.10) consists
of a manifold which connects the sample amount at one end and the capillary
at the other end. The sample is contained in an ampoule of 6-mm outside
diameter (o.d.) with a sealed tip of 2-mm o.d. The ampoule is connected
to the apparatus with flexible vacuum tubing that will permit the breaking
of the ampoule tip under vacuum. To transfer the sample, the ampoule is
kept at room temperature, while the capillary is cooled with liquid nitrogen.
When enough sample has condensed in the capillary, it is sealed and excess
liquid is returned into the sample ampoule by cooling the ampoule which
can be resealed. An alternative technique to measure the quantity of very
low boiling liquids is to handle them as gases (see Sec. 1.8).

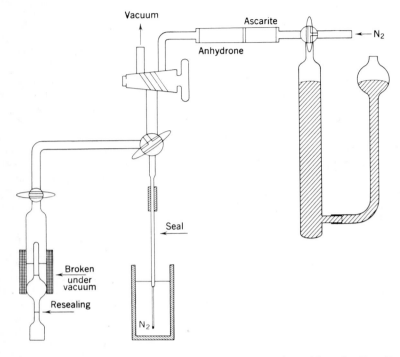

FIG. 1.10. Distilling apparatus for the transfer of low-boiling liquids or gases. (From Ref. 80, courtesy of J. Wiley & Sons, Inc.)

Aluminum capsules [81] are used to weigh liquid samples for combustion in the automated C, H, N analyzers. Küber [82] has constructed a machine to fill the aluminum capsules in which air can be replaced by another gas. Kennedy [83] has pointed out, however, that aluminum tubes are extremely time-consuming, besides being reactive toward liquid samples containing active halogens or acidic functions.

In analytical procedures which involve decomposition of the organic substance in liquid media (e.g., micro Kjeldahl procedure for nitrogen, Sec. 3.3, or determination of metals by acid digestion, Sec. 9.6), volatile samples are conveniently weighed in microweighing bottles. As illustrated in Fig. 1.11, the ground-glass stopper of the microweighing bottle has an opening so that a hook can go through it to pull the stopper apart after the closed bottle is submerged in the solvent in the reaction flask.

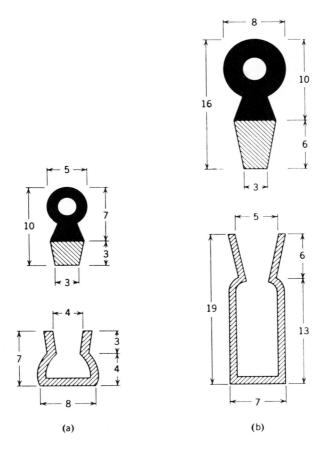

FIG. 1.11. Weighing bottles (scale in mm). (a) Micro. (b) Semi-micro. (Courtesy Microchem. J.)

1.7 EXPERIMENTAL: MEASUREMENT OF LIQUID VOLUME

Burets, pipets, and syringes are instruments for the measurement of liquid volumes. A buret may be defined as a device to deliver variable volumes of fluid at the specified level of precision. In contrast, a pipet is usually employed to deliver or to contain a fixed volume. Therefore, while some pipets are graduated, measurement by the graduation is not expected to be very precise. Utilization of syringes in the organic analysis laboratory is a recent development. Depending on the intended use, a chemical syringe is called either a syringe-buret or a syringe-pipet.

It is understandable that burets come in different sizes and shapes.
Generally speaking, the burets for organic elemental analysis are used to
measure standardized solutions in titrimetric experiments. Since 0.01 N
solutions are frequently employed and the volume of titrant is usually in the
range of 2 to 7 ml, microburets with a capacity of 10 ml are most convenient. Some examples are illustrated in Fig. 1.12. The Pregl and Machlett
burets have an automatic zero device; hence measurements are always
started at the zero line, thus eliminating one possible error in observation
and recording. It should be noted that the rubber tubing and pinch clamp in
the buret for delivery of an alkali solution is preferably replaced by a Teflon stopcock. Mention may also be made that current Pregl burets are
graduated in 0.05 ml and not in 0.02 ml. The Machlett type permits storage of the stock solution under an inert atmosphere, but it is not as stable

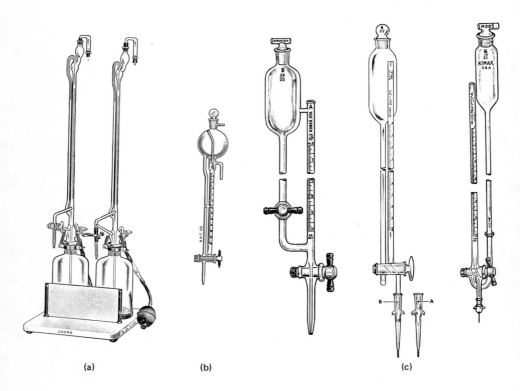

(a)   (b)   (c)

FIG. 1.12. Microburets. (a) Pregl type. (b) Machlett type. (c) Koch
type. (Courtesy A. H. Thomas Co.)

or convenient as the Pregl type. The Koch type has a small reservoir and no automatic zero mechanism; however, it is more flexible and less expensive. The British Standards Institution has recently specified the maximum and minimum scale lengths for burets [84].

Figure 1.13 shows the operation of a Metrohm piston-type buret. The titrant is delivered from precision bore glass tubes by a Teflon piston, advanced by a screw drive. The buret capacity is 10 ml, with an accuracy of ±0.02 ml and a reproducibility of ±0.01 ml. In the manual model (Fig. 1.13A), a thumbwheel on the graduated drum protruding from the front of the housing advances a piston to deliver the buret contents. The shape of housing permits the fingers of either hand to curve behind while the thumb moves the wheel for delivery. Each full revolution (indicated by a unit advance of the pointer on a coarse scale) delivers 1 ml of liquid. Drum scale division is read directly in 0.01-ml graduations. The motor-driven model (Fig. 1.13B) uses the same glass part and piston as the manual

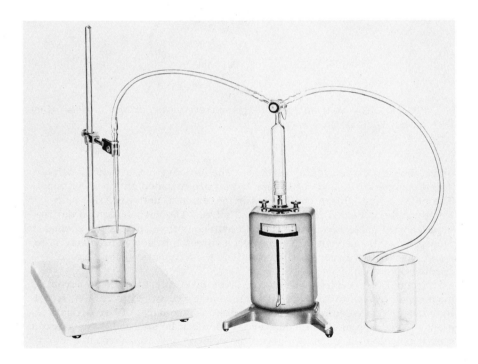

FIG. 1.13A. Metrohm piston-type buret, manual model. (Courtesy A. H. Thomas Co.)

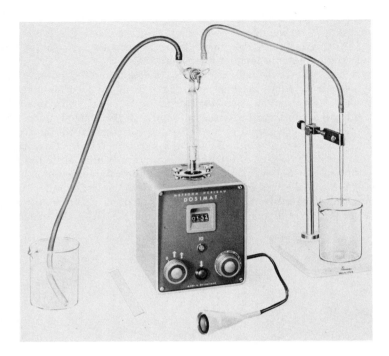

FIG. 1.13B. Metrohm piston-type buret, motor driven model. (Courtesy A. H. Thomas Co.)

drive, but has direct digital readout. The delivery mode switch, delivery speed control knob, and refill push button are mounted on the front panel. The mode switch allows continuous or incremental delivery. Delivery speed can be varied from 0.01 to 0.33 ml/s. The push-button on the remote control cable provides either a continuous flow, or, in the second mode, a precise aliquot of 0.01 ml each time the button is pressed. The panel push-button reverses the motor for the automatic filling and self-zeroing operation.

Examples of syringe-burets are given in Fig. 1.14. These burets have capacities up to 2 ml. In the micrometer buret (Fig. 1.14A), a Teflon plunger displaces titrant from precision-bore buret of borosilicate glass. Accuracy of ±1 micrometer division can be attained. The micrometer mechanism is made of plastic; the plunger operates through a Viton

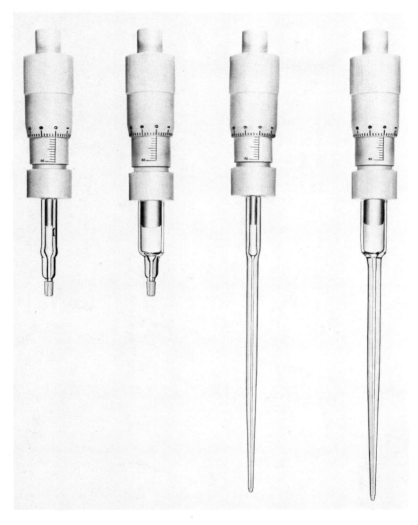

FIG. 1.14A. Gilmont micrometer burets. (Courtesy A. H. Thomas Co.)

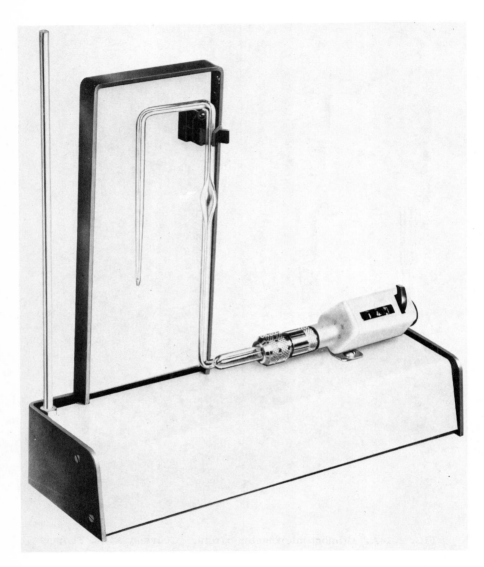

FIG. 1.14B. Manostat digital readout buret. (Courtesy A. H. Thomas Co.)

O-ring seal. Berger and DeForest [84a] have described a simple technique to employ the micrometer buret for photometric titrations using a
Spectronic 20 spectrometer. In the manostat digital readout buret (Fig.
1.14B), the titrant is delivered indirectly by the operation of a stainless
steel plunger which displaces mercury in the reservoir.

Since syringes to deliver 10 μl of liquid with an accuracy of ±1% are
now commercially available, they may be utilized to measure organic
liquids for analysis. Burroughs and Goodrich [85] have shown that the
densities of liquids can be accurately determined by means of 10-μl syringes. Hence, for routine multiplicate analyses in certain experiments
(e.g., closed-flask combustion, Sec. 5.3), it is expedient to determine
the density of the liquid material and measure aliquots using the microliter syringe instead of weighing each sample. Anderson and Reichelt
[85a] have fabricated a simple automatic dispensing apparatus based on a
disposable 2-ml syringe; it can be operated continuously, and delivers
volumes that are only a fraction of the total liquid flowing through it.

1.8 EXPERIMENTAL: MEASUREMENT OF GAS VOLUME

The need for measuring a volume of gas for organic elemental analysis
arose with the development of fluoro and perfluoro compounds [86]. As
the molecular weight of the compound increases through the substitution
of hydrogen by fluorine, the boiling point decreases, and the analyst is
faced with the problem of handling gaseous samples. There are very few
published procedures for measurement of gases for organic elemental
analysis. If applicable, the gas may be measured in a syringe and injected into the combustion train through a rubber septum attached to an
appropriate side tube.

Olson and Knafla [87] have constructed an apparatus to measure gas
samples for the determination of nitrogen by the micro Dumas method
using a modified Coleman instrument (Sec. 3.5). As illustrated in Fig.
1.15, the gas bulb is 3.4 ml in volume and 15 mm in length, and is fitted
with valves E and F which have Teflon needles. A and B are T-bore
stopcocks. The stopcock C is for purging the system with carbon dioxide.
The gas sample is mixed with carbon dioxide which enters through valve
E. The sample carbon dioxide mixture is then swept into the combustion
tube through valve F.

Balis et al. [88] have determined carbon, hydrogen, and chlorine in
gaseous compounds using the combustion train shown in Fig. 1.16. The
sampling vessel is filled with the gas sample (see inset) and then introduced into the gas holder F by the following operation: (1) The mercury

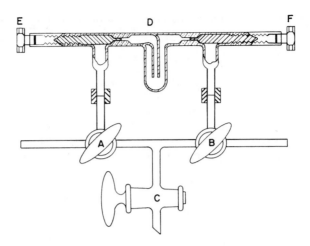

FIG. 1.15. Apparatus for handling gas samples. (From Ref. 87, courtesy Microchem J.)

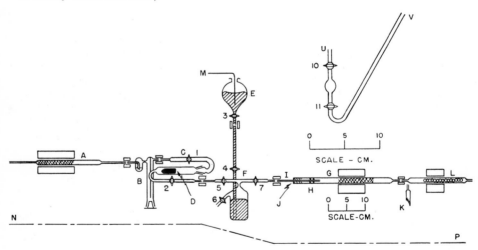

FIG. 1.16. Diagram of train arranged for combustion of highly volatile liquids. A, quartz preheater (platinum foil filling at 900°C); B, Kraissl bubble counter and absorption tube; C, sample tube for volatile liquids; D, glass-enclosed iron weight; E, separatory funnel filled with mercury; F, gas holder (135-ml volume when sealed with 1-cm depth of mercury); G, quartz combustion tube (platinum-foil filling at 900°C); H, platinum-foil baffle; I, quartz capillary; J, rubber stopper; K, air jet; L, quartz tube (silver filling at 600°C); M, wire to ground; NP, alignment of apparatus as seen from above. (From Ref. 88, courtesy Anal. Chem.)

is drawn from the gas holder F by evacuating the funnel E, while admitting oxygen through stopcock 5; (2) F is evacuated through stopcock 6; (3) the long arm V of the sampling vessel is filled with mercury to stopcock 11, and the short arm U is then connected by rubber tubing to stopcock 6 of the gas holder F; (4) the gas sample is expanded into F; (5) the last portion of gas in the sampling vessel is pushed into F by filling the bulb with mercury from V; (6) the gas holder F is finally filled with oxygen to atmospheric pressure. In the course of combustion, the gas sample in gas holder F is forced through by allowing mercury to flow from the separatory funnel E to the gas holder F. M is a ground wire; if there is no grounding, tiny bubbles of mercury will continually leap up from the mercury surface in F, and there is the possibility of an explosion hazard owing to the accumulation of static charge.

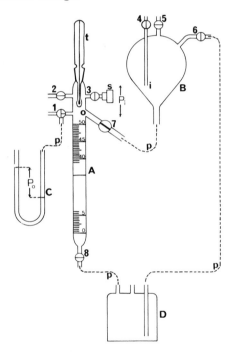

FIG. 1.17. Constant-pressure gas buret. A; buret; B, Mariotte bottle; C, open-tube water manometer; D, recovery tank for displacement liquid; 1-8; stopcocks (1-5, all-glass greased stopcocks; 6-8, PTFE-glass stopcocks; 7, stopcock with 15-mm hole length and 4-mm hole diameter, with hydrodynamic resistance to water of 0.02 mmHg s/ml); t, thermometer; i, lowest end of the withdrawal tube in the Mariotte bottle; o, orifice for entry of displacement liquid into the buret; s, sleeve with screw for micrometric driving of a gasketed plunger; p, PVC tubes (the tube between B and A was 70 cm long, 0.08 cm in diameter, and had hydrodynamic resistance to water of 0.05 mm Hg s/ml). (From Ref. 89, courtesy Anal. Chem.)

In the above two procedures, a fixed volume of the gaseous material is taken for analysis. If the analytical method calls for variable volumes of gas sample, the constant pressure gas buret described by Forina [89] may be employed. Based upon the principle of the Mariotte bottle, this apparatus (Fig. 1.17) is a volume-calibrated container, for the delivery of gas under a constant pressure, into a vessel or through one or more solutions, with the overcoming of some dynamic and hydrostatic pressures.

REFERENCES

1. N. D. Cheronis and T. S. Ma, Organic Functional Group Analysis, Wiley, New York, 1964.
2. T. S. Ma, Rec. Chem. Progr. $\underline{26}$:113 (1965).
3. T. S. Ma and V. Horak, Microscale Manipulations in Chemistry, Wiley, New York, 1976.
4. Encyclopedia Britannica, Encyclopedia Britannica Inc., Chicago, 1973.
5. A. G. Debus (ed.), World Who's Who in Science, Marquis Co., Chicago, 1968.
6. P. L. Heissner (ed.), Handbuch der allgemeinen und technischen Chemie, Vol. V, Part 2, Carl Gerold, Wien, 1829, p. 198.
7. P. G. Stecher (ed.), The Merck Index, 8th Ed., Merck, Rahway, N.J., 1968, p. 1094; M. Windholz (ed.), The Merck Index, 9th Ed., Merck, Rahway, N.J., 1976, p. 1266.
8. L. Carius, Annalen $\underline{116}$:1 (1860); $\underline{136}$:129 (1865); $\underline{145}$:301 (1868); Berichte $\underline{3}$:697 (1870).
9. T. S. Ma, Analysis of Carboxylic Acids and Esters, in S. Patai (ed.), The Chemistry of Carboxylic Acids and Esters, Wiley, New York, 1969, p. 872.
10. H. Roth, Quantitative Organic Microanalysis of Fritz Pregl (translated by E. B. Daw), Blackiston, Philadelphia, 1937.
11. T. S. Ma and G. Zuazaga, Ind. Eng. Chem., Anal. Ed. $\underline{14}$:280 (1942).
12. M. Vecera, Organic Elemental Analysis (in Czech), SNTL, Prague, 1967.
12a. F. Ehrenberger and S. Gorbach, Methoden der organischen Elementar- und Spurenanalyse, Verlag Chemie, Weinheim, 1973.
13. K. Hozumi (ed.), Organic Micro Quantitative Analysis (in Japanese), Nangoto, Tokyo, 1969.
14. T. Mitsui, Pure Appl. Chem. $\underline{10}$:45 (1965).
15. G. Ingram, Organic Elemental Analysis, Reinhold, New York, 1962.
15a. J. P. Dixon, Modern Methods in Organic Microanalysis, van Nostrand, London, 1968.
16. A. Steyermark, Quantitative Organic Microanalysis, 2nd Ed., Academic, New York, 1961.

17. A. P. Terentev, Organic Analysis (in Russian), Moscow University Press, Moscow, 1966.
18. N. E. Gel'man, Pure Appl. Chem. 25:15 (1971).
19. T. S. Ma and M. Gutterson, Anal. Chem. 42:105R (1970).
20. T. S. Ma and M. Gutterson, Anal. Chem. 46:437R (1974).
20a. U. S. Pharmacopoeia XIX, U.S. Pharmacopoeia Convention, Inc., Rockville, Md., 1975, p. 626.
21. W. Schöniger, Mikrochim. Acta 1955:123.
22. T. S. Ma and M. Gutterson, Anal. Chem. 36:150R (1964).
23. K. Ubik, J. Horacek, and V. Pechanic, Collect. Czech. Chem. Commun. 37:102 (1972).
24. K. Ubik, Microchem. J. 17:556 (1972).
25. K. Ubik, Microchem. J. 18:29 (1973).
26. S. Ebel, Z. Anal. Chem. 264:16 (1973).
27. F. Salzer, Microchem. J. 16:145 (1971).
28. T. S. Ma and M. Gutterson, Anal. Chem. 44:445R (1972).
29. R. Scheidl and V. Toome, Microchem. J. 18:42 (1973).
29a. D. Fraisse and R. Richard, Microchem. J. 21:170 (1976).
30. G. M. Maciak, R. A. Byers, and P. W. Landis, Microchem. J. 18: 8 (1973).
30a. E. C. Olson, Upjohn Co., Kalamazoo, Mich., 1975.
30b. H. Wagner and R. Kübler, Ciba-Geigy A.G., Basel, Switzerland, 1976.
30c. A. Dirscherl, Hoffmann-LaRoche A.G., Basel, Switzerland, 1976.
30d. R. W. Frei and W. Pfirter, Sandoz Ltd., Basel, Switzerland, 1976.
30e. W. Merz, Talanta 21:481 (1974).
31. E. Haberli, Mikrochim. Acta 1973:597.
31a. R. Stoffel and K. H. Gade, Pharm. Z. 119:1286 (1974).
32. S. A. Liebman, D. H. Ahlstrom, T. C. Creighton, G. D. Pruder, N. R. Averitt, J. L. Walker, and E. J. Levy, Anal. Chem. 44:1411 (1972); 45:1360 (1973).
33. S. A. Liebman, D. H. Ahlstrom, C. D. Nauman, G. D. Pruder, N. R. Averitt, and E. J. Levy, Research/Development 1972, December 24.
34. W. Simon and J. T. Clerc, Pure Appl. Chem. 25:35 (1971).
35. B. F. A. Griepink and G. Dijkstra, Z. Anal. Chem. 257:269 (1971).
36. F. W. McLafferty, Interpretation of Mass Spectra, 2nd Ed., Benjamin, New York, 1973.
37. R. I. Reed, Application of Mass Spectrometry to Organic Chemistry, Academic, London, 1966.
38. G. W. A. Milne (ed.), Mass Spectrometry: Techniques and Applications, Wiley, New York, 1971.
39. G. Ege, Zahlentafeln Zur Massenspektrometrie und Elementaranalyse, Verlag Chemie, Weinheim, 1970.

39a. W. Walisch, O. Jaenicke, and A. Siewert, Talanta 18:175 (1971); 22:167, 345 (1975).

40. C. G. Hammar and R. Hessling, Anal. Chem. 43:298 (1971).

41. H. C. E. van Leuven, Anal. Chim. Acta 49:364 (1970).

42. N. E. Vanderborgh and W. T. Ristau, Anal. Chem. 45:1529 (1973).

43. C. Merritt, private communication, June 1974.

44. J. B. Stothers, Carbon-13 NMR Spectroscopy, Academic, New York, 1972.

45. A. R. Abraham, The Analysis of High Resolution NMR Spectra, Elsevier, Amsterdam, 1971.

46. D. W. Mathieson, Nuclear Magnetic Resonance for Organic Chemists, Academic, London, 1967.

47. G. C. Levy and G. L. Nelson, Carbon-13 Nuclear Magnetic Resonance for Organic Chemists, Wiley, New York, 1972.

48. E. Ricci, Anal. Chem. 43:1866 (1973).

49. K. Asai and D. Ishii, J. Chromatogr. 69:355 (1972).

49a. G. H. Morrison and R. A. Nadkarni, Radiochem. Radioanal. Lett. 24:103 (1976); private communication, November, 1976.

49b. R. L. Hutson, J. J. Reidy, K. Springer, H. Daniel, and H. B. Knowles, IEEE Trans. Nuclear Sci. NS-22, No. 3 (1975); private communication, July 1976.

50. R. Levy (ed.), Erreurs en Microanalyse Organique Elementaire (reprinted from Pure Appl. Chem.), Butterworths, London, 1972.

51. I. Meyer and C. Geissler, Nahrung 17:507 (1973).

51a. J. Bosset and G. Steiger, Mitt. Geb. Lebensmitt. 65:470 (1974).

52. Office International du Cacao et du Chocolate, Int. Choc. Rev. 27:281 (1972).

53. R. A. Lalancette and A. Steyermark, J. Ass. Offic. Anal. Chem. 57:26 (1974).

54. R. Lalancette, A. Steyermark, D. M. Lukaszewski, and P. L. Kostrzewski, J. Ass. Offic. Anal. Chem. 56:888 (1973).

55. R. E. Denney and P. A. Smith, Analyst 99:166 (1974).

56. L. H. Scroggins, J. Ass. Offic. Anal. Chem. 56:892 (1973); 57:22 (1974); 58:146 (1975).

57. R. A. Lalancette, A. Steyermark, R. A. Lee and D. M. Lukaszewski, J. Ass. Offic. Anal. Chem. 56:897 (1973).

58. F. J. Johnson, J. Ass. Offic. Anal. Chem. 55:979 (1972); 56:1084 (1973).

58a. R. Nadkarni, Radiochem. Radioanal. Lett. 19:127 (1974).

58b. J. F. Uthe, H. C. Freeman, J. R. Johnston, and P. Michalik, J. Ass. Offic. Anal. Chem. 57:1363 (1974).

58c. Metallic Impurities in Organic Matter Sub-Committee, Analyst 100:54 (1975).

59. The Joint Panel of the Microchemical Methods Group and the Analytical Standards Sub-Committee of the Analytical Methods Committee of the Society for Analytical Chemistry, Analyst 97:740 (1972).

60. J. P. Martin and J. H. Thompson, Analyst 97:405 (1972).

61. T. S. Ma and M. Gutterson, Anal. Chem. 34:111R (1962).

62. T. S. Ma, "Quantitative Microchemical Analysis," in F. J. Welcher (ed.), Standard Methods of Chemical Analysis, 6th Ed., Vol. 2, van Nostrand, Princeton, 1963, p. 357.

63. K. O. Hodgson, F. Mares, D. F. Starks, and A. Stretweisser, Jr., J. Amer. Chem. Soc. 95:8652 (1973).

64. A. E. Crease and P. Legzdins, J. Chem. Soc. 1973:1501.

65. H. Schneider and E. Nold, Z. Anal. Chem. 269:113 (1974).

66. A. Colombo and R. Vivian, Microchem. J. 18:589 (1973).

67. Fisher Scientific Co., Pittsburgh, Pa., Cat. No. 21-13-5 and 21-131-10; Fisher Scientific Lab. Rep. 11, No. 3 (1973).

68. Pamphlet, Foss America, Inc., Fishkill, N.Y., 1974.

69. R. K. Patterson, Anal. Chem. 45:605 (1973).

70. R. E. Lee, Jr., and J. Hein, Anal. Chem. 46:931 (1974).

71. Appareil de Microanalyse (pamphlet), Ateliers Herrmann-Moritz, Chassant, France, 1970.

72. R. Herrmann, Chim. Anal. 50:89 (1968).

73. Pamphlet, W. C. Heraeus, Hanau, Germany, 1970.

74. M. Marzadro, J. Zavattiero, and F. A. Mazzeo, Ann. 1st Super Sanita, 8:9 (1972); M. Marzadro, Microanalisi Quantitativa Organica, Edzioni Scientifiche Italiane, Napoli, 1958.

75. V. B. Fish, J. Chem. Educ. 46:A323 (1969).

76. R. T. E. Schenck and T. S. Ma, Mikrochimie 40:236 (1953).

77. V. Nuti, Farmaco 28:528 (1973).

77a. B. Cousin and C. Muller, Talanta 21:1287 (1974).

78. T. S. Ma and K. W. Eder, J. Chinese Chem. Soc. 15:112 (1947).

79. G. Ingram, in I. M. Kolthoff and P. J. Elving (eds.), Treatise on Analutical Chemistry, Part II, Vol. 11, Wiley, New York, 1965, p. 375.

80. O. Schwarzkopf and F. Schwarzkopf, in M. Tsutsui (ed.), Characterization of Organometallic Compounds, Part I, Wiley, New York, 1969, p. 41.

81. A. C. Thomas and C. D. Robinson, Mikrochim. Acta 1971:1.

82. R. Küber, Mikrochim. Acta 1974:213.

83. C. I. Kennedy, Microchem. J. 17:325 (1972).

84. British Standards Institution, BS 846 (1962), Amendment No. 2, January 31, 1975.

84a. S. A. Berger and P. R. DeForest, Mikrochim. Acta 1974:689.

85. J. E. Burroughs and C. P. Goodrich, Anal. Chem. 46:1614 (1974).

85a. B. L. Anderson and K. L. Reichelt, Anal. Biochem. 59:517 (1974).

86. T. S. Ma, Microchem. J. 2:91 (1958).

87. P. B. Olson and R. T. Knafla, Microchem. J. 13:362 (1968).

88. E. W. Balis, H. A. Liebhafsky, and L. B. Bronk, Ind. Eng. Chem., Anal. Ed. 17:56 (1945); H. Liebhafsky, private communication, March 1978.

89. M. Forina, Anal. Chem. 46:1622 (1974).

Carbon and Hydrogen

2.1 GENERAL CONSIDERATIONS

A. Basic Principles and History of Carbon and Hydrogen Analysis

As a matter of convenience, carbon and hydrogen are generally determined together (for exceptions see Sec. 2.1F). The basic principle is very simple: The organic compound is decomposed by oxidation, whereupon the carbon is converted to carbon dioxide and hydrogen to water, as depicted by the following equation:

$$C_x H_y + (2x + \frac{1}{2}y)O = xCO_2 + \frac{1}{2}yH_2O$$

The mechanism of the oxidation process, however, is complicated and has not been fully elucidated. It is evident that the cleavage of the carbon and hydrogen bonds depends on the structure of the molecule, and that the formation of carbon dioxide as well as water may not follow the same paths in all cases.

From the viewpoint of quantitative analysis, the crucial questions are: (1) Can the decomposition reaction be brought to completion as expected? (2) Can the water and carbon dioxide thus produced be totally collected and accurately measured? Fortunately, affirmative answers to both questions were obtained about 150 years ago, when Berzelius (1817) and Liebig (1831) performed the oxidation in the presence of solid oxidants and determined water and carbon dioxide, respectively, by weighing the products after collecting them in separate receivers. Thus the foundation of organic chemistry was laid, for it became possible to deduce the formula of an organic compound by its carbon and hydrogen analysis.

The next stage of development took place around 1910 when Pregl demonstrated that the determinations could be carried out with 3-5 mg of organic compounds by using a Kuhlmann microbalance of 20-g capacity and 1-μg sensitivity. Previously, the sample size ranged from 200 to 500 mg. Since most organic compounds are difficult to prepare, and many are unstable on standing, the advantage of the Pregl method was quickly recognized, culminating in his award of the Nobel Prize in 1923. It is fair to say that the Pregl approach of performing quantitative organic analysis at the milligram region has played an important role in the rapid expansion of biochemistry and synthetic organic chemistry during the past 50 years.

B. Current Practice

Generally speaking, the currently preferred methods for carbon and hydrogen determination are based on the Pregl-Lieb procedure [1], in which the organic compound is driven by an oxygen stream to come into contact with some solid oxidant (called the "packing") at about 700°C. Copper oxide, used by Liebig, remains the favorite oxidizing agent, although many other reagents [1a, b] have been proposed, the most popular being Körbl's mixture, obtained by heating silver permanganate [2]. The latest reagent recommended is barium chromate [2a]. Belcher and Ingram [3] have advocated oxidation in an empty tube in a rapid stream of oxygen at 900°C or higher. Subsequently Ingram [4] has recommended decomposition under a closed-system condition, provided that the reaction chamber contains enough oxygen. This technique is known as "ignition combustion," "flash combustion," or "static-state oxidation." It depends on the nearly instantaneous complete oxidation of the sample. On the other hand, Korshun and co-workers [5, 6] and Kuck [7], also using the empty tube, consider it advantageous to partly oxidize the organic compound first, and they have incorporated the prepyrolysis step in their procedure. Needless to say, the empty tube methods do not permit gradual combustion, which is deemed essential for the complete decomposition of low-boiling hydrocarbons and some refractory polymeric substances. Francis [7a] has recommended the use of ceramic tubes in place of the expensive quartz tubes for flash combustion.

It should be realized, especially for the organic chemist who synthesizes new compounds or isolates unknown substances from nature but does not perform the analysis himself, that the determination of carbon and hydrogen is not free from difficulties and complications. For instance, hygroscopic samples may absorb atmospheric moisture on transit, leading to high hydrogen values; pyrophoric substances tend to decompose prematurely; steroids possessing angular methyl groups may yield low carbon and hydrogen results because of the formation of methane which escapes oxidation. A common complication is the presence of heteroelements which are discussed in the next section.

C. Compounds Containing Heteroelements

When the sample submitted for carbon and hydrogen determination also contains other elements, provision should be made to circumvent the possible interference of such elements during the decomposition of the organic compound or in the measurement of water and carbon dioxide. For instance, whereas organic nitrogen compounds may form nitrogen gas upon heating, they frequently produce oxides of nitrogen which interfere with the determination. One method to remove this interference is based on the absorption of the oxides of nitrogen in a suitable reagent which has no effect on water and carbon dioxide. Either lead dioxide or manganese dioxide can be used; the former is placed in the combustion tube while the latter is packed in a separate tube situated between the water and carbon dioxide absorption tubes (see Sec. 2.6, Fig. 2.9G). Another method is to pass the mixture containing water vapor, carbon dioxide, and oxides of nitrogen through copper at 500°C which converts all oxides of nitrogen to molecular nitrogen. It is interesting to note that the second method was the standard technique developed by Liebig, discarded in the Pregl-Lieb procedure, and revived half a century later in the automated C,H,N analyzers currently on the market.

The interference of sulfur is circumvented by means of a metal such as silver which forms silver sulfate. The same metal is also capable of removing chlorine, bromine, and iodine by forming the corresponding halide. On the other hand, silver is inefficient for eliminating the interference due to the presence of fluorine, for which magnesium oxide should be used.

Surprisingly, some compounds with high oxygen content, like carbohydrates, are difficult to analyze; the oxygen in the molecule does not seem to facilitate quantitative oxidation. The reason is that these compounds tend to char upon heating, resulting in a layer of carbon which interferes with the decomposition of the material below it. The same difficulty may be encountered in the analysis of organometallic compounds. With few exceptions (e.g., mercury) the metallic element or its inorganic product remains behind in the container and may prevent part of the organic moiety from being oxidized. If the sample contains alkali metals, stable carbonates may be formed, resulting in low yields of carbon dioxide.

D. Measurement of Water and Carbon Dioxide

Up until the recent acceptance of automated C,H,N analyzers, determination of water and carbon dioxide has been invariably carried out gravimetrically, the amount of product measured being in, or above, the milligram region. For carbon and hydrogen analysis below this level, it is not possible to weigh the products accurately because of the difficulty in handling the absorption tubes in an environment which is not free from moisture and carbon dioxide (see Chap. 11 for further discussion). On the other hand, gravimetric finishes are not favored in the automatic devices because of the difficulty in design and because of the time-consuming feature of gravimetry. Many approaches for water and carbon dioxide measurements have been proposed, as can be seen in Sec. 2.2A, Table 2.1.

E. Automated Apparatus

Automated apparatus for carbon and hydrogen determination became popular in the 1960s, and a number of instrument manufacturers have produced commercial models. One group employs the gravimetric finish. In the Coleman [8] apparatus, the electric furnaces, packed combustion tube, and absorption tubes are arranged in vertical positions. The Thomas [9] apparatus uses a vertical empty tube and combustion furnace but a horizontal arrangement of absorption tubes. The Sargent [10] and Heraeus [10a] models keep the horizontal combustion train of the Pregl-Lieb procedure but incorporate a synchronous motor to drive the short burner at speeds which are adjustable from 5 to 15 mm per min. Another group of automated apparatus retain the feature of a horizontal combustion train and successive absorption of water and carbon dioxide which are measured by thermal conductivity. In the Perkin-Elmer [11] apparatus the sample, which can vary from 0.5 to 5 mg, is decomposed in an oxygen atmosphere in a static state. The thermal conductivity of the gas mixture is first taken, then measured after the water has been removed by passing through magnesium perchlorate, and again after the carbon dioxide has been absorbed in Ascarite. In the Technicon [12] apparatus, the recommended sample size is between 0.3 and 0.6 mg. Combustion is carried out in an dynamic state. Water is first absorbed in silica gel to allow the measurement of the thermal conductivity of carbon dioxide. Then water is recovered by heating the absorption tube and measured. The same principle is used in the Heraeus Ultra Analyzer [13]. In still another group of instruments, gas chromatographic columns are utilized to separate the gas mixture. In the F & M apparatus of Hewlett-Packard [14], the sample (less than 0.5 mg) is covered with catalyst-oxidant and inserted into the combustion tube containing copper oxide at 1050°C. In the Carlo Erba [15] apparatus, the sample is placed in a revolving magazine from which it is dropped into copper oxide at 1050°C; the magazine can hold 23 samples with a weight range of 0.5 to 1.8 mg. It is of interest to note that these five above-mentioned automatic instruments contain a section of metallic copper in the combustion train to reduce the oxides of nitrogen, if present, and are designed for the simultaneous determination of carbon, hydrogen, and nitrogen by means of one weighing of the sample. Contrastingly, the Aminco [16] apparatus is constructed for carbon and hydrogen determination only. While gasometry and thermal conductivity are employed in the finishes, the water formed is reacted with calcium hydride,

$$2H_2O + CaH_2 = 2H_2 + Ca(OH)_2$$

and the resultant hydrogen gas is measured in the same thermister detector used for carbon dioxide. Nitrogen oxides, if present, are removed by manganese dioxide. Several automated apparatus have been constructed by Japanese instrument firms though they have not yet been put into the world market.

Every automated apparatus has its special features and limitations. Prices of the different categories vary, and the maintenance cost depends on the availability of repair service. The chief advantage of automation is speed. For instance, the Thomas [9] gravimetric device can handle 40 samples in a day; the Carlo Erba [15] recording instrument is able to make 46 determinations in 8 hr, plus 23 more if the analyst replaces the revolving magazine before he leaves the laboratory. Therefore the first consideration in such investment is the frequency of carbon and hydrogen analyses expected in the establishment. Secondly, the variety of organic compounds submitted for analysis should be considered. Since automation alleviates certain human errors, it is generally true that the automated apparatus excel in "repeatability" (i.e., statistical evaluation of multiplicate analyses using identical procedure by the same person on one compound). Thus duplicate determinations are superfluous unless the sample is suspected to be nonhomogeneous. On the other hand, if successive samples fed into the automatic machine differ widely in physical and chemical characteristics or elemental compositions, it is possible that the whole series of determinations become erratic and useless. It may be mentioned that the automated instruments are seldom recommended in labor-intensive localities, and that the manually operated apparatus has the advantages of being inexpensive, flexible, and easy to maintain. For these reasons, detailed procedures of both types are given in this book (see Secs. 2.3-2.7).

F. Determination of Carbon or Hydrogen Only

Whereas the general practice is simultaneous analysis of carbon and hydrogen in the organic sample, there are occasions in which only carbon or only hydrogen determination is possible or needed. For example, when deuterium or tritium is used in tracer studies, only water is of concern in the combustion products. Conversely, tracer techniques employing Carbon-14 involve only the determination of carbon dioxide, and the decomposition can be performed in the closed flask (see Chap. 5). For carbon dating, the organic material is oxidized to carbon dioxide, which is then converted to acetylene and finally to benzene to be measured in the radioactivity counter [17]. Apelgot et al. [17a] have determined Carbon-14 or tritium in biological samples without decomposition.

When the organic material is in an aqueous solution, it is impossible to determine both the carbon and hydrogen contents of the organic compound. The decomposition is effected by wet oxidation (see Table 2.1). Determination of the "active hydrogen" in an organic compound is not considered as elemental analysis and is treated in books on organic functional group analysis [18].

2.2 REVIEW OF RECENT DEVELOPMENTS

A. Summary of Recent Publications

The literature on carbon and hydrogen analysis which has appeared since about 1961 is summarized in Table 2.1. The reader who wishes to cover the literature prior to that year is referred to the excellent article by Ingram and Lonsdale [19]. The methods and techniques listed in Table 2.1 are arranged roughly in an order according to the preference and frequency of application in the analytical laboratories. Some publications not cited in the table may be found in Chap. 10 which deals with simultaneous determination of other elements in addition to carbon and hydrogen. It should be noted that the references listed are not exhaustive. They are adequate, however, to exemplify the variety of methods and procedures. Gel'man and Bunz [19a] have compiled a bibliography covering the papers which appeared from 1953 to 1971.

A wide range of decomposition methods have been studied during the past decade. The consensus of the investigators, however, still seems to favor the combustion tube packed with solid reagents. Many new findings have been incorporated in the commercial automatic instruments. According to the Russian reports, the empty tube method has been successful both in manual operation and in automated devices. This and several other methods have their advantages, but they probably will not be accepted as the standard method for general purposes.

TABLE 2.1.   Summary of Recent Literature on Carbon and Hydrogen Analysis.

Decomposition method	Product determined	Finish technique	Reference
Combustion tube with packing	CO_2, H_2O	Gravimetric	20-57a
	CO_2, H_2O	Gasometric; gas chromatography	58-70
	CO_2, H_2O	Manometric	71
	CO_2, H_2O	Titrimetric	72-76
	CO_2, H_2O	Conductimetric	77, 78
	CO_2	IR spectrometry	79
	CO_2	Spectrophotometry	80
	H_2O	Coulometry	81, 82
	^3H_2O, $^{14}CO_2$	Radiometric	83-84a
Empty tube	CO_2, H_2O	Gravimetric	85-90c
	CO_2, H_2O	Titrimetric	91, 92
	CO_2, H_2O	Gasometric	93
	HD	Thermal conductivity	93a
	CO_2	IR spectrometry	94
	H_2O, CO_2	Coulometry	94a
Closed flask	CO_2	Titrimetric	95-98
	$CaCO_3$	Gravimetric	99
	$CaCO_3$	Radiometric	99
	^3H_2O, $^{14}CO_2$	Radiometric	99a
Wet oxidation	CO_2	Manometric	100, 101
	CO_2	Radiometric	102
Na_2O_2 bomb	CO_2	Radiometric	103

TABLE 2.1. (Continued)

Decomposition method	Product determined	Finish technique	Reference
CuO in ampoule	H_2O, CO_2	Gas chromatography	103a
Mg fusion	$H_2 \rightarrow H_2O$	Gravimetric	104
Combustion with S	H_2SO_4	Titrimetric	105
Combustion in NH_3	HCN	Titrimetric	106

B. The Problem of Difficult Samples

Considerable attention has been given in recent years to the search for effective reagents and techniques to handle difficult samples. There are several reaons for this. (1) Since lead chromate in the original packing of the Pregl-Lieb procedure had to be abandoned when quartz combustion tubes were employed, an effort was made to find another oxidizing agent to complement copper oxide. (2) After World War II the number of synthetic organic polymers increased by leaps and bounds, and numerous natural polymeric substances were investigated. These materials cannot always be manipulated in the combustion train by general volatilization of the sample in the microboat. (3) With the development of fluorine chemistry, the analyst suddenly faced a series of compounds in which the cleavage of the C-F bond is extremely difficult. (4) The current interest in organometallic compounds has produced many analytical samples, all of which leave residues on decomposition and many of which are sensitive to the atmosphere. (5) It is essential for the instrument manufacturer to find a long-life and safe combustion tube packing before the commercial automated apparatus can be marketed.

In the late 1950s many analysts evaluated the thermal decomposition product of silver permanganate proposed by Körbl [2] as the packing material. It is an efficient oxidant and contains the built-in reagent to remove sulfur and halogens. Unfortunately, it does not have a fixed composition, hence cannot be standardized, and it may become explosive in inexperienced hands. As can be seen in Table 2.2, the current emphasis is on oxides of cobalt, tungsten, vanadium, and cerium. Combinations of several solid reagents are generally recommended for the combustion tube packing. For instance, Yeh [30] has described a universal filling which consists of 2 cm of gold wire, 1 cm of silver wire, 5 mm of platinized asbestos, and 15-cm section of a mixture containing nine parts MgO and

TABLE 2.2. Recent Literature on Combustion of Difficult Samples for
Carbon and Hydrogen Analysis.

Nature of sample	Recommended treatment	Reference
General	Cobalto-cobaltic oxide packing at 650-700°C	21, 22, 26, 33, 54, 107-109
	CO_3O_4, CO_2O_3, Ag, Fe_2O_3, CuO	110
	$AgMnO_4$	24
	Au, Ag, platinized asbestos, MgO, silver tungstate, zirconium oxide	30
	Pt gauze, Ag gauze, SiO_2 grains	110a
	$Ag + MnO_2$	111
	$KMnO_4$	112
	$AgMnO_4$ at 500°C, CO_3O_4 at 910°C, $AgMnO_4$ at 490°C successively	113
	$Ag_2WO_4 + CeO_2 + MgO$	114
S, halogens	Silver vanadate	38
F, P, B	$CeO_2 + CuO$	43
Perfluoro	Silver vanadate at 500°C	115, 116
	Mix sample with CuO	48
Sn, halogens	Mix sample with Mg + CuO	117
B	1300°C combustion	118
	Tungstic oxide at 1000°C	119
B, Si polymer	$MnO_2 + Cr_2O_3 + WO_3$	120
Tl	Cover sample with silica	121
Se	Mix sample with CuO + MgO	122
Ge	Cover sample with WO_3	123
Si	CO_3O_4 on pumice	124

TABLE 2.2. (Continued)

Nature of sample	Recommended treatment	Reference
Si, S, metals	Add Aerosil 200 to sample	124a
P, Zn, alkali metals	Cobaltic oxide + tungstic oxide	125
Refractory	Tungstic oxide + silver tungstate + AgMnO$_4$	53
	CO$_3$O$_4$ + Al$_2$O$_3$ + Ag + CeO$_2$	126
Polycyclic, triterpenoids	Cover sample with CO$_3$O$_4$	127
Liquid	Sealed aluminum pan	128
	Glass capillary	129
	Thermoelectric cooling device	130
Hygroscopic	Aluminum capsule	131
	Quartz container	132
Air-sensitive; easily flammable	Special device	133, 134
Explosive	Mix sample with CuO or CO$_3$O$_4$	135

one part Ag$_2$WO$_4$/ZrO$_2$ (1:4), and finally 2 mm of platinized asbestos. Binkowski and Levy [126] have reported that a combustion tube packed with cobalto-cobaltic oxide on aluminum oxide, silver on pumice, and cerium dioxide are effective for the analysis of compounds containing fluorine and phosphorus.

The difficulty of complete combustion of long-chained and condensed-ring compounds is well known. When the Pregl-Lieb procedure is employed (see Sec. 2.6), the sample should be spread out in the microboat and the combustion temperature of the short burner gradually increased. In the case of other procedures, the study of Gawargious [127] on triterpenoids which contain about 30 carbon atoms and fused tetra- and penta-cyclic structures is of interest. Gawargious found it necessary to (1) modify the CO$_3$O$_4$ packing of Vecera et al. [22] by incorporating 1% NiO,

(2) introduce two silica spirals in the empty tube method of Korshun and Klimova [5], and (3) cover the sample with Co_3O_4 powder in the Belcher-Ingram [3] procedure.

Handling of liquid samples is a problem in practically all automatic analyzers. Culmo and Fyans [128] have described a technique that involves the use of an aluminum pan in which the material is hermatically sealed using a sealing press. Pointing out the drawback of aluminum capillary tubes (they are time-consuming, easily fractured, and reactive toward compounds containing active halogens or acidic functions), Kennedy [129] has proposed the use of glass capillaries of 10-mm length (l) and 0.5-mm inside diameter (i.d.), sealed at either one end or both ends. It is worthy of note that in the Pregl-Lieb procedure, the movement of the liquid sample can be conveniently controlled. Trutnovsky [130] has constructed a thermoelectric cooling device to be attached to the portion of the combustion tube where the sample is situated. This device, while useful, blocks the view of the sample.

Correct results of carbon and hydrogen determination of hygroscopic, explosive, and air-sensitive substances depend heavily on the proper technique of handling the sample. Thus Pella [133] has described a procedure in which the sample is shielded from the stream of oxygen by housing it in a quartz capsule that can be shifted from outside. The combustion is initiated by a furnace which is advanced at a constant speed. The reader is referred to the manual on microchemical manipulations [136] for general techniques suitable for handling these types of organic compounds.

Ieki and Daikatsu [137] have designed a sophisticated scheme to study the course of decomposition during carbon and hydrogen analysis by combining the determination with differential thermal analysis. The organic compound is combusted slowly on a thermocouple detector in a stream of air and the carbon and hydrogen values are determined gravimetrically. During the heating process, the thermal characteristics of the sample, such as dehydration, fusion, vaporization, sublimation, explosion, decomposition, and oxidation, are indicated as the exothermic peak or endothermic negative peak in a differential thermogram. The differential thermocouples are installed under the quartz vessel holding the sample and in the two absorption tubes. Anomalous analytical results may be interpreted by referring to data concerning the type of thermal decomposition exhibited by the sample.

C. Removal of Interferences

Table 2.3 summarizes the recent publications on the reagents investigated for the prevention of interferences caused by the presence of heteroelements such as alkali metals, and for the removal of combustion products that

TABLE 2.3. Recent Literature on Removal of Interference in Carbon and Hydrogen Analysis.

Element present	Reagent employed	Reference
N	$MnO(OH)_2$ gel, external	138
	$MnO_2 + PbO_2$, external	139
	$CrO_3 + H_2SO_4$ on Sil-O-Cel, external	113
S, halogens	Ag, MgO, Al_2O_3, CeO_2, etc.	140
Halogens	Cu, external	141
	Ag granules	142
	CeO_2	143
Br	Barium silicate at 750°C	144
Cl	Stronium silicate at 600°C	145
F	NaF + Ag, external	146, 146a
	MgO pellets	147
	Vermiculite $(MgO + Al_2O_3)$	148
	HCl on silica gel	148a
F, P	MgO	64, 149
P	Silica insert	149a
	Ag on pumice	150
	Quartz spoon	150a
Se, halogens	Ag on pumice	55
Metals	Cover sample with silver tungstate	151
Hg	Ag sponge in cool zone	152
Si	MgO	153
Si, F	MgO, asbestos	94
As, Sn, F	Pb_3O_4, external	39

interfere with the measurement of water and carbon dioxide. As mentioned previously (Sec. 2.1E), metallic copper is the reagent of choice in the commercial C, H, N analyzers since it can convert the otherwise interfering oxides of nitrogen into a measurable entity.

Padowetz [140] has designed an apparatus to perform comparative studies on reagents useful for the removal of sulfur and halogens in carbon and hydrogen analysis. The diagram of the apparatus is shown in Fig. 2.1. The reagent under investigation is packed for a length of 20 mm in the otherwise empty combustion tube which is electrically heated. The gas flowing through the tube is monitored by a flowmeter. The heteroelement is added to the gas current in exactly defined portions. Sulfur is introduced as sulfur dioxide which is injected into the system by means of a syringe. Halogens are used in the form of aqueous solutions of hydrogen halides. Measured amounts of the solution are transferred into a platinum microboat which is then vaporized by the movable furnace. The control tube is connected to the end of the combustion tube as shown. It is filled with glass beads which are wetted with dilute hydrogen peroxide solution. The unabsorbed portion of sulfur dioxide or hydrogen halide is retained in

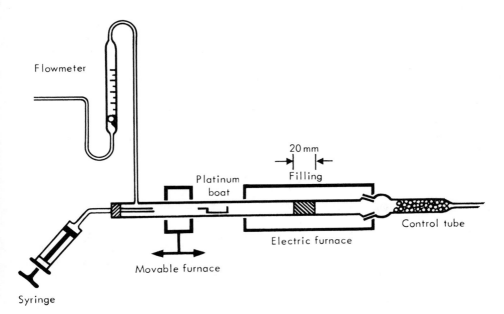

FIG. 2.1. Apparatus of Padowetz for comparative studies on the removal of halogens and sulfur in carbon and hydrogen analysis. (From Ref. 140, courtesy Microchem. J.)

the control tube and determined by acidimetric titration. It has been found that (1) silver vanadate and silver with aluminum oxide have a higher absorption capacity than silver alone for sulfur dioxide, (2) magnesium aluminum oxide $(3Mg \cdot Al_2O_3)$ is considerably more effective than magnesium oxide or silver tungstate on magnesium oxide for removing hydrogen fluoride, and (3) silver with aluminum oxide is more efficient than silver manganate, vanadate, or tungstate, or barium silicate for the retention of hydrogen chloride.

Binkowski and co-workers [140a] have studied the formation of P oxides from phosphorus compounds during combustion. For the complete absorption of P_2O_5, it should be done in two stages.

D. Methods of Finish

Judging from the number of papers listed in Table 2.1, many microanalysts presently use gravimetric finish for the determination of water and carbon dioxide. The future tendency, however, appears to be leaning toward nongravimetric procedures. Salzer [154] has reviewed the physicochemical measuring methods used in automated apparatus for carbon and hydrogen determination. Gas chromatography has been extensively utilized [154a-c]. Davies [155] has reviewed the nongravimetric methods for determining carbon dioxide. The methods can be enumerated as follows: (1) thermal conductivity, (2) conductimetry, (3) coulometry, (4) nonaqueous titrimetry, (5) potentiometric titrimetry, (6) colorimetric titrimetry, (7) colorimetric spectrophotometry, (8) manometry, (9) infrared spectrometry, (10) mass spectrometry, and (11) photoactivation analysis.

Nakamura et al. [156] have proposed a method in which both carbon and hydrogen are measured in the form of water with coulometric finish. The organic compound is mixed with Co_3O_4 and combusted in a nitrogen atmosphere. After reduction of the nitrogen oxides with copper, water is absorbed in silica gel and carbon dioxide is conducted through lithium hydroxide at 215°C where it is converted to water:

$$CO_2 + 2LiOH = H_2O + Li_2CO_3$$

The water passes through a Pt/P_2O_5 electrolytic cell. The current is integrated in millivolts per second, and the weight of hydrogen is related to the peak height. After the determination of carbon dioxide, the water in silica gel is released by heating at 250°C and measured in the same electrolytic cell. A similar method has been described by Karlsson [156a].

In contrast to the above, Campiglio [76] has proposed a method in which both elements are measured as carbon dioxide. After combustion of the sample, water is initially absorbed on $CaCl_2$, while carbon dioxide is

absorbed in dimethylformamide in the presence of monoethanolamine and determined by titration with 0.02 $\underline{N}$ tetra-n-butylammonium hydroxide. Then
water is converted to carbon dioxide by reaction with N,N'-carbonyldimid-
azole,

and the carbon dioxide thus formed is titrated by using the same standardized titrant. Another method for the conversion of water vapor into carbon
dioxide has been reported by Roemer et al. [76a].

E. Modification and Evaluation of Automated Apparatus

While automated apparatus for the determination of carbon and hydrogen
have passed their developmental stage, publications have continued to appear suggesting modifications and improvements. Sels and Demoen [157]
have found that erroneous results for carbon in the Coleman apparatus are
caused by the difference between the true and apparent weights of the carbon
dioxide absorption tube which is related to the amount of carbon dioxide
evolved during analysis. The proposed remedy comprises (1) adjusting the
sample size to generate a constant amount of carbon dioxide, (2) adding
self-sealing valves to the absorption tube and weighing it exactly 3 min
after the end of the combustion cycle, and (3) modifying the absorption
train in order to connect it directly to the combustion tube. Subsequently
the same workers [52] designed special absorption tubes with programmed
weighing, and recommended the use of helium as the sweep gas. Binkowski
put forward an automated procedure using gravimetry [157a], while
Rousseau [158] employed coulometry. Culmo [158a] has described improvements for the Perkin-Elmer apparatus which include the filling of combustion, reduction, and absorption tubes; sample carrier design; and handling
techniques. Binkowski and Levy [126] have modified the Technicon apparatus by raising the temperature to 1050-1100°C for the combustion tube
which is packed with Co_3O_4 on Al_2O_3, Ag on pumice, and CeO_2. In view
of the preference of thermal conductivity detectors in the automated apparatus, Hozumi and co-workers [159-161] have studied the method for
counterbalancing, and the effects of temperature and flow rate, and they
have elucidated the analytical errors statistically.
　　There were more than 10 commercial models of automated apparatus
available in 1972. The manufacturers usually claim that their instruments
are suitable for the determination of carbon and hydrogen in all types of
organic samples. The senior author has visited analytical laboratories in

several countries where these instruments were installed or under test. The general response has been favorable. Most microanalysts confided, however, that slight changes and adjustments had to be made in order to obtain correct results. It should be noted that these instruments were designed by engineers who might not be familiar with the stringent requirements of quantitative carbon and hydrogen analysis. Furthermore, unlike the Pregl-Lieb procedure in which the analyst assembles the apparatus (thus adjusts it properly) and can check it step by step, the commercial automatic analyzers are delivered assembled. A new machine that gives good answers with benzoic acid does not guarantee correct values for unknown organic compounds.

Only a few reports have been published concerning the evaluation of the automatic analyzers. Childs and Henner [56] have made a direct comparison of the Perkin-Elmer, Hewlett-Packard, and the classical procedures for a period of 6 months with eight analysts. A total of 19 compounds of varied characteristics and compositions were analyzed, but no liquid sample was included. The study indicates that the Pregl-Lieb procedure, although slower, gives the most reliable results. Acceptable results, however, can be achieved by the two automated apparatus. According to Gel'man [94], her automated C,H apparatus which is based on "wide tube combustion" with infrared determination of carbon dioxide and coulometric determination of water, has been tested on more than 3000 samples, including polymers and fluorine- or phosphorus-containing compounds. This apparatus is modified from a previous design [93] in which water and carbon dioxide were measured by optical-acoustic analyzers. A report on the assessment of the Technicon apparatus has been published by its manufacturer [162]. It includes the reports of five experts who tested this instrument during the period from 1964 to 1966. In 1969 Manser [162a] used the Technicon apparatus to analyze 250 different samples containing heteroelements N, O, S, Cl, I, F, P, Cr, Co, and Sn, respectively, during a period of 14 days, averaging 4-5 determinations per hour, and obtained values of C within $\pm 0.2\%$ and values of H within $\pm 0.1\%$. In our opinion, the Technicon device allows more flexibility than the other commercial instruments, though it is not so compact. Probably for economic reasons, however, the manufacturer has not made further developments on this automated apparatus.

At the International Symposium on Microchemical Techniques held in University Park, Pennsylvania, August 1973, the performances of three commercial C,H,N analyzers (Hewlett-Packard, Perkin-Elmer, and Carlo Erba) were reported by microanalysts who had used the respective instruments for several years. The following observations made by Fildes, while referring to a particular instrument, apply generally to all types of automated machines which employ rapid combustion and calibration techniques. (1) It is essential to use matching standards to obtain the calibration constants. (2) It is difficult to handle volatile, very low-melting, or readily sublimable compounds. (3) The instrument is not satisfactory for the analysis of materials which can easily loose water or solvent of crystallization during the equilibration stage of each run. (4) Difficulties are encountered with samples containing more than 90% carbon or more than 9% hydrogen.

Generally speaking, volatile liquids, unstable solids, and compounds with high carbon or hydrogen contents are preferably analyzed by means of the Pregl-type apparatus. It is advisable, therefore, to keep the manual combustion train as a standby when the microanalytical laboratory is expected to receive a wide range of research samples to be analyzed. On the other hand, the automated apparatus becomes indispensable when it is desired to computerize organic elemental analysis (see Sec. 1.3B).

2.3 EXPERIMENTAL: DETERMINATION OF CARBON, HYDROGEN, AND NITROGEN USING THE PERKIN-ELMER MODEL 240 ELEMENTAL ANALYZER

A. Principle

The organic sample is combusted in pure oxygen under static conditions in a Pregl-type system. Interfering combustion products such as halogens and sulfur and phosphorus oxides are retained in the combustion tube by appropriate reagents. The resultant CO_2, H_2O, N_2, and oxides of nitrogen are swept into a Dumas-type reduction tube by a stream of helium where the oxides of nitrogen are reduced to N_2 and the excess O_2 is removed. The final measurement of C (as CO_2), H (as H_2O), and N (as N_2) present in the sample is accomplished by three pairs of thermal conductivity cells connected in series. A trap between the first pair of cells absorbs any water from the gas mixture before it enters the second cell so that the signal is proportional to the amount of water removed. Likewise, another trap between the second pair of cells removes CO_2 and the signal is proportional to the CO_2 removed. The last pair analyzes N_2 by comparing the remaining sample gas plus He with pure helium. The instrument is calibrated with a pure material containing C, H, and N before samples are run.

B. Apparatus

1. Perkin-Elmer Model 240 Elemental Analyzer (Perkin-Elmer Corp., Norwalk, Connecticut).
2. Microbalance - balance capable of weighing to the nearest 1 μg.
3. Perkin-Elmer Volatile Sample Sealer Accessory.

4. Platinum boats (for solids).
5. Recorder (0-1 mV).
6. Reaction tubes:

   For combustion: 11 mm o.d. x 9 mm i.d. x 25 in. ℓ, quartz
     (12 in. active length).
   For reduction: 11 mm o.d. x 9 mm i.d. x 14-1/4 in. ℓ, Vycor
     (12 in. active length).

7. Scrubbers and traps: 11 mm o.d. x 9 mm i.d. x 8 in. ℓ, Pyrex
   (7-1/2 in. active length).

C. Reagents

In combustion tube:

   Silver tungstate on magnesium oxide.
   Silver oxide + silver tungstate on Chromosorb P 60/80 mesh.
   Silver vanadate.
   Silver gauze, platinum gauze.
   Quartz wool.

In reduction tube:

   Copper wire, 60/80 mesh.
   Quartz wool.
   Silver gauze.

In water trap:

   Magnesium perchlorate.

In CO_2 trap (serves also as the He, O_2 scrubber):

   Colorcarb (NaCa hydrate + Ascarite).
   Magnesium perchlorate.

Other:

   Helium, best grade.
   Oxygen, best grade.

D. Procedure

Preliminary Adjustments of the Apparatus

Fill the combustion and reduction tubes, traps, and scrubbers as described in Fig. 2.2C. Set the combustion furnace temperature at 950°C and the reduction furnace at 650-700°C. Adjust the helium pressure to 18-18.5 lb and the oxygen pressure to 25.5 lb.

Before actually starting a run, check to see that the furnaces are up to their proper temperatures and that the helium and oxygen regulators are at the proper settings. Turn the recorder on. Flush the detectors with helium by manually setting the programmer index wheel (PIW) to position no. 9. After a few minutes, flush the entire system by manually setting the PIW to position no. 1. After 5 min switch to position no. 0 and put the main power switch on "detect." In this position all power is applied. The instrument is now ready for operation.

To condition the instrument, place an unweighed sample (approximately 2 mg) of acetanilide into a Pt boat and insert into the open end of the ladle. Remove the sample fitting plug and slide the ladle into the combustion tube. Replug the combustion tube using the plug to push the ladle further in. Push the "start" button. This starts the program.

When the inject bulb lights, introduce the sample into the combustion zone with the aid of a magnet. Press the "combust" button. The analysis is fully automatic from then on.

The first signals that come out on the recorder chart are respectively the nitrogen, carbon, and hydrogen "zero" readings. The course and fine controls have previously been set so that these zero values will be on scale. After the zero readings have come out, remove the ladle from the hot zone so that it will be cool for the next run. The second set of signals coming out will be the nitrogen, carbon, and hydrogen "reads." Adjust the attenuations so that they are on scale with good measurable deflections.

Analysis

The procedure just described is identical to that used for weighed samples, determining boat and ladle blanks, and for the determination of the sensitivity factors. After the instrument has been conditioned, the empty boat and ladle are run through the entire program as described above in order to obtain the boat and ladle blanks. Two successive blank runs giving good reproducibility are sufficient. The sensitivity factors, i.e., the number of microvolts equivalent to 1 μg of nitrogen, carbon, and hydrogen, respectively, are then determined by weighing accurately approximately 2 mg of pure standard (e.g., NBS 141 acetanilidé) and putting it through the previously described cycle. Two successive runs of a pure standard giving good reproducibility are sufficient for determination of the sensitivity factors. With the boat and ladle blanks and the sensitivity factors established, the instrument is now ready for the routine determination of samples.

E. Calculation

To calculate the percentage of N, C, and H in an organic sample using the Model 240 Elemental Analyzer, the sensitivity factors (i.e., the microvolt

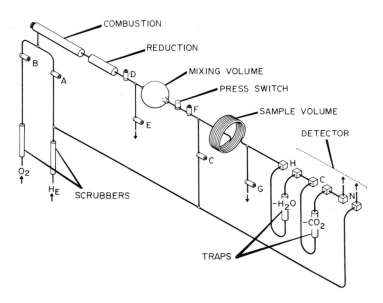

FIG. 2.2A. Simplified diagram of combustion train and analytical system of Perkin-Elmer Model 240 Elemental Analyzer. (Courtesy Perkin-Elmer Corp.)

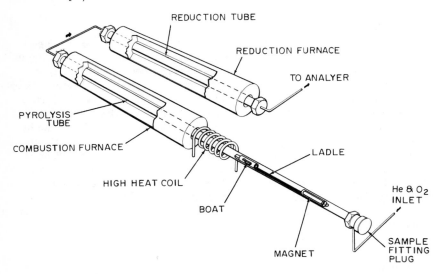

FIG. 2.2B. Combustion train of Perkin-Elmer Model 240 Elemental Analyzer. (Courtesy Perkin-Elmer Corp.)

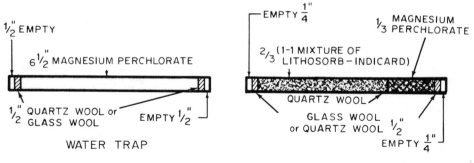

WATER TRAP

CO_2 TRAP; He,O_2 SCRUBBER

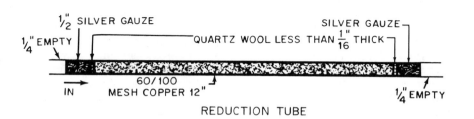

REDUCTION TUBE

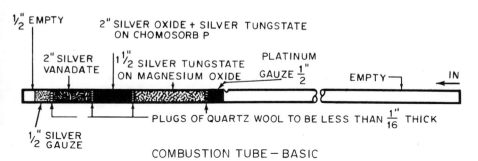

COMBUSTION TUBE — BASIC

FIG. 2.2C. Recommended makeup for combustion and reduction tubes, traps, and scrubbers. (Courtesy Perkin-Elmer Corp.)

FIG. 2.2D. Computerized Perkin-Elmer Elemental Analyzer. (Courtesy Perkin-Elmer Corp.)

equivalent to 1 μg of N, C, and H) must first be determined. These values are obtained by running a weighed sample of a pure standard through the apparatus in the same manner as an unknown sample would be run. The total signal in microvolts (less the zero readings and ladle and boat blanks) divided by the weight in micrograms of each element (N, C, H) gives the respective sensitivity factors.

To obtain the total signal for N and H from the recorder chart which has a 0-1000 μV range and is numbered from 0 to 100, multiply the reading by 10 and then multiply by the attenuation. For example, if the N attenuation is 4 and recorder chart reads 35 then the total signal for N is 35 x 10 x 4 = 1400 μV.

The hydrogen signal is obtained in the same manner.

For C, obtaining the signal from the recorder is slightly different, because the C signal is not attenuated but suppressed. Each position suppresses 10 000 μV x number. For example, position 2 is 20 000 μV. The recorder reading is therefore multiplied by 100 and added to the attenuation position x 10 000. For example, if the C position is set at 3 and the recorder reads 65 then the total C signal is 65 x 100 + 30 000 = 36 500 μV.

The following table of data illustrates how the sensitivity factors for each element (N, C, H) are determined. A weighed sample of pure acetanilide (2.055 mg) is run through the cycle and the following data obtained. The ladle and boat blanks have previously been determined by running them through the cycle several times without any sample. These values are constant for a particular instrument.

		N	C	H
A	attenuation	4	3(30 000)	16
R	reads	475	4300	645
Z	zeros	265	2120	282
LB	ladle and boat blank	100	50	100
C	A x R for N and H			
	A + R for C	1900	34 300	10320
B	total blank (Z + LB)	365	2170	382

From the above data the net signals are as follows:

For N,

$$4 \times 475 = 1900$$

$$1900 - (265 + 100) = 1535 \ \mu V$$

For C,

$$30\,000 + 4300 = 34\,300$$

$$34\,000 - (2120 + 50) = 32\,130 \; \mu V$$

For H,

$$16 \times 645 = 10\,320$$

$$10\,320 - (282 + 100) = 9938 \; \mu V$$

The weight of N, C, and H in the standard is obtained by multiplying the fraction of N, C, and H in acetanilide by the sample weight.

Example:

Weight N = 2.055 x 0.1036 = 212.90 μg
Weight C = 2.055 x 0.7109 = 1460.9 μg
Weight H = 2.055 x 0.0671 = 137.89 μg

The sensitivity factors for each element are then determined by dividing the net signal by the respective element weight.

For example:

Sensitivity factor for N is $\dfrac{1535 \; \mu V}{212.90 \; \mu g}$ = 7.21 $\mu V/\mu g$

Sensitivity factor for C is $\dfrac{32\,130 \; \mu V}{1460.9 \; \mu g}$ = 21.99 $\mu V/\mu g$

Sensitivity factor for H is $\dfrac{9938 \; \mu V}{137.89 \; \mu g}$ = 72.07 $\mu V/\mu g$

The sensitivity factor for N, for example, says that each 7.21 μV of signal represents 1 μg of N.

Final Calculation

$$\%N = \frac{\text{Net signal } (\mu V) \times 100}{\text{Sensitivity factor } (7.21) \times \text{sample weight } (\mu g)}$$

$$\%C = \frac{\text{Net signal } (\mu V) \times 100}{\text{Sensitivity factor } (21.99) \times \text{sample weight } (\mu g)}$$

$$\%H = \frac{\text{Net signal } (\mu V) \times 100}{\text{Sensitivity factor } (72.07) \times \text{sample weight } (\mu g)}$$

F.   Comments

(1) Volatile liquid samples are conveniently handled by use of a pan and cover kit (P.E. No. 219-0062) made of aluminum. The clean pan and cover

are accurately weighed.   The pan is placed in the Volatile Sample Sealer
Accessory (P. E. No. 219-0061).   Approximately 2 mg of liquid sample is
added and the cover is sealed to the pan.   The sealed sample is then reweighed.   It is advisable to seal the pan in an atmosphere of helium or
oxygen in order to reduce the nitrogen blank to a minimum and keep it
reproducible.

(2) The sensitivity factors remain remarkably constant during the life
of a particular pair of combustion and reduction tubes.   When a difference
in the normal carbon or nitrogen sensitivity factor is observed it usually
indicates that either the combustion or reduction tube is depleted.   Both
should be replaced.   The condition of the water and CO_2 traps can be as-
certained by inspection.

(3) Special procedures are required for the determination of N in inorganic nitrates ($NaNO_3$, KNO_3), of C in certain boron polymers, and for
trace determinations of H or C.

(4) Computerization of the Perkin-Elmer Elemental Analyzer has been
reported by several groups of investigators [163], and the system is commercially available [163a].

(5) For routine samples we have found that the Perkin-Elmer instrument is more accurate than the Pregl apparatus.

(6) Alicino [164] has made the apparatus more automatic by the instal-
lation of an Infotronic CRS-30 and a novel sample-entry device.   He also
has modified the composition of the combustion-tube filling.   Slightly less
than the recommended amounts of tube filling are employed in order to
accommodate a longer length (30-40 mm) of platinum gauze of a slightly
heavier mesh than that recommended in the Perkin-Elmer manual.

(7) The Japanese instrument [165] Yanagimoto Model MT-2 (Fig. 2.3)
uses differential thermal conductometry, while Hitachi Model 026 uses
gas chromatography even though it has the external appearance of the
Perkin-Elmer Model 240 Elemental Analyzer.

FIG. 2.3.   Yanaco C, H, N Analyzer.   (Courtesy Yanagimoto Co.)

2.4 EXPERIMENTAL: DETERMINATION OF CARBON, HYDROGEN AND NITROGEN USING THE HEWLETT-PACKARD (F & M) C,H,N ANALYZER

This section was written by Leonard C. Klein [166] of FMC Corporation, Princeton, New Jersey, who had used the following method for more than 5 years. The Hewlett-Packard F & M) instrument is modified as indicated under Sec. 2.4B.

A. Principle

Organic compounds are combusted at 1000°C in the presence of an oxygen donor. Carbon, hydrogen, and nitrogen are converted to carbon dioxide, water, and nitrogen oxides, respectively. The oxides of nitrogen are then reduced to nitrogen gas. The gases are separated by chromatography and measured by thermal conductivity. The conductivity signals are integrated and the data related to percent carbon, hydrogen, and nitrogen.

B. Apparatus

  Hewlett-Packard (F & M) Carbon, Hydrogen, Nitrogen Analyzer Model
     185 (see Fig. 2.4A) (Hewlett-Packard, Avondale, Pennsylvania).
  Infotronics Integration Model CRS-108 with Victor Printer.
  Ratio Electrobalance - Cahn Cat. 1605.

Modifications and Instrument Conditions

  Analyzer

       1/4 in. x 10 ft Porapak Q column, 100-200 mesh.
       Oxidation furnace temperature 1000-1100°C.
       Reduction furnace temperature 500-550°C.
       Column oven temperature 90-110°C.
       Shell oven temperature 10°C below column temperature.
       Helium flow rate approximately 100 ml/min.

  Integrator and Balance

     The weight dial on the balance is tied to the gain control of the operational amplifier in the integrator (see Fig. 2.4B). This modification eliminates the sample weight in the calculation.

  Integrator Settings

       Tracking rate up: 60 μV/min.
       Tracking rate down: 200 μV/min.

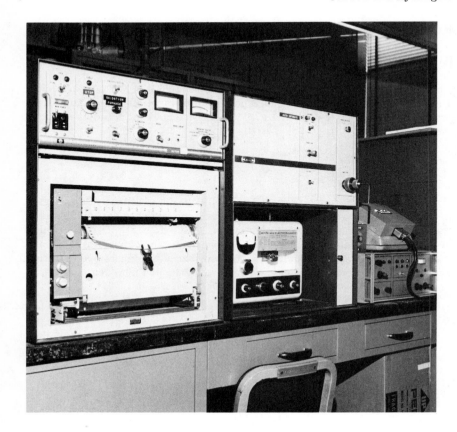

FIG. 2.4A. Hewlett-Packard C,H,N Analyzer and Cahn electrobalance. (Courtesy Leonard C. Klein.)

Input noise rejection: 7-max.
Peak rate: minimum 10 s per peak.
Slope sensitivity: 2.
Control set on readout.

C. Chemicals

In oxidation furnace - copper oxide.
In reduction furnace - cuprin (copper).
Oxidation catalyst - 2 parts MnO_2, 1 part $CrO_2 \cdot SiO_2$, . Prepared
by processing MnO_2, CrO_3, and Celite 454 via a prescribed process [167].

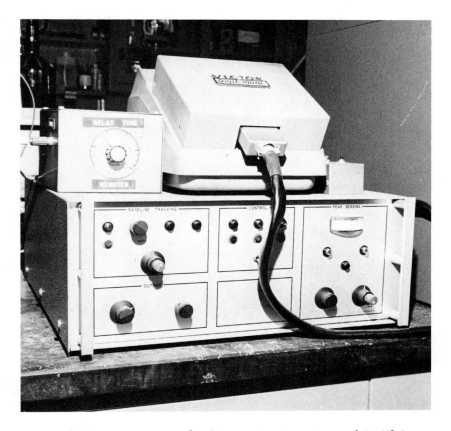

FIG. 2.4B. Integrator and printer. (Courtesy Leonard C. Klein.)

D. Procedure

Analysis

Place a sample boat on the "A" pan of the balance and a tare boat on the "C" pan. Set the balance on the "1" mg range and zero the balance. Put sufficient sample in the boat to obtain a 0.5-0.8 mg weight and complete the weighing process. (Rebalance the beam and weight dial.) Transfer the sample (boat) to the injection rod and cover with catalyst. Insert the injection rod into the instrument, positioning the sample in the cool end of the combustion tube. Allow the instrument to reequilibrate (4-6 min) and inject the sample by first activating the furnace bypass and integrator delay and then cracking the injection part long enough to inject the sample.

The nitrogen, carbon dioxide, and water peaks are automatically integrated and the areas printed out. When the last peak has been recorded the integrator is reset. Record the three peak areas.

Calibration

Determine the calibration constants (K_E) by measuring the areas produced by known compounds such as acetanilide and calculating the constants using formula I in Sec. E below.

Using the constants, calculate the C, H, and N content of unknowns using formula II.

E. Calculations

$$\text{I.} \quad K_E = \frac{\%E \text{ in known compound}}{\text{Peak area}}$$

II. %E of unknown = Peak area of element x K_E

F. Comments

(1) Blanks are usually unnecessary unless the hydrogen content of the sample is below 3%. If the catalyst is of good quality, the lower limit could be 1-2%.

(2) The constants should be obtained by running known compounds whose C, H, and N content are significantly different, e. g., acetanilide and phenyl thiourea.

(3) It is advisable to run an unweighed sample with both ladles at the beginning of the day.

(4) It may be necessary to run known compounds whose elemental content is similar to the unknown if the carbon or nitrogen is unusually high, i.e., N > 25-30%, C > 75-80%.

(5) A disadvantage of this ratio system is that the weight dial on the balance may not be changed until the hydrogen peak has been integrated. Since only the weight dial is connected to the integrator, the next sample can go through the weighing process with the weight dial step completed just after the hydrogen peak is integrated.

The advantages of this system are that the weight is not read and good duplicates are readily discernible by observing the closeness of the corresponding peak areas.

(6) According to Inouye [167a], analysis of difficult-to-combust fluoro compounds is accomplished by utilizing potassium perchlorate as an auxiliary oxidizing agent.

2.5 EXPERIMENTAL: DETERMINATION OF CARBON, HYDROGEN, AND NITROGEN USING THE CARLO ERBA ELEMENTAL ANALYZER

This section is based on the report of Dr. E. Pella and B. Colombo [168], Istituto Carlo Erba per Ricercha Teropeutiche, Milan, Italy.

A. Principle

The organic samples are packed into light-weight containers of tin, and dropped at preset times into a vertical quartz tube, heated to 1050°C, through which a constant flow of helium is maintained. When the samples are introduced, the helium stream is temporarily mixed with pure oxygen. Flash combustion takes place, primed by the oxidation of the container. Quantitative combustion is achieved by passing the gases over Cr_2O_3.

The mixture of the combustion gases is transferred over copper at 640°C to eliminate excess of oxygen; then, without stopping, it is introduced into the chromatographic column, at 130°C and filled with a vinyl-ethylbenzene-divinylbenzene polymer as stationary phase. The individual components are separated, eluted in the order N_2, CO_2, H_2O, and measured by a thermal conductivity detector. The response signal feeds a potentiometric recorder and in parallel an integrator with digital printout.

B. Apparatus

Analyzer

Carlo Erba Elemental Analyzer Model 1104 (Carlo Erba, Milan, Italy). The complete apparatus is shown in Fig. 2.5; the schematic diagram is given in Fig. 2.6. This instrument automatically determines carbon, hydrogen, and nitrogen. Combustion of the sample, separation of the combustion gases, and measurement by thermal conductivity are all carried out in the dynamic mode. The reference side is used in the instrument as part of another permanent analytical circuit for oxygen determination. There are thus two analytical circuits, one for C,H,N determination, and the other for oxygen. They do not work simultaneously, but alternately as analytical and reference circuits. The oxygen circuit is considered here only as a reference circuit.

The C,H,N circuit consists of two gas sources, for helium and oxygen, an oxygen dispenser valve for periodic injection, an automatic sampler, an independently heated **vertical** oxidation reactor, a vertical U-shaped reduction reactor, a gas chromatographic column and a detector, both housed in a thermostatically controlled oven, and a recording-measuring system.

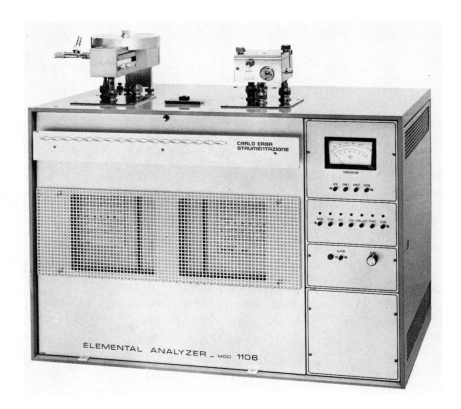

FIG. 2.5. Carlo Erba Elemental Analyzer. (Courtesy E. M. Becker Co.)

Combustion Train

The combustion train (Fig. 2.7) comprises the oxidation reactor and the reduction reactor. The oxidation reactor consists of a quartz tube, the size and packing of which are shown in Fig. 2.7a. The tube is arranged upright so that the samples may be introduced from the top. The vertical position makes the packing more stable and the tube may be easily removed from above. A quartz crucible is placed in the combustion zone to protect the tube from damage during flash combustion (Fig. 2.7a) and to collect the spent containers.

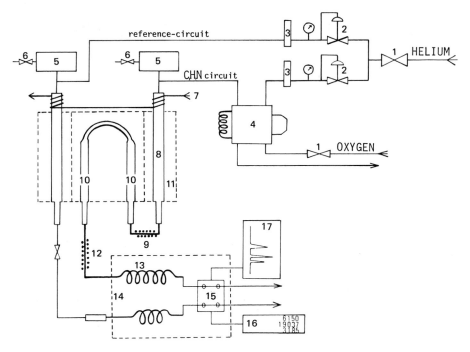

FIG. 2.6.   Scheme of the Carlo Erba apparatus.   1, pressure regulator; 2, flow controller; 3, purification tubes; 4, oxygen injection valve; 5, sample dispenser; 6, sweeping valve; 7, water cooler; 8, oxidation reactor; 9, heated connecting tubing; 10, reduction reactor; 11, oven for oxidation reactor; 12, heated tubing; 13, C,H,N chromatographic column; 14, thermostatically controlled oven; 15, thermal conductivity detector; 16, digital integrator-printer; 17, recorder. (Courtesy Mikrochim. Acta.)

The lower part of the tube is tapered in order to avoid dead volume after the packing. The oxidation and reduction reactors are linked, as shown in Fig. 2.6, by a silver tube (4 x 2 mm), heated to 120°C and connected to the circuit by Viton O-rings and metal gasket connections.

The reduction reactor consists of a quartz U-tube (Fig. 2.7b) standing upright, and heated to 640°C. It is loaded from both ends, and packed by means of a vibrator. It is connected to the chromatographic column via a stainless steel tube (2 x 1 mm), heated to 120°C throughout its length.

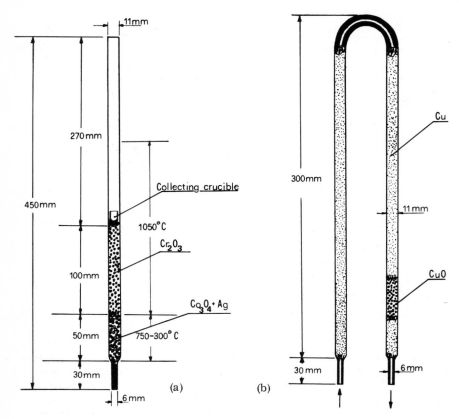

FIG. 2.7. The Carlo Erba combustion train. (a) Oxidation reactor.
(b) Reduction reactor. (Courtesy Mikrochim. Acta.)

Chromatographic Column

A stainless steel column is used, diameter 6 x 5 mm, length 33 m,
filled and tightly packed with Porapak QS (80-100 mesh) pretreated according to the manufacturer's instructions. After packing, the column is
coiled, connected to the two parts of the circuit and fixed into the heating
block. It is then heated to a preselected constant temperature at about
130°C. The filling causes a 0.5 kg/cm^2 pressure drop, with a flow rate
of 25 ml/min.

Detection and Measuring System

The detector used is a conventional catharometer with four WX rhenium-tungsten 9225 filaments as sensing elements (Gow-Mac Instruments Co., Madison, New Jersey), assembled in a Wheatstone bridge with a voltage source stabilized at 20 V; the bridge current is 120 mA. It is placed inside a cylindrical aluminum block, together with which the chromatography columns are wound. After 60 min operation, a constant temperature of $130 \pm 0.05°C$ and a bridge supply voltage constant to ± 0.05 mV are reached.

The detector is equipped with an automatic sensitivity switch, which after the N_2 signal has been registered lowers the sensitivity by a factor of 4 for the CO_2 signal and then returns it to the initial value.

The electric signal from the detector feeds the potentiometric recorder and an electronic digital integrator with data printout, in parallel with the recorder. The recorder is used essentially for controlling the combustion process and to permit the time control of the successive operations. The electronic integrator is connected directly to the detector bridge without automatic attenuation. The integrator is a simplified voltage-to-frequency converter, in view of the analytical system used. It contains no drift correctors or slope detectors, and no valley sensors, since the baseline is steady and drift-free. There is, however, an automatic zero device. As the integrator receives the signal directly from the detector, errors caused by the recorder, such as pen delay, variations in chart speed, and excessive peaks, are avoided. On attenuation 8, the integrator speed is 12 000 pulses/min full scale. Total accuracy is to within 0.1%.

C. Reagents

Helium R for G.C., purity 99.998% (SIAD, Bergamo, Italy).
Oxygen, purity 99.999% (SIAD, Bergamo, Italy).
Tin foil for containers, thickness 0.01 mm (Carlo Erba, Milan).
Cr_2O_3, 25-60 mesh (Carlo Erba, Milan): For its preparation, 200 g
   of $Cr(NO_3)_3 \cdot 9H_2O$, analytical grade, are dissolved in 1500 ml of
   distilled water. The stirred solution is treated slowly with 150 ml
   of 32% ammonia solution. After precipitation of the hydroxide, the
   solution is boiled for about 15 min. The precipitate is filtered off
   on a Gooch crucible and washed several times with distilled water.
   It is then dried for 12 h in an oven at 120°C, and heated at 1100°C
   in an oxygen stream. The dry pulverized reagent is then compressed
   under 1500 kg/cm^2 pressure in a tableting machine, broken up,
   and granulated to 0.7-0.25 mm.

Co_3O_4/Ag, 25-60 mesh: Prepared according to Horacek and Körbl [169]. Before use, it is heated in an oxygen stream for 2 h at 750°C.

Cu, containing Ag, 25-45 mesh: Commercially available CuO is ground and sieved to 0.4-0.7 mm, then 100 g are treated with saturated silver nitrate solution (7.5 ml). The mixture is heated, with stirring, in an evaporating dish until the nitrate completely decomposes. The reagent obtained is reduced in a stream of carbon monoxide at 300°C and then heated for 2 h at 750-800°C.

Porapak QS 80-100 mesh (Waters Assoc. Inc., Framingham, Massachusetts): Before use the product is purged for 5 h at 230°C in a stream of helium.

D. Procedure

Preliminary Operations

The following preliminary operations should be carried out before each analytical cycle: (1) the detector is switched on 1 h before starting the analysis; (2) the recorder is switched on (the integrator remains on all the time); (3) the recorder pen is zeroed and the baseline checked at attenuation 1; (4) the attenuation switch is set on 8; (5) the integrator signal is zeroed to the baseline; (6) the manual control is switched over to automatic. At this point the analytical cycle may be started.

Routine Analysis

For routine analysis a sample of between 0.5 and 1.0 mg is weighed. In this case the 5 ml oxygen dosing loop is used, and the attenuation values are 8, 32, and 8. An electronic microbalance is used; larger samples may be weighed on a normal microbalance.

Solid samples are weighed into cylindrical containers (height 6 mm, diameter 6 mm), made of tin foil (6 mg). The containers are washed with solvent and dried before use. Tin containers are also suitable for samples containing alkali metals, boron, phosphorus, or any substance that does not burn easily. When the substances have been weighed, the containers are closed with tweezers and folded over, then placed in the numbered holes of the sampler drum. Tin containers can be mechanically sealed with pliers. This procedure can be applied to volatile solid samples. Silver capillaries (height 6 mm, diameter 2 x 1 mm) are used for sampling volatile liquids. The liquid is introduced by means of a microsyringe; then the capillaries are mechanically sealed and weighed. If the weight is constant, perfect sealing has been obtained. (Ellison [168a] has described a crimping device for solids and a sealing device for liquids.)

One analytical cycle consists of 23 individual combustions, two or three of which will be of standard compounds. The drum is placed in the sampler, the cover is screwed down, and the sampler body is purged. After 2 min the sweeping valve is closed, and the carrier gas enters the analytical train. At the circuit inlet, the positive pressure of about 0.8 kg/cm^2 is restored. The carrier-gas flow-rate (25 ml/min) at the outlet of the C,H,N and reference channels is checked.

The following general steps should then be taken, before starting up the analytical cycle: (1) the programmer is set for 7 min analysis time; (2) the "sample in delay" timer is set at about 15 s; (3) the integrator autozero is set at about 30 s before the N_2 peak; (4) the auxiliary timers are set for automatic sensitivity switching before and after the CO_2 peak. These times also apply for printing out the integral counts for N_2 and CO_2. The printout of the H_2O peak takes place about 1 min after the end of analysis.

E. Calculations

The printed counts (I_x) are multiplied by the calibration factors (f_x) and divided by the sample weight (E) in μg; then %x is given by

$$\%x = \frac{I_x f_x}{E}$$

The calibration factors are determined by combustion of standard substances and represent the amount of element in micrograms corresponding to 100 counts. Correction for blanks is needed only in working out the hydrogen content. A typical blank figure is 20 counts, equivalent to 1.4 μg of H_2O.

2.6 EXPERIMENTAL: DETERMINATION OF CARBON AND HYDROGEN BY GRAVIMETRY

A. Principle

The organic material is oxidized in a current of oxygen in a combustion tube containing copper oxide at about 800°C, whereupon carbon is converted quantitatively to carbon dioxide, and hydrogen to water vapor. The respective products are then retained in suitable absorbents and weighed. The method described below is very close to the procedure developed by Pregl and Lieb (see Sec. 2.1B).

B. Apparatus

A commercial assembly for gravimetric determination of carbon and hydrogen is illustrated in Fig. 2.8. The schematic diagram of the combustion train is shown in Fig. 2.9. It should be noted that some parts of the setup may be modified or omitted with advantage. For example, when pure oxygen is available, the oxygen cylinder A can be directly connected to the constant-pressure device B via a simple reduction valve (gas regulator); the rubber tubing and clamp is replaced by a glass stopcock; and the oxygen purifier C also becomes superfluous.

C. Reagents

Oxygen, USP medicinal grade.
Copper oxide, wireform.
Silver thread or silver wool.
Asbestos (long fibers), or quartz wool.
Manganese dioxide pellets or lead peroxide, specially prepared for retention of nitrogen oxides.
Anhydrone, for H_2O absorption.
Ascarite, for CO_2 absorption.
Kronig cement.

D. Procedure

Packing the Combustion Tube

A combustion tube suitable for the analysis of organic compounds containing nitrogen, chlorine, bromine, iodine, and sulfur is shown in Fig. 2.10; it should be packed as follows. Insert a silver wire about 10 mm longer than the tip through the capillary tip. Push a wad of silver thread toward the tip of the combustion tube from the open end. Then pad asbestos fibers to a length of about 5 mm. (This serves as a choking plug to control the speed of the gas stream.) Next introduce a section of lead peroxide about 50 mm long (this section should be heated at $180 \pm 10°C$ during combustion). Add a little asbestos as partition, followed by the roll of silver thread, 50 mm long, and another asbestos partition. Now introduce the copper oxide, 180 mm long, asbestos partition, and finally the roll of silver. Close the combustion tube with a cork stopper which fits snugly and squarely at the opening (see Sec. 2.6F, Note 1).

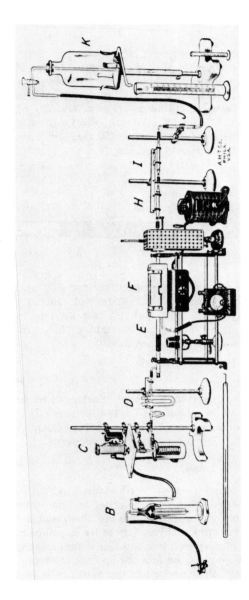

FIG. 2.8. The assembly for carbon and hydrogen determination. (Courtesy A. H. Thomas Co. From Ref. 20, STANDARD METHODS OF CHEMICAL ANALYSIS, Vol. 2A, 6th Ed., edited by Frank J. Welcher, 1963 by Litton Educational Publishing, Inc., reprinted by permission of Van Nostrand Reinhold Company.)

FIG. 2.9. Schematic diagram of C,H combustion train. 1, Oxygen cylinder; 2, pressure regulator; 3, bubble counter U-tube; 4, combustion tube; 5, H_2O absorber; 6, NO absorber; 7, CO_2 absorber; 8, guard tube; 9, Mariotte bottle; 10, graduated cylinder. (From Ref. 20, STANDARD METHODS OF CHEMICAL ANALYSIS, Vol. 2A, 6th Ed., edited by Frank J. Welcher   1963 by Litton Educational Publishing, Inc., reprinted by permission of Van Nostrand Reinhold Company.)

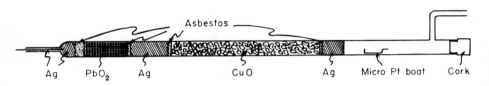

FIG. 2.10. Packing of microcombustion tube for C,H analysis using lead peroxide to remove nitrogen oxides. (From Ref. 20, STANDARD METHODS OF CHEMICAL ANALYSIS, Vol. 2A, 6th Ed., edited by Frank J. Welcher   1963 by Litton Educational Publishing, Inc., reprinted by permission of Van Nostrand Reinhold Company.)

If manganese dioxide is used for removal of nitrogen oxides, the section of PbO_2 is omitted and the section of CuO is lengthened. If MgO or quartz chips are placed in the combustion tube besides CuO, the setup is suitable for the analysis of fluorine-containing compounds.

Packing the Absorption Tubes

Pack the absorption tube for water as illustrated in Fig. 2.11. Shape the cotton wad at the terminal joining the combustion tube in tapered form, as shown, so that the water will spread along the absorbent instead of collecting at the end. Seal the stopper by means of Krönig cement. The carbon dioxide absorber is prepared in a similar manner except that the tapered cotton wad is placed near the stopper and the absorption tube is first filled with 10 mm of Anhydrone before the Ascarite is added.

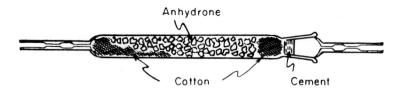

FIG. 2.11. Packing of the water absorber. (From Ref. 20, STAND-
ARD METHODS OF CHEMICAL ANALYSIS, Vol. 2A, 6th Ed., edited by
Frank J. Welcher   1963 by Litton Educational Publishing, Inc., re-
printed by permission of Van Nostrand Reinhold Company.)

When manganese dioxide is employed to retain nitrogen oxides (Note 2),
a third absorption tube is placed between the water absorber and the carbon
dioxide absorber. Fill one-half (near the water absorber) of this tube with
manganese dioxide pellets and the other half with Anhydrone.

Assembling the Combustion Train

Follow the sequence of arrangements indicated in Figs. 2.8 and 2.9 to
assemble the combustion train. Fill the constant-pressure device B with
water containing a little sodium hydroxide to prevent the atmospheric carbon dioxide from passing into the gas stream. Fill the bubble counter of
unit D with concentrated sulfuric acid until the tip of the gas inlet tube is
about 3 mm immersed in the liquid. Pack one side (joining the combustion
tube) of the U-tube in D with Anhydrone and the other side with Ascarite,
putting some cotton at both ends. Seal the U-tube with Krönig cement. Fill
the guard tube J with Anhydrone and the Mariotte bottle with water.
Connect all glass apparatus bud-to-bud by means of soft thick-walled
rubber tubing specially impregnated with paraffin. Lubricate the rubber
connections sparingly with glycerol. Join the stopper of the Anhydrone
tube H with that of the Ascarite tube I (or to the MnO_2 tube if it is used),
the other end of I to the guard tube J, and then the Anhydrone tube with the
capillary tip of the combustion tube. Adjust the level of the delivery tube
of the Mariotte bottle so that 100 ml of water leaves the reserve in 20 min.

Checking the Combustion Train

Before the analysis, check the system by letting oxygen into the
constant-pressure device B while the stopcock above the Mariotte bottle K
remains closed. If bubbles continue to pass through the bubble counter,
there is a leak in the combustion train. On the other hand, when the stopcock is opened and no bubble appears, the system is clogged up somewhere.

The Course of Analysis

Pass a current of oxygen through the combustion train. Heat the section of copper oxide at about 850°C and that of lead peroxide at 180°C. Meanwhile, weigh out the sample. Wearing clean gloves (or using a piece of chamois), remove the absorption tubes H and I while the stopcock above the Mariotte bottle is closed. Disconnect the absorption tubes and wipe their tips with a tissue paper (Note 3). Place the absorption tubes near the microbalance for 5 min, then place the Anhydrone tube on the hook of the balance pan and take the weight on the 10th minute; similarly weigh the Ascarite tube on the 15th min (Note 4).

Now replace the absorption tubes in the combustion train. Shut off the oxygen inlet to the bubble counter. Remove the cork to open the combustion tube. Transfer the microboat containing the sample into the mouth of the combustion tube by means of the forceps and push it inside with the aid of a platinum wire hook until it is about 50 mm from the silver roll. (For a liquid sample contained in a sealed capillary, place the platinum foil, Fig. 1.9, at the neck of the combustion tube, cut off the tip of the capillary and place the open capillary on the foil, then push the foil toward the silver roll until the capillary tip is 50 mm away from the silver roll.)

Close the combustion tube with the cork, and check for leaks and also smooth passage of the gas stream. Then start combustion by placing the burner under the microboat (or platinum foil). Slowly advance the burner toward the silver roll. Solid sample usually melts first and then volatilizes or sublimes. Liquid sample will come out of the capillary tip and settle in the combustion tube. In order to prevent incomplete combustion, it is important to control the rate at which the organic compound enters the oxidizing mixture. A sample of 3-6 mg ordinarily requires about 10 min.

After the sample has disappeared, collect 100 ml of water from the Mariotte bottle. Then turn off the stopcock, remove the absorption tubes, and wipe and weigh them as described above. The increases in the weights of the respective tubes give the amounts of water and carbon dioxide produced from the sample.

E. Calculations

$$\%H = \frac{\text{Weight of } H_2O \times 0.1119 \times 100}{\text{Weight of sample}}$$

$$\%C = \frac{\text{Weight of } CO_2 \times 0.2729 \times 100}{\text{Weight of sample}}$$

F. Notes

Note 1: The corkstopper of the combustion tube is shaped from a cork (free from pores) that is wider than the combustion tube, by slowly and carefully squeezing the cork into the neck of the tube. The cork stopper is better than a rubber stopper or a ground-glass cap. If there is a sudden gas expansion inside the combustion tube, the rubber stopper or glass cap might cause an explosion, while the cork is just pushed open and pressure is released. The cork stopper has the same life as the combustion tube.

Note 2: Currently, microanalysts favor the use of the external scrubber containing manganese dioxide to retain oxides of nitrogen, primarily because quartz combustion tubes are employed. Quartz tubes can be heated above 900°C but have a short life in the presence of lead. If Supremax combustion tubes are available, lead peroxide can be used, together with lead chromate [1], and the combustion temperature is kept at about 800°C.

Note 3: The expert may prefer to use a piece of chamois instead of the gloves, and to wipe the absorption tubes with moistened flannel cloth and chamois, using cotton on knurled metal rods for the tips. The novice, however, usually finds it difficult to use this technique.

Note 4: An experienced operator can weigh the absorption tube immediately. Since he performs all weighing in the same time span, the errors cancel out.

2.7 EXPERIMENTAL: RAPID MULTIPLE GRAVIMETRIC DETERMINA-
TION OF CARBON AND HYDROGEN

A. Principle

The principle utilized in this method is identical to that described in Sec. 2.6. By using a large-bore furnace and smaller combustion tubes, it is possible to place two to four tubes in the furnace at one time and thus to analyze two to four samples simultaneously.

B. Apparatus

The assembly is shown in Fig. 2.12, and the schematic diagram of the combustion train is given in Fig. 2.13.

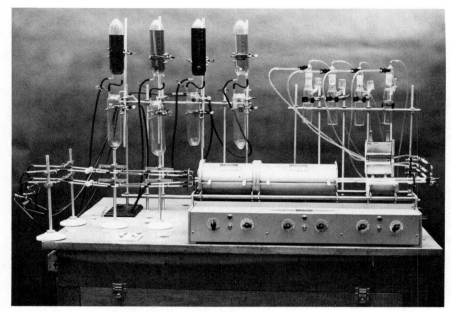

FIG. 2.12. Assembly for multiple C, H analysis.

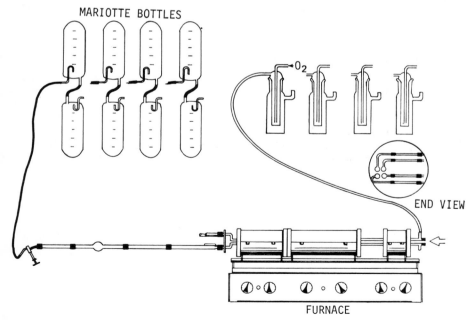

FIG. 2.13. Schematic multiple combustion train. (Courtesy Mikrochim. Acta.)

Combustion furnace (Hevi-Duty, Milwaukee, Wisconsin): The furnace is provided with three independent heating units which slide on rails, and each is controlled by its own rheostat. The heating units are hinged so that the top halves may be raised to observe the progress of the combustion, to provide for rapid cooling, and to position the combustion tubes more readily. The cores of the heating units have an inner diameter of about 32 mm, and the lengths are about 10, 20, and 30 cm. The 10-cm heating unit is movable and is used as the sample burner. The 30-cm unit is stationary and positioned over the copper oxide filling. The 20-cm unit is also stationary and is positioned over the silver wool filling. The furnace is modified as follows: (1) the nickel alloy trough, which extends through all three sections of the furnace is removed in order to provide more space for the combustion tubes; (2) the 25.4 mm of asbestos insulation between the sample burner and the long burner is removed and replaced by about 4 mm of asbestos sheet to insure continuity of temperature between the two heating units.
Pressure regulators.
Modified Mariotte bottles [170], 500-ml capacity.
Absorption tubes and tube fillings (see Sec. 2.6).
Combustion tube extensions (Fig. 2.14).
Combustion tubes and tube fillings (Fig. 2.15).

C. Reagents

Oxygen, USP medicinal grade (Linde Co., New York, New York).
MgO on quartz chips, prepared by adding quartz chips to a suspension of Fisher reagent MgO in water. The supernatant is poured off, and the remaining material is dried at 140°C for 1 h. The dried reagent is then placed on a 30 mesh sieve, and all material which does not pass through is used.
Quartz chips, quartz wool, and silver wool: Quartz chips (2-4 mm in diameter) and quartz wool may be obtained from P. W. Blackburn, Inc., Dobbs Ferry, New York. The silver wool is obtained from Baker Platinum Div., Newark, New Jersey.
Copper oxide in wire form is obtainable from J. T. Baker, Phillipsburg, New Jersey.

D. Procedure

Raise the tops of the three heating units. Place a piece of asbestos insulation on top of the lower half of the sample burner and place the four combustion tubes in the furnace (two on top, two on bottom). Hold the tubes rigidly in place by fitting the open ends and capillary ends of the tubes

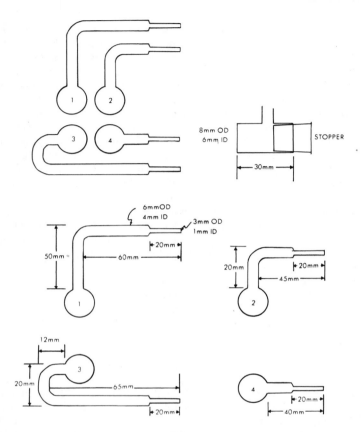

FIG. 2.14. Combustion tube extensions. (Courtesy Mikrochim. Acta.)

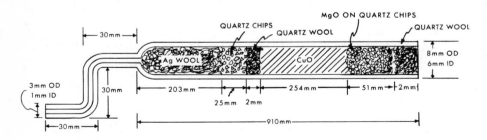

FIG. 2.15. Combustion tube fillings. (Courtest Mikrochim. Acta.)

through small squares of Masonite through which four appropriately sized holes are drilled. Secure the Masonite with wire to both ends of the combustion furnace. Lower the tops of the silver burner and of the long burner over the combustion tubes. Weigh four samples of about 3-10 mg and place them in their respective combustion tubes 6 cm from the long burner. Connect the purified oxygen supply to the open ends of the combustion tubes by means of the combustion tube extensions (Fig. 2.14), and connect the previously weighed absorption tubes to the capillary ends of the combustion tubes. (These capillary ends are bent to insure spacing between the absorption tubes for ease of manipulation. Each combustion tube has its own individual oxygen supply, pressure regulator, guard tubes, absorption tubes, and Mariotte bottle.)

Adjust the oxygen flow rate to approximately 30 ml/min. Remove the asbestos insulation from the lower half of the sample burner and lower the top of the sample burner over the combustion tubes. Position the sample burner 9 cm from the long burner, which is 3 cm behind the sample. (The temperature of the sample burner is 900-950°C. The temperature of the silver burner is between 450° and 500°C.) After 2 min, move the sample burner 3 cm closer to the long burner and in two more 2-min intervals (moving 3 cm at a time), rest the sample burner against the long burner, where it should remain for an additional 6 min. (The long burner is at a temperature of 900-950°C.) The total elapsed combustion time is about 12 min. A check of the oxygen flow rate may be made by observing the Mariotte bottle, where about 360 ml of water should be displaced. At the completion of the combustion period remove the two upper water absorption and carbon dioxide absorption tubes. Wipe the absorption tubes with a chamois cloth and remove the static charges with the aid of a Tesla coil. Then weigh the tubes immediately. After weighing, return them to the combustion train and repeat the procedure with the lower absorption tubes. (To speed up the procedure, four new samples may be weighed out during the combustion period so that, after the absorption tubes are returned to the train, the next series of four combustions may be started without any delay. Before starting each new series, it is, of course, necessary to invert the Mariotte bottles so that the water is in the upper halves.)

E. Calculations

The calculations in this method are the same as those given in Sec. 2.6.

F. Comments

In addition to the obvious advantages of speed and economy of space, this method has other advantages: (1) the cost of the combustion apparatus and replacement parts is much lower than that of existing automatic and semi-automatic furnaces designed for microcombustion analysis; (2) since four different tubes are being employed simultaneously, the operator's efficiency is not much lowered by contamination or breakage of one tube due to an accident or because of the explosive nature of one sample; (3) if the operator is analyzing samples which require different combustion tube fillings he can change the filling in one tube, while the other tubes can be filled in the normal fashion for routine-type determinations.

REFERENCES

1. F. Pregl (assisted by H. Lieb), Die Quantitative Organische Mikro-analyse, 3rd Ed., Springer, Berlin. 1930.

1a. V. Pechanec, Collect. Czech. Chem. Commun. $\underline{38}$:2917 (1973).

1b. A. A. Abramyan, S. M. Atashyan, R. A. Megroyan, A. A. Kocharyan, and A. S. Tevosyan, Arm. Khim. Zh. $\underline{27}$:740, 832 (1974).

2. J. Körbl, Collect. Czech. Chem. Commun. $\underline{20}$:948 (1955); Mikrochim. Acta $\underline{1956}$:1705.

2a. A. Campiglio, Mikrochim. Acta $\underline{1976(I)}$:651.

3. R. Belcher and G. Ingram, Anal. Chim. Acta $\underline{4}$:118 (1950).

4. G. Ingram, Analyst $\underline{86}$:411 (1961).

5. M. O. Korshun and V. A. Klimova, Zh. Anal. Khim. $\underline{2}$:274 (1947).

6. M. O. Korshun, N. E. Gel'man, and N. S. Sheveleva, Zh. Anal. Khim. $\underline{13}$:695 (1958).

7. J. A. Kuck (ed.), Methods in Microanalysis (translated from Russian), Gordon and Breach, New York, 1964.

7a. H. J. Francis, Jr., private communication, 1976.

8. Coleman Model 33 Carbon-hydrogen Analyzer (1968), Coleman Instruments, Maywood, Ill.

9. Thomas Model 35 Carbon-hydrogen Analyzer (1970), A. H. Thomas Co., Philadelphia.

10. Sargent C-H Combustion Apparatus (1968), Sargent Co., Chicago.

10a. Heraeus Micro Analyzer, Standard Model (1970), W. C. Heraeus GMBH, Hanau, Germany.

11. Perkin-Elmer Model 240 Elemental Analyzer (1970), Perkin-Elmer Corp., Norwalk, Conn.

12. Technicon CHN Microanalyzer (1970), Technicon Instruments Co. Ltd., Surrey, England.

13. Heraeus Automatic CHN-O Ultra Analyzer (1970), W. C. Heraeus GMBH, Hanau, Germany.

14. Hewlett-Packard Model 185B CHN Analyzer (1972), Hewlett-Packard, Avondale, Pa.

15. Carlo Erba Elemental Analyzer Model 1102 (1972), Carlo Erba Scientific Instruments, Rodano, Italy.

16. Aminco Carbon and Hydrogen Analyzer (1970), American Instrument Co., Silver Springs, Md.

17. J. E. Noakes, S. M. Kim, and L. K. Akers, Geochim. Cosmochim. Acta 31:(1)94 (1967).

17a. H. W. Scharpenseel and F. Pietig, Geoderma 2:273 (1968).

17b. S. Apelgot, R. Chemama, M. Frilley, G. Tham, and M. Guggiari, Bull. Cancer 60:41 (1973).

18. N. D. Cheronis and T. S. Ma, Organic Functional Group Analysis, Wiley, New York, 1964, p. 409.

19. G. Ingram and M. Lonsdale, in I. M. Kolthoff and P. J. Elving (eds.), Treatise on Analytical Chemistry, Part II, Vol. II, Wiley, New York, 1965, p. 297.

19a. N. E. Gel'man and E. N. Bunz, Determination of Carbon and Hydrogen in Organic Compounds, Science Publishing House, Moscow, 1974.

20. T. S. Ma, in F. J. Welcher (ed.), Standard Methods of Chemical Analysis, 6th Ed., Vol. 2, Van Nostrand, Princeton, 1963, p. 366.

21. M. Marzadro, Ann. Chim. (Rome) 49:911 (1959).

22. M. Vecera, F. Vojtech, and L. Synek, Collect. Czech. Chem. Commun. 25:93 (1960).

23. W. Pfab, Z. Anal. Chem. 187:354 (1962).

24. C. C. Chang, H. C. Huang, and J. P. Chang, Acta Chim. Sinica 28:75 (1962).

25. T. Mitsui, K. Yoshikawa, and K. Furuki, Mikrochim. Acta 1962:385.

26. C. Meyer and G. Vetter, Chem. Tech. (Berlin) 13:104 (1961).

27. W. Knobloch, F. Knobloch, and G. Mai, Mikrochim. Acta 1961:576.

28. W. Simon, P. F. Sommer, and G. H. Lyssy, Microchem. J. 6:239 (1962).

29. N. Oda, G. Tsuchihashi, and S. Ono, Microchem. J. 8:69 (1964).

30. C. S. Yeh, Microchem. J. 7:303 (1963).

31. G. J. Kakabadse and B. Manohim, Analyst 88:816 (1963).

32. R. C. Rittner and R. Culmo, Mikrochim. Acta 1964:631.

33. M. Vecera, Mikrochim. Acta 1964:196.

34. H. J. Francis, Jr., Microchem. J. 7:462 (1963).

35. S. Mizukami and T. Ieki, Microchem. J. 7:50 (1963).

36. J. P. Chang and C. C. Chang, Acta Chim. Sinica 29:307 (1963).

37. G. Kainz and H. Horwatitsch, Mikrochim. Acta 1963:720.

38. M. Ebeling and L. Malter, Microchem. J. 7:179 (1963).

39. S. L. Wang, Acta Chim. Sinica 30:211 (1964).

40. D. G. Newman and C. Tomlinson, Mikrochim. Acta 1964:1023.

41. J. Lakomy, L. Lehar, and M. Vecera, Collect. Czech. Chem. Commun. 28:3271 (1963).

42. J. F. Cannelongo and J. J. Kobliska, Microchem. J. 7:421 (1963).
43. G. J. Kakabadse and B. Manohin, Mikrochim. Acta 1965:1136.
44. E. A. Reich, M. A. Carroll, and J. E. Jarembo, Microchem. J. 11:264 (1966).
45. H. Pieters and W. J. Buis, Microchem. J. 8:383 (1964).
46. T. F. Holmes and A. Lauder, Analyst 90:307 (1965).
47. V. Pechanec and J. Horacek, Collect. Czech. Chem. Commun. 30:1082 (1965).
48. P. B. Olson, Microchem. J. 13:75 (1968).
49. V. V. Mikhailov and T. I. Tarasenko, Zavod. Lab. 15:1380 (1967).
50. H. Trutnovsky, Mikrochim. Acta 1968:97.
51. F. Sels and P. Demoen, Mikrochim. Acta 1970:48.
52. F. Sels and P. Demoen, Mikrochim. Acta 1970:516.
53. J. E. Fildes, Mikrochim. Acta 1970:978.
54. V. Pechanec and J. Horacek, Collect. Czech. Chem. Commun. 35:2749 (1970).
55. N. I. Shakhova, I. A. Zavlokina, and R. S. Shish, Zh. Anal. Khim. 24:282 (1969).
56. C. E. Childs and E. B. Henner, Microchem. J. 15:590 (1970).
57. T. Teki and K. Daikatsu, Microchem. J. 17:93 (1972).
57a. A. A. Abramyan and M. A. Balyan, Arm. Khim. Zh. 26:383 (1973).
58. O. E. Sundberg and C. Maresch, Anal. Chem. 32:274 (1960).
59. A. M. Vogel and J. J. Quattrone, Jr., Anal. Chem. 32:1754 (1960).
60. M. Vecera, Collect. Czech. Chem. Commun. 26:2298 (1961).
61. M. Vecera, Z. Anal. Chem. 208:15 (1965).
62. I. A. Revel'skii, R. T. Borodulina, V. G. Klimova, and T. M. Sovakova, Dokl. Akad. Nauk. SSSR 159:861 (1964).
63. F. Scheidl, Microchem. J. 13:155 (1968).
64. W. Walisch and K. Schafer, Mikrochim. Acta 1968:765.
65. A. P. Mischenko and M. F. Rodicheva, Zh. Anal. Khim. 22:1536 (1967).
66. G. Kainz and E. Wachberger, Z. Anal. Chem. 220:15 (1966).
67. H. Trutnovsky, Microchem. J. 16:266 (1971).
68. V. G. Berezkin, B. M. Luskina, S. Syavtisillo, and A. P. Terent'ev, Zh. Anal. Khim. 23:1254 (1968).
69. Perkin-Elmer Corp., British Patent 1,153,821 (1966).
70. E. Bailey and W. Brown, Talanta 16:469 (1969).
71. W. Ihn, W. Herb, and I. Noack, Mikrochim. Acta 1963:1132.
72. N. E. Gel'man and V. Y. Van, Zh. Anal. Khim. 15:487 (1960).
73. G. F. Anisimova and V. A. Klimova, Zh. Anal. Khim. 18:412 (1963).
74. A. Campiglio, Farmaco, Ed. Sci. 22:196 (1967).
75. W. Merz, Anal. Chim. Acta 48:381 (1969).

76.   A. Campiglio, Farmaco, Ed. Sci. 24:748 (1969).
76a. F. G. Roemer, M. A. Jonker, and B. F. Griepink, Z. Anal. Chem.
      273:287 (1975).
77.   S. Greenfield and R. A. D. Smith, Analyst 88:886 (1963).
78.   G. Kainz, K. Zidek, and G. Chromy, Mikrochim. Acta 1968:235.
79.   J. A. Kuck, J. W. Berry, A. J. Andreath, and P. A. Leutz,
      Anal. Chem. 34:403 (1962).
80.   N. Oda, S. Ono, and H. Matsumori, Japan Analyst 18:854 (1969).
81.   G. F. Anisimova and V. A. Klimova, Zh. Anal. Khim. 23:411
      (1968).
82.   M. N. Chumachenko and N. B. Levina, Zh. Anal. Khim. 23:1250
      (1968).
83.   M. F. Barakat and A. H. Zahran, Z. Anal. Chem. 239:93 (1968).
84.   D. F. Fenton, J. P. Mackey, and P. B. Creedon, Chem. Ind.
      (London) 1970:956.
84a. H. Frohofer, Z. Anal. Chem. 271:203 (1974).
85.   G. Kainz and F. Scheidl, Mikrochim. Acta 1963:902.
86.   N. F. Egorova and T. E. Pokrovskaya, Zh. Anal. Khim. 19:366
      (1964).
87.   H. J. Francis, Jr. and E. J. Minnick, Microchem. J. 8:245 (1964).
88.   W. I. Awad, Y. A. Gawargious, and S. S. M. Hassan, Mikrochim.
      Acta 1967:847.
89.   G. Dugan and V. A. Aluise, Anal. Chem. 41:495 (1969).
90.   E. Schwarz-Bergkampf, Mikrochim. Acta 1967:1001.
90a. E. Kozlowski, Chem. Anal. (Warsaw) 18:677 (1973); 19:99, 835
      (1974).
90b. Y. Baba, Microchem. J. 21:75 (1976).
90c. H. J. Francis, Jr., reported at the Federation of Analytical Chem-
      istry and Spectroscopy Societies 3rd Annual Meeting, Philadelphia,
      November 1976, Abstract No. 271.
91.   H. Malissa and E. Pell, Microchem. J., Symp. Ser. 2:371 (1962).
92.   M. Pecar, Acta Pharm. Jugoslav. 16:151 (1966).
93.   N. E. Gel'man, P. I. Bresler, B. N. Ruzin, N. V. Grek, N. S.
      Sheveleva, and A. A. Mel'nikova, Dokl. Akad. Nauk SSSR 161:107
      (1965).
93a. D. Fraisse, D. Girard, and R. Levy, Talanta 20:667 (1973).
94.   N. Gel'man, Talanta 16:464 (1969).
94a. D. Fraisse, Bull. Soc. Chim. Fr. 1972:3631; 1973:514; Talanta
      18:1011 (1971).
95.   R. S. Juvet and J. Chiu, Anal. Chem. 32:130 (1960).
96.   G. Gutbier and W. Ihn, Mikrochim. Acta 1966:24.
97.   M. Leclercq and A. Mace de Lepinay, Mem. Poudres 46:129 (1964).
98.   G. Ingram, Analyst 86:411 (1961).

99. C. P. Lloyd-Jones, Analyst 95:366 (1970).

99a. M. Higuchi, M. Oota, and A. Mukade, Radioisotopes (Tokyo) 23:523 (1974).

100. T. S. Ma, in F. J. Welcher (ed.), Standard Methods of Chemical Analysis, 6th Ed. , Vol. 2, Van Nostrand, Princeton, 1963, p. 373.

101. J. Binkowski and B. Bobtanski, Chem. Anal. (Warsaw) 9:515 (1964).

102. G. R. Watson and J. P. Williams, Anal. Biochem. 33:356 (1970).

103. L. A. Ford, J. Ass. Offic. Anal. Chem. 53:86 (1970).

103a. O. Mlejnek, Chem. Zvesti 27:421 (1973).

104. M. Jurecek, V. Cervinka, and K. Cejka, Collect. Czech. Chem. Commun. 34:1162 (1969).

105. S. Mlinko and M. K. Hermann, Mikrochim. Acta 1967:872.

106. S. Mlinko, Mikrochim. Acta 1963:456.

107. M. Vecera, D. Snobl, and L. Synek, Mikrochim. Acta 1961:872.

108. B. Arventiev, M. Leonte, and H. Offenberg, Am. Stiint. Univ. Al.I. Cuza, Iasi Sect. I 6:183 (1960).

109. G. Kainz and H. Horwatitsch, Z. Anal. Chem. 176:175 (1960).

110. J. Horacek, J. Körbl, and V. Pechanec, Mikrochim. Acta 1960:294.

110a. V. P. Grigoryan and B. D. Ryzhkov, Zavod. Lab. 40:797 (1974).

111. S. Mizukami and T. Ieki, Microchem. J. 7:293 (1963).

112. A. A. Abramyan, A. A. Kocharyan, and R. A. Megroyan, Izv. Akad. Nauk. Arm. SSR, Khim. Nauk 19:849 (1966).

113. W. J. Kirsten, Mikrochim. Acta 1964:487.

114. E. D. Lomadze, Zh. Anal. Khim. 27:2266 (1972).

115. A. I. Lebedeva, N. A. Nikolaeva, and V. A. Orestova, Izv. Akad. Nauk SSR, Ctd. Khim. Nauk 1961:1350.

116. A. M. G. Macdonald and G. G. Turton, Microchem. J. 13:1 (1968).

117. V. S. Bazalitskaya and M. K. Dzhamaletdinova, Zavod. Lab. 33:427 (1967).

118. D. E. Butterworth, Analyst 86:357 (1961).

119. S. Mizukami and T. Ieki, Microchem. J. 7:485 (1963).

120. E. Celon and S. Bresadola, Anal. Chem. 40:972 (1968).

121. A. I. Lebedeva, N. A. Nikolaeva, V. A. Orestova, and E. V. Shikhman, Izv. Akad. Nauk. SSR, Sor. Khim. 1964:475.

122. V. S. Bazalitskaya, Izv. Akad. Nauk Kaz. SSR, Ser. Khim. 1969:86.

123. H. Pieters and W. J. Buis, Microchem. J. 8:383 (1964).

124. L. Uhle, Z. Anal. Chem. 231:194 (1967).

124a. W. Wojnowski and A. Olszewska-Borkowska, Chem. Anal. (Warsaw) 20:209 (1975).

125. Y. A. Gawargious and A. M. G. Macdonald, Anal. Chim. Acta 27:119 (1962).

126. J. Binkowski and R. Levy, Bull. Soc. Chim. Fr. 1968:4289.

127. Y. A. Gawargious, Talanta 16:1112 (1969).

128. R. Culmo and R. Fyans, Mikrochim. Acta 1968:816.
129. C. I. Kennedy, Microchem. J. 17:325 (1972).
130. H. Trutnovsky, Mikrochim. Acta 1968:371.
131. A. D. Campbell, Mikrochim. Acta 1968:833.
132. W. J. Kirsten, private communication, July 1969; Mikrochim. Acta 1966:105.
133. E. Pella, Mikrochim. Acta 1964:943.
134. P. P. Wheeler and A. C. Richardson, Mikrochim. Acta 1964:609.
135. A. D. Campbell, L. S. Harns, R. Monk, and D. R. Petrie, Mikrochim. Acta 1968:836.
136. T. S. Ma and V. Horak, Microscale Manipulations in Chemistry, Wiley, New York, 1976.
137. T. Ieki and K. Daikatsu, Microchem. J. 17:93 (1972).
138. M. Vecera and D. Snobl, Collect. Czech. Chem. Commun. 25:2013 (1960).
139. G. Kainz and J. Mayer, Mikrochim. Acta 1963:543.
140. W. Padowetz, Microchem. J. 14:110 (1969).
140a. J. Binkowski, S. Gizinski, R. Kaminski, and W. Reimschussel, Mikrochim. Acta 1976(I):623.
141. A. S. Zabrodina and S. Y. Levina, Zh. Anal. Khim. 17:644 (1962).
142. T. Mitsui, O. Yamamoto, and K. Yoshikawa, Mikrochim. Acta 1961:521.
143. U. Bartels, Chem. Tech. (Berlin) 19:696 (1967).
144. E. I. Margolis and V. N. Bibileishvili, Vestn. Mosk. Univ., Ser. Khim. 1963:46.
145. E. I. Margolis and G. G. Lyamina, Vestn. Moskov Univ., Ser. Khim. 1963:66.
146. P. R. Wood, Analyst 85:764 (1968).
146a. Y. Tomida, T. Ando, and W. Funasaka, Japan Analyst 22:1465 (1973).
147. A. D. Campbell and A. M. G. Macdonald, Anal. Chim. Acta 26:275 (1962).
148. J. F. Alicino, Microchem. J. 9:22 (1963).
148a. S. W. Bishara, M. E. Attia, and H. N. A. Hassan, Mikrochim. Acta 1974:819.
149. F. Kasler, Microchem. J. 13:430 (1968).
149a. E. Kozlowski and B. Kobylinska-Mazurek, Chem. Anal. (Warsaw) 18:1161 (1973).
150. J. Binkowski and M. Vecera, Mikrochim. Acta 1965:842.
150a. J. Binkowski, Chem. Anal. (Warsaw) 18:989 (1973); 19:1050 (1974).
151. E. Kissa and M. Soepere-Yllo, Mikrochim. Acta 1967:287.
152. V. Pechanec and J. Horacek, Collect. Czech. Chem. Commun. 27:232 (1962).

153. Y. A. Gawargious and A. M. G. Macdonald, Anal. Chim. Acta
      27:300 (1962).
154. F. Salzer, Microchem. J. 16:145 (1971).
154a. V. Rezl and J. Janak, Chromatogr. Rev. 17:233 (1973).
154b. G. A. Butrinova and V. G. Zizin, Zavod. Lab. 40:147 (1974).
154c. V. Rezl and B. Kaplanova, Mikrochim. Acta 1975 I:493.
155. D. H. Davis, Talanta 16:1055 (1969).
156. K. Nakamura, K. Ono, and K. Kawada, Ann. Sankyo Res. Lab.
      22:54 (1970).
156a. R. Karlsson, Mikrochim. Acta 1974:963.
157. F. Sels and P. Demoen, Mikrochim. Acta 1969:530.
157a. J. Binkowski, Chem. Anal. (Warsaw) 19:879 (1974).
158. J. C. Rousseau, private communication, 1973.
158a. R. Culmo, Mikrochim. Acta 1969:175.
159. K. Hozumi and H. Tamura, Japan Analyst 16:1193 (1967).
159a. K. Hozumi, O. Tsuji, and H. Kushima, Microchem. J. 15:481
      (1970).
160. H. Tamura and K. Hozumi, Japan Analyst 19:60 (1970).
161. M. Shimitzu and K. Hozumi, Japan Analyst 19:1041 (1970).
162. Techniçon Instruments Co., Report on Customer Experience with
      the Technicon CHN-Analyzer, Technicon International Division,
      Domont, France, 1967.
162a. W. Manser, private communication, August 1969.
163. Papers presented at the International Symposium on Microchemical
      Techniques, University Park, Pa., August 1973. See also D. E.
      Harrington and W. R. Bramstedt, Microchem. J. 21:60 (1976);
      K. Hakariya, H. Saito, and T. Akehata, Bull. Tokyo Inst. Tech.
      122:23 (1974).
163a. Control Equipment Corp., Lowell, Mass., Announcement, October
      1974.
164. J. F. Alicino, Microchem. J. 18:350 (1973).
165. K. Hozumi, private communication, August 1972.
166. L. C. Klein, private communication, November 1973.
167. Pamphlet, Preparation of 185 Oxidant (Catalyst), Hewlett-Packard,
      Avondale, Pa., 1972.
167a. K. Inouye, private communication, November 1973.
168. E. Pella, private communications, August 1973 and October 1974;
      E. Pella and B. Colombo, Mikrochim. Acta 1973:697.
168a. M. Ellison, Chem. Ind. (London) 1974:787.
169. J. Horacek and J. Körbl, Chem. Ind. (London) 1958:101.
170. F. O. Fischer, Anal. Chem. 21:827 (1949).

Nitrogen

3.1 GENERAL CONSIDERATIONS

A. Basic Principles

The element nitrogen occurs in carbon compounds in a variety of bondings, for example, N-C, N-H, N-O, N-Cl, N-Br, N-P, N-S, N-N. In this chapter we discuss the principles and procedures by which the total quantity of nitrogen in an organic substance can be determined. The methods for quantitative analysis of nitrogen in specific linkages such as N-N or N-O are covered in treatises on organic functional group analysis [1].

The determination of nitrogen holds a unique position in organic elemental analysis. Whereas determinations of carbon, hydrogen, and other elements are usually performed on pure organic compounds, a vast majority of nitrogen analyses are carried out on mixtures. It is also apparent that nitrogen determination in mixtures cannot be done by direct physical methods such as nuclear magnetic resonance.

Because the determination of nitrogen has the widest applications in chemical analysis, the number of nitrogen analyses routinely performed probably exceeds the total number of determinations of all other elements. As a matter of fact, in some analytical laboratories, such as those for agricultural products and clinical materials, nitrogen may be the only element of concern.

There are three methods to convert nitrogen in the organic compound into a measurable form. The principles involved are as follows:

(1) The Dumas method [1a] (1830) is based on the destruction of the organic compound by oxidation, followed by the reduction of oxides of nitrogen by means of copper to molecular nitrogen:

$$\text{Organic N compound} \xrightarrow{[O]} \text{N oxides} \xrightarrow{Cu} N_2$$

(2) The Kjeldahl method [2] (1883) utilizes the action of concentrated sulfuric acid on certain nitrogen compounds resulting in the formation of ammonium bisulfate from which ammonia is quantitatively released by means of caustic alkali:

$$\text{Amino N compound} \xrightarrow{H_2SO_4} (NH_4)HSO_4 \xrightarrow{NaOH} NH_3$$

(3) The ter Meulen method [3] (1924) depends on the catalytic hydrogenation of the organic compound concomitant with pyrolytic cleavage of the C-N bonds, to produce ammonia:

$$\text{Organic N compound} \xrightarrow{H_2} NH_3$$

It is interesting to note that the Kjeldahl method is by far the most frequently used procedure, although its mechanism is the least understood. Some textbooks have ascribed the sulfuric acid action yielding ammonia as "wet oxidation." In our opinion, this term is misleading. While oxidation does occur in the reaction mixture, producing carbon dioxide and water, ammonia cannot be a product of oxidation. The discovery of Kjeldahl was a happy coincidence because he worked on proteins in which nitrogen exists only in N-C and N-H bondings that can be converted to ammonia by hydrolytic cleavage. In some cases, when the organic nitrogen is in the amide form, ammonia can be quantitatively recovered without the sulfuric acid treatment. Thus urea can be determined as ammonia by the use of the hydrolytic enzyme urease:

$$\underset{\underset{O}{\overset{\|}{}}{H_2N-C-NH_2}} \xrightarrow{\text{urease, } H^+} 2NH_4^+ + CO_2$$

B. Current Practice

The Micro Kjeldahl Procedure

The current micro Kjeldahl procedure (see Sec. 3.3) was developed by Ma and Zuazaga [4] over 35 years ago. It involves sulfuric acid digestion of the sample, steam distillation of ammonia from caustic alkali into boric acid solution, and direct titration of the ammonia by means of 0.01 $\underline{N}$ HCl.

This procedure has several advantages over the Pregl method which was adapted from the original Kjeldahl procedure that included absorption of ammonia in a definite amount of standardized acid followed by back-titration of the unconsumed acid. (1) The simplified operation of the present micro Kjeldahl procedure is more rapid. (2) It requires only standardized 0.01 $\underline{N}$ acid and no 0.01 $\underline{N}$ alkali which is difficult to keep. (3) There is no danger of spoiling a determination because of an insufficient quantity of acid to absorb the ammonia. (4) In case the contents of the receiving flask are sucked back into the distilling flask, distillation can be resumed without deleterious effect. Ma and Zuazaga [4] have found that quantitative recovery of 0.06-1.00 mg of nitrogen can be accomplished by steam distillation for 2 min at the rate of 5 ml condensate per min, and that the titration endpoint is best detected by the use of a mixed indicator containing five parts bromocresol green and one part methyl red. Other mixed indicators have been proposed during the past three decades. The ideal indicator, however, would be a single compound; this is still to be discovered. This micro Kjeldahl procedure is the standard method recommended by the Association of Official Analytical Chemists [5].

The popularity of the Kjeldahl method for nitrogen determination is primarily due to the fact that it can be performed with simple and inexpensive equipment (see Sec. 3.3) and is easy to operate. It is particularly suited for the analysis of agricultural and biochemical materials since they comprise mostly organic nitrogen compounds of the amino and amido types which are amenable to sulfuric acid digestion. Polymeric substances such as plastics and proteins may require high temperature and supplementary oxidants like hydrogen peroxide to effect complete destruction of the organic matter to produce the maximum amount of ammonia.

Experimental directions for the micro Kjeldahl procedure are given in Sec. 3.3. A commercial automatic micro Kjeldahl digestion and spectrophotometric determination apparatus is available from Technicon Industrial Systems, Tarrytown, New York. Another automated assembly involving flask digestion (0.5-1 g sample), steam distillation, and titration is marketed by Foss Electric Ltd., York, England.

The Modified Micro Kjeldahl Procedures

By modified micro Kjeldahl procedure we mean that the digestion process requires certain changes in order to effect quantitative conversion of the nitrogen in the organic sample to ammonia. The nitrogen compounds which require modification of the procedure can be classified into two categories as follows.

Heterocyclic Nitrogen Compounds. Heterocyclic aromatic compounds with nitrogen in the ring (e.g., nicotinic acid) require the presence of

mercury in the digestion mixture for the cleavage of the C-N-C bond. This
was reported by Phelps [6] and confirmed by Shirley and Becker [7]. Lake
and co-workers [8] have found that the presence of mercury and a digestion
time of 1 h at 370°C are necessary conditions. Since these conditions are
considerably more drastic than the regular micro Kjeldahl digestion process, this type of compounds are known as "refractory" samples.

Compounds with the N-O or N-N Linkage. For many years the Kjeldahl
method was employed only for the determination of nitrogen in amino compounds in which the nitrogen atom is in the oxidation state of -3. Since it
is possible to reduce organic compounds containing the N-O or N-N linkage
to amines, a number of investigators have attempted to modify the Kjeldahl
process in order to analyze these compounds. For instance, aromatic
nitro compounds [9] can be quantitatively reduced in the micro Kjeldahl
flask followed by the regular digestion and determination. A mixture of
phosphorus and hydriodic acid works well with the N-N linkage.

In view of the simplicity and convenience of the micro Kjeldahl method,
its extension to nonaminoid compounds is always welcome in the analytical
laboratory. It should be noted, however, that the modified micro Kjeldahl
procedures still await further study, as discussed in Sec. 3.1C.

The Micro Dumas Apparatus

After Pregl [10] adapted the Dumas method to the micro (3-5 mg)
scale, the combustion and measuring assembly remained practically unal-
tered for four decades until Shelberg [11] proposed the post-treatment tube
and Gustin [12] designed the new gasometer. In the classical method, the
organic nitrogen compound is oxidized in a combustion tube packed with
copper oxide and a section of metallic copper which serves to reduce the
oxides of nitrogen to molecular nitrogen. The latter is conducted into the
gasometer (called nitrometer or azotometer) by a current of pure carbon
dioxide. The nitrometer is filled with potassium hydroxide solution which
absorbs carbon dioxide and acidic gases. Because carbon dioxide disso-
ciates at the temperature employed in combustion to form carbon monoxide
and oxygen,

$$2CO_2 \xrightleftharpoons{800°C} 2CO + O_2$$

a cooling period is necessary before the gas mixture can be driven into the
nitrometer. In the Shelberg [11] arrangement, the combustion tube contains
only copper oxide while a short tube packed with copper and copper oxide
maintained at 400°C is placed between the combustion tube and the nitrometer. The Gustin [12] gasometer comprises an absorption chamber in which

the potassium hydroxide solution can be agitated by magnetic stirring to insure complete removal of all caustic-soluble gases. The absorption chamber is connected to a mercury-filled measuring capillary or to a measuring syringe attached to a micrometer screw. The latter has a shaft extension to engage the digital dial such that measurements are read directly to 0.001 ml. Taking advantage of these two innovations, Gustin [12] constructed an automated micro Dumas analyzer which was the prototype of the commercial model manufactured by Coleman Instrument Co.

The automated apparatus is presented in detail in Sec. 3.5, and the classical micro Dumas nitrogen combustion train in Sec. 3.4. Determination of nitrogen by the Dumas principle using the automatic C, H, N analyzers is described in Secs. 2.3-2.5.

C. Interferences and Difficulties

Unlike the determination of carbon and hydrogen, there are very few interferences due to heteroelements in the determination of nitrogen. In the Dumas method, fluorine-containing compounds may plug up the nitrometer inlet due to the hydrolysis of silicon tetrafluoride with the water condensate. This can be prevented by incorporating magnesium oxide in the packing. Incomplete combustion of aliphatic chains or polychlorinated compounds may produce volatile hydrocarbons or chlorocompounds which are not absorbed by potassium hydroxide. In the Kjeldahl process, the only element that interferes is platinum which, probably due to the formation of $Pt(NH_3)_x$ complexes, causes low results. Hence, if a platinum container is used to measure the sample, it should not remain in the digestion flask.

Prevention of leaks and insurance of complete absorption of carbon dioxide are the main concerns in the micro Dumas procedure. Heterocyclic nitrogen compounds and polymeric substances are difficult to combust. As an alternative to CuO, it is recommended that this type of sample be mixed with vanadium pentoxide or cobalt oxide. The latter is also employed for analyzing organometallic compounds. Abramyan and co-workers [12a] have recommended CuO in the presence of $KMnO_4$ for compounds containing the N-alkyl groups.

Useful as the Kjeldahl method is, its difficulties should be kept in mind. Since the reaction mechanism of sulfuric acid digestion has not been elucidated, the process is empirical. Wilfarth [13] found in 1885 that the presence of mercury facilitated digestion, and Gunning [14] in 1889 added potassium sulfate. It has been established that mercury has a definite catalytic effect on the decomposition of the organic material while potassium sulfate merely serves to raise the temperature of the medium during heating. For the latter purpose, the reaction can be performed by heating with sulfuric acid alone in a sealed tube. The temperature should be kept below 450°C in order to prevent oxidation of ammonia. The influence of halogens and some metallic elements in sealed-tube digestions has been studied [14a].

For the purpose of complementing the action of sulfuric acid to break down carbon skeletons, oxidizing agents have been proposed as additives among which may be mentioned hydrogen peroxide, perchloric acid, and permanaganate. The use of the last two oxidants requires extreme caution in order to avoid explosions.

When Kjeldahl digestion is carried out in an open vessel, the consensus among analysts is that a "catalyst" should be added. Numerous papers have been written on this subject, although there are no definite conclusions. It may be mentioned that these reagents are not catalysts in the real sense, since the amounts used are always larger than the organic sample to be analyzed and they are never recovered unchanged. Among the 40 elements tested, mercury is considered the best, followed by selenium. In our experience, selenium serves well for the determination of simple aliphatic and aromatic amines, amides, and α-aminoacids. Some workers have reported low nitrogen values when selenium is used at high temperatures and prolonged heating. It should be noted that, in the micro Kjeldahl procedure, nitrogen loss may occur if the temperature of the digest is too high or the time too long, because part of the ammonium sulfate solution may be evaporated to dryness at the edges with the concomitant thermal decomposition of the salt.

It goes without saying that the success of the Kjeldahl digestion depends on keeping the nitrogen atom in the oxidation state of -3 after cleavage of the C-N bond. Mercury probably helps by forming the stable nonvolatile $HgNH_2^+$ complex. For the same reason, it is important to destroy this complex (e.g., by adding thiosulfate) prior to the distillation of ammonia.

The nitrogen atoms in the pyridine and pyrrole nuclei, which generally are considered as belonging to the amino type, are resistant to hydrolytic cleavage unless mercury is present in the digestion mixture [15]. Probably a strong Hg-N bond is formed as the initial step. On the other hand, some heterocyclic compounds such as barbiturates [4] give a quantitative yield of nitrogen with the regular micro Kjeldahl procedure.

It is apparent that a nitrogen atom not at the oxidation state of -3 has to be reduced to this state before it can form ammonia. In organic chemistry, nitro compounds contain nitrogen atoms at the highest oxidation state. Thus, the ultimate reduction product of the aromatic nitro function is the amino function, and the intermediate stages are nitroso, azoxy, azo, and hydrazo functions:

$$2\ -NO_2 \xrightarrow{[H]} 2\ -NO \xrightarrow{[H]} \underset{\underset{O}{\diagdown\diagup}}{-N-N-} \xrightarrow{[H]} -N=N- \xrightarrow{[H]} \overset{H\ H}{\underset{|\ \ |}{-N-N-}} \xrightarrow{[H]} 2-NH_2$$

It is interesting to note that workers have reported quantitative conversion of nitro [16,17] and nitroso [18] compounds to ammonia by the regular micro Kjeldahl digestion. The explanation is that the reaction between sulfuric acid and organic matter produces sulfur dioxide and carbon, both of which are reducing agents. Nevertheless, for analytical purpose, it is recommended to add a reduction step prior to the sulfuric acid digestion. A number of reducing agents have been suggested, including organic substances like sucrose and cellulose. Using a large amount of organic reagent in the digestion mixture, however, tends to weaken the sulfuric acid. Raney nickel may serve as the reductant; but Tanabe [18a] has found that only one N in semicarbazones or phenylhydrazones is converted into NH_3.

It should be remembered that compounds containing the N=N linkage may lose a molecule of nitrogen upon heating. Even for a relatively stable compound like azobenzene, erratic results are obtained unless the reduction in acetic acid solution with zinc and hydrochloric acid is carried out slowly at room temperature [19]. Similarly, some compounds containing the nitrile function lose HCN on heating.

The modified micro Kjeldahl procedure may be applied to the determination of nitrogen in nitrates and nitrites. Nitrates are reduced in acid solution with chromium metal, and the ammonia is subsequently distilled and titrated. Nitrites, on the other hand, must be oxidized first with hypochlorite in alkaline medium to nitrates and then reduced, in order to prevent the loss of nitrogen oxides upon addition of acid.

Whereas open-chain N-N structures are amenable to the Kjeldahl technique after reduction with nascent hydrogen, this method cannot be extended to heterocyclic compounds containing the same linkage. For instance, while phenylhydrazine can be determined by the modified micro Kjeldahl procedure, its derived cyclic compounds (e.g., pyrazolones and tetrazolium salts) yield low results with or without the reduction step [20]. It appears that these cyclic structures always liberate some molecular nitrogen upon cleavage.

When the regular or modified micro Kjeldahl procedure is used to determine nitrogen in pure compounds, the results can be checked against the theoretical percentage of nitrogen. On the other hand, if the Kjeldahl method is used to analyze mixtures for nitrogen, the procedure should be standardized against a known sample of similar composition, and the experimental conditions should be closely controlled.

3.2 REVIEW OF RECENT DEVELOPMENTS

A. Summary of Recent Publications

Gustin and Ogg [21] have written an excellent account of the determination of nitrogen covering the literature up to 1962. The reader is also referred to the extensive discussion of the Dumas method by Kirsten [22], the Kjeldahl method by Bradstreet [23], and the review by Fleck [23a]. Recent publications are summarized in Table 3.1. It can be seen that the Dumas and Kjeldahl methods received about the same attention in the past decade. The absence of research papers on the ter Meulen method is significant. Unless an effective and long-life catalyst can be found, practicing analysts will shun the hydrogenation technique for quantitative analysis of nitrogen in organic materials. A patent [66] has been granted, however, on a device for nitrogen determination using nickel on magnesium oxide in humidified hydrogen, followed by coulometric determination of the resulting ammonia.

TABLE 3.1. Summary of Recent Literature on the Determination of Nitrogen.

Decomposition method	Product determined	Finish technique	Reference
Combustion train	N_2	Gasometric	24–56a
	N_2	Gas chromatographic	57–63a
	$^{15}N_2$	Mass spectrometry	64, 65
	NH_3	Coulometric	66
	$^{15}N_2$	Electrodeless discharge	66a
Kjeldahl digestion	$(NH_4)^+$	Titrimetric	67–87
	$(NH_4)^+$	Spectrophotometric	88–92a
	$(NH_4)^+$	Coulometric	93, 93a
	$(NH_4)^+$	Gravimetric	94
	N_2	Gasometric	95
Sealed tube	$(NH_4)^+$	Titrimetric	96–97
	^{15}N	Mass spectrometry	97a
Alkali fusion	CN^-	Titrimetric	98, 99
Hydrogenation	N_2	Thermal conductivity	99a
Neutron activation		Radiometric	111a
Chemical ionization	^{15}N	Mass spectrometry	112a

B. Methods of Decomposition

In the classical Dumas procedure, decomposition is carried out in an atmosphere of carbon dioxide, and oxygen gas is avoided because it complicates the collection of nitrogen. Recently, however, combustion in oxygen has been advocated, especially for materials difficult to combust or samples of low nitrogen content. For instance, Merz [55] has determined nitrogen in foodstuffs, fertilizers, resins, leather, etc. by combustion in pure oxygen in a vertical quartz tube at 850°C, while Thürauf and Assenmacher [63] have recommended a combustion temperature of 1000°C and Dindorf and Hoecker [53] favored 1100°C. The use of high temperature is necessary in fast combustions. However, Baba [63a] recommended 850°C for rapid flushed oxygen combustion. Kainz and Kasler [100] found considerable amounts of methane in the nitrometer with automatic combustion whereas no methane was present using the Pregl technique. By raising the temperature of the automatic combustion to 850-900°C, all traces of methane were eliminated.

   Cobalt oxide is preferred by most analysts for mixing with difficult samples in place of copper oxide. Silver manganate and manganese dioxide have been used by Wang and Jen [36] and Liu and Chou [37], while vanadium pentoxide has been proposed by Pippel and Romer [101] and Gelman et al. [101a] have employed PbO. For the analysis of nitrogen-containing fluorocarbons using the automated micro Dumas apparatus, Olson and Knafla [50] have used the packing sequence of CuO--NaF--CuO; for perfluorinated amines, a section of CuO is replaced by copper to minimize the formation of CF_4 resulting from attack of the $-NF_2$ group on carbon. For the determination of heterocyclic nitrogen compounds, Saran and co-workers [56] have recommended a combustion tube filling of the classical micro Dumas procedure that comprises 4 cm of platinized asbestos, 8.5 cm of CuO, 4.5 cm of Cu, and 8 cm of CuO.

   The digestion mixtures for the Kjeldahl method continued to be scrutinized. Baker [71] studied the influence of various metal catalysts and concluded that a mixture of HgO (20 mg), K_2SO_4 (2.25 g) and H_2SO_4 (1.5 ml) was the best. Rexroad and Gathey [71a] found that 0.04 g of $CuSO_4$ had the same effect as 0.7 g of HgO. Marten and Catanzaro [81] proposed a mixture containing 90% H_2SO_4, 2% $HClO_4$, 0.5% $HgSO_4$, and 0.02% SeO_2. These workers claimed that their combination avoided the formation of SO_3 which was responsible for the oxidation of some ammonia to molecular nitrogen. On the other hand, Terent'ev and Luskina [77] have recommended a mixture which contains CrO_3, $CuSO_4$, $K_2S_2O_8$, SeO_2, and H_2SO_4. Glowa [77a] has proposed to replace HgO with the nontoxic ZrO_2.

   Davidson et al. [92] reported that automatic digestion was not satisfactory; they have developed a system which employs manual digestion followed by automatic determination of ammonia. In order to perform digestion on a large number of samples simultaneously, Reardon [102] has constructed a rack that accommodates 60 test tubes to be heated on a hot-plate 18 x 12 in. (see Sec. 3.3). Faithfull [103] has described an aluminum block (18 x 4 x 4 in.) with 76 holes (17 x 88 mm); with two such units heated on an electric plate, 304 samples can be processed in a day.

C. Methods of Finish

Most papers on the micro Dumas method published during the past decade measured the volume of nitrogen produced. The conventional nitrometer is used with the classical procedure (see Sec. 3.4). As Teflon plugs became available, the problem of frozen stopcocks has been eliminated. Mitsui [41] has described an improved nitrometer which has a large capacity for caustic solution and can be used for 150 determinations before refilling. As shown in Fig. 3.1, it operates without a leveling bulb, the volume of gas being corrected to 1 atmosphere by a calibration graph. Gas chromatography and thermal conductivity measurement are the preferred finish for rapid and automatic determinations. Thürauf and Assenmacher [104] have designed an apparatus in which the combustion gases are conducted through potassium hydroxide solution and the remaining elementary nitrogen is measured by a thermal conductivity cell using helium as the carrier gas. The direct current voltage, proportional to the quantity of nitrogen, at the outlet of the thermal conductivity cell is recorded by an integrator. The constant gas flow necessary for the measurement is secured by a novel pressure equalizing system. Nitrogen quantities down to 1 μg can be detected. One determination takes 4 min. Automated procedures for micro Dumas determination of nitrogen in proteins [104a] and foodstuffs [105a] have been proposed.

Titrimetric finish is the normal procedure for micro Kjeldahl determinations that include the distillation step. Automatic analyzers in general omit distillation and use a spectrophotometric finish though it is not as accurate as the other methods. Cedergreen and Johansson [93] have proposed to determine the ammonia after digestion coulometrically in the presence of potassium bromide at pH 8.60 with a platinum–foil indicator-electrode and a saturated calomel reference electrode. Bromine is generated quantitatively at the anode:

$$2Br^- = Br_2 + 2e^-$$

In alkaline solution the bromine disproportionates to hypobromite, which reacts with ammonia so that 1 mole of ammonia corresponds to the generation of three redox equivalents,

$$3BrO^- + 2NH_3 = N_2 + 3Br^- + 3H_2O$$

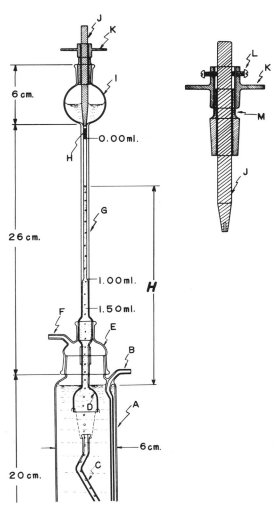

FIG. 3.1A. Micronitrometer of Mitsui.
(From Ref. 41, courtesy Microchem.
J.)

FIG. 3.1B. Micro Dumas assembly in the Shionogi Research
Laboratory, Osaka, Japan,
1975, using a modified Mitsui
micronitrometer.

Selenium cannot be used in the digestion process because of coulometric interference.

At a time when gravimetry had been losing favor in quantitative analysis, it was interesting to find a paper which describes the determination of ammonia after Kjeldahl digestion by weighing ammonium tetraphenylborate [94]. Mercury and potassium interfere. Mercury reacts nonstoichiometrically with sodium tetraphenylborate and must be removed, either by formation of the soluble tetraiodomercurate complex with sodium iodide before the precipitation stage, or by precipitation of HgS. Correction for potassium is made by boiling an aliquot of the sample to expel ammonia and precipitating potassium tetraphenylborate alone. This method of finish is not recommended unless the distillation apparatus is not available and nondistillation techniques cannot be utilized.

D. Comparison of Procedures

Several groups of investigators have made a comparison of procedures for the determination of nitrogen for various purposes. The procedures studied and materials analyzed are tabulated in Table 3.2. From time to time the Association of Official Analytical Chemists has carried out collaborative studies [105] on the Kjeldahl method using specific samples and experimental directions.

TABLE 3.2. Comparison of Procedures for the Determination of Nitrogen.

Procedures compared	Materials tested	Reference
Classical Dumas vs C,H,N analyzers	Pure compounds	106
Kjeldahl vs. classical Dumas	Biological materials	68
Kjeldahl vs. automated Dumas	Biological materials	107
Kjeldahl vs. automated Dumas	Fertilizers	108
Kjeldahl vs. automated Dumas	Agricultural materials	109
Kjeldahl vs. automated Dumas	Feeds	105
Kjeldahl vs. C,H,N analyzer	Petroleum	110
Kjeldahl vs. automated Kjeldahl	Milk products	82, 82a
Kjeldahl vs. automated Kjeldahl	Foods	83a
Enzymatic vs. automated Kjeldahl	Blood urea	111
Neutron activation vs. Kjeldahl	Biological materials	111a

The general conclusion from the recent studies is that the new automatic instruments for the determination of nitrogen in organic materials satisfy the requirements of quantitative analysis. Childs and Henner [106] have found that, while the classical micro Dumas procedure gives the most reliable results for pure compounds, acceptable results can be achieved by either the Perkin-Elmer C,H,N analyzer or the Hewlett-Packard C,H,N analyzer. Studies [68] on biological materials using the classical Dumas and the Kjeldahl procedures have shown both to be of equal precision. Comparison of the Coleman automated Dumas apparatus and Kjeldahl procedure by Stitcher and co-workers [107] indicates close agreement in results obtained by the two methods. The Coleman analyzer has been tested extensively in agricultural analysis. Morris and co-workers [108,109] have found that the Kjeldahl procedure is more precise, but that the automated Dumas procedure will allow analysis of fertilizers for nitrogen well within the tolerances for control purposes. Ebeling [105] has reported the results of a collaborative study comparing the automatic Dumas procedure and the Kjeldahl method for nitrogen determination in feeds and recommended that the automatic Dumas method be adopted as official first action by the Association of Official Analytical Chemists.

In the analysis of heterogeneous solid plant samples, the particle size has a definite effect on the nitrogen values. For instance, Hamlyn and Gasser [112] have found that the coefficient of variation in measuring the percentage of nitrogen in spruce seedlings decreases with fineness of grinding and that the percent nitrogen in finer fractions of material ground to pass a 1 mm sieve increases with decreasing size. No normally-ground material is satisfactory using the Kjeldahl method with 50-mg samples. When the Dumas procedure is employed, difficulty may be encountered in transferring the sample into the combustion tube because the finely ground particles (e.g., hay) adhere electrostatically to the sides of the quartz tube.

The comparative study on petroleum samples by Smith et al. [110] is of interest because the percent nitrogen is at low levels and because nitrogen compounds in petroleum are usually of the refractory type. These investigators have found the automatic C,H,N instrument useful for the analysis of nitrogen in petroleum ranging from 0.05 to 0.1% nitrogen, a borderline level that presents difficulty for conventional methods.

Carpenter and LaFleur [111a] have compared the determination of nitrogen in biological materials by the nuclear track technique and by the Kjeldahl method. The overall elapsed time is the same, but there are fewer manipulations involved in the neutron activation method. Also there is no differentiation due to the chemical state of the nitrogen in the sample, therefore some of the uncertainties of the Kjeldahl analysis are removed.

3.3 EXPERIMENTAL: DETERMINATION OF NITROGEN BY THE MICRO
     KJELDAHL METHOD

A. Principle

When a compound containing aminoid nitrogen is heated with concentrated
sulfuric acid, the organic nitrogen is converted to ammonium sulfate. The
reaction is catalyzed by the presence of certain metals and facilitated by
raising the reflux temperature through the addition of potassium sulfate.
Some nonaminoid substances such as nitro and nitroso compounds can be
analyzed by the micro Kjeldahl method after being subjected to reduction
with zinc and hydrochloric acid prior to sulfuric acid digestion. Hydra-
zides can be reduced by means of red phosphorus and hydriodic acid (see
Sec. 3.3D).

     After the organic nitrogen is quantitatuvely converted to ammonium
sulfate, the reaction mixture is transferred to a distilling apparatus where
sodium hydroxide is introduced to liberate ammonia. The latter is absorbed in boric acid solution and determined by direct titration with
standardized 0.01 $\underline{N}$ HCl.

B. Apparatus

The equipment for micro Kjeldahl determination is simple and easily constructed as described below. Any commercial apparatus or other home-
made device can be used with slight modification of the experimental
procedure.

     Digestion Flasks

     These can be made from glass tubing or 150-mm (6-in.) test tubes
with the bottom blown out to form a bulb of about 25-mm diameter and 6-ml
capacity. For the procedure which includes a prereduction step, the bulb
should have a capacity of 15 ml.

     Digestion Stand

     A digestion stand for six flasks can be made from a hard asbestos
board, 10 x 35 cm, with six openings of 22 mm diameter drilled in it (Fig.
3.2). The board is set on a metal frame with four legs 10 cm high. A
metal wire supports the necks of the digestion flasks. The burner is made
from copper tubing 45 cm long, situated 2.5 cm below the openings of the
asbestos board. Six holes, each about 1 mm in diameter, are drilled along
the copper tubing which is connected to a Bunsen burner, or propane gas
cylinder with a regulator. If the digestion is not carried out in the hood, a
fume duct, which is connected to the water tap through a glass aspirator,
is added to the stand.

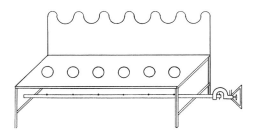

FIG. 3.2. Simple digestion rack. (Courtest <u>Anal. Chem.</u>)

Distillation Apparatus

The distilling apparatus illustrated in Fig. 3.3 comprises two compact units joined glass-to-glass by means of a short rubber tubing B. The whole apparatus is conveniently clamped onto an iron stand and occupies an area 30 x 40 cm. The steam generator A is fabricated from a 1-liter flask. When it is two-thirds filled with water before distillation begins, enough steam will be generated for 8-12 determinations. The plug for the funnel E is made of perfluoro plastic (Teflon) resistant to strong alkali.

C. Reagents

   Sulfuric acid, concentrated, nitrogen-free.
   Selenium powder, or selenized Hengar granules.
   Copper sulfate-potassium sulfate mixture (1:1).
   Mercuric oxide.
   Sodium hydroxide solution, 30%.
   $NaOH/Na_2S_2O_3$ solution (required when mercuric oxide is used):
      Dissolve 120 g of NaOH and 20 g of $Na_2S_2O_3 \cdot 5H_2O$ in 400 ml of
      distilled water.
   Boric acid solution, 2%.
   Indicator solution: Mix 50 ml of 0.2% bromocresol green with 10 ml
      of 0.2% methyl red, both in 95% ethanol.
   Standardized HCl, 0.01 <u>N</u>.

For reduction:

   Zinc powder.
   Acetic acid.
   Methanol.
   Concentrated hydrochloric acid.
   Red phosphorus.
   Hydriodic acid, 57%.

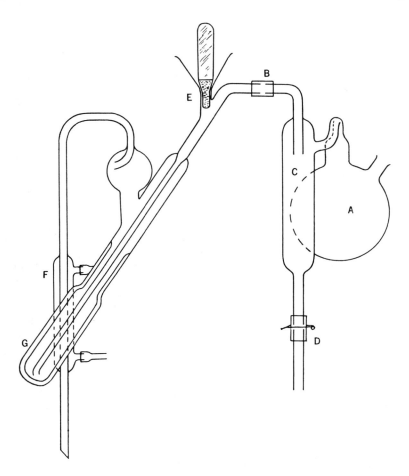

FIG. 3.3. Simple micro Kjeldahl distillation apparatus. (From Ref. 1, courtesy J. Wiley & Sons, Inc.)

D. Procedure

Digestion for Amino Compounds

Accurately weigh 2-5 mg of sample (containing 0.14 to 1.4 mg of nitrogen) and transfer it to the bottom of the digestion flask. Use a long-handle weighing tube (Fig. 1.7) for solids, and a porcelain or glass microboat for semisolids and heavy oils. Weigh volatile liquids in a one-chamber capillary; cut the tip and handle and introduce all three into the digestion flask; use a glass nail to crush the capillary under sulfuric acid; keep the glass nail in the flask.

Add about 10 mg of selenium powder (see Sec. 3.3F, Note 1) and 40 mg of copper sulfate-potassium sulfate mixture, followed by 1 ml of concentrated sulfuric acid. Place the digestion flask on the digestion stand and boil the reaction mixture gently with a flame about 2 cm high. The reaction is usually complete in 10 min. Upon cooling, add 2 ml of distilled water and cool the solution again.

Reduction and Digestion of Nitro Compounds

Accurately weigh 3-8 mg of the compound (corresponding to about 0.5 mg nitrogen) and transfer it to the bottom of the digestion flask. Add 1 ml of acetic acid to dissolve the sample; warm the flask, if necessary, to effect solution. Upon cooling, introduce 100 mg of zinc powder and 1.5 ml of methanol. Now add two drops of concentrated hydrochloric acid to generate hydrogen gently. When the evolution of gas slows down, introduce more hydrochloric acid. Toward the end, heat the flask over a small flame to keep the evolution of hydrogen proceeding smoothly. Then add two drops of sulfuric acid and gently boil the reaction mixture to remove the volatile solvent (but not to dryness). Upon cooling, add 1.5 ml of sulfuric acid and again heat the flask until the solution darkens. Add 700 mg of potassium sulfate and 25 mg of selenium powder, followed by 0.5 ml of sulfuric acid. Then digest the reaction mixture by boiling gently for 1 h. Allow the contents of the digestion flask to cool; before the mixture completely solidifies, carefully introduce 3 ml of distilled water along the wall of the flask.

Reduction and Digestion of Hydrazides

Weigh an appropriate amount of sample in a porcelain boat or micro weighing bottle (Fig. 1.11) and place in a micro Kjeldahl flask. Into the flask add a small amount of red phosphorus on the tip of a microspatula, one selenized Hengar granule, and 1 ml of hydriodic acid, and heat the mixture with a low flame for 15 min. Cool, and wash down the walls with 2 ml of distilled water; add one more selenium granule and 2 ml of concentrated sulfuric acid and boil gently. With the aid of a Bunsen burner, drive off the excess iodine by playing the burner on the walls of the micro Kjeldahl flask. Continue for about 30 min until all the iodine is driven off and the reaction mixture turns gray-white. Remove the flask from the heat, cool, add $CuSO_4 \cdot K_2SO_4$ and HgO catalysts and digest for 15 min.

Distillation and Titration

Bring the water in the steam generator A (Fig. 3.3) to boiling with a Bunsen burner flame about 10 cm high. Insert the plug to funnel E and close pinch clamp D. (With cold water running through condenser F, the distillation rate should be about 5 ml per min.) Now move the burner away

from the steam generator A, whereupon the liquid in the distilling flask G
is sucked back into the trap C. Fill funnel E with distilled water and mo-
mentarily lift up the plug to drain the water into flask G. Replace the
burner under the steam generator A for about 20 s and again move it away.

Meanwhile, deliver 5 ml of 2% boric acid and 0.05 ml of mixed indicator into a 50-ml conical borosilicate flask. When the distilling flask
has been automatically emptied, replace the Bunsen burner under the
steam generator A, open pinch clamp D to drain the liquid in trap C into
a beaker. Leave the pinch clamp D on the glass tubing, through which the
steam escapes. Now place the conical flask containing boric acid under the
condenser F and support the flask in an oblique position so that the tip of
the condenser is completely immersed in the liquid.

Smear a trace of Vaseline on the lip of the digestion flask to prevent
the contents of the flask from running over to the outside wall. Remove
the plug of the funnel E and pour the contents of the digestion flask into the
distilling flask G. Quickly rinse the digestion flask twice with 2-ml portions of distilled water and pour the rinsings into the distilling flask G.
Now, introduce the sodium hydroxide solution (8 ml for amino compounds,
15 ml for nitro compounds) through the funnel E and replace the plug.
Move the pinch clamp D to close the rubber tubing, whereupon steam enters
the distilling flask G and stirs up its contents. Ammonia which is liberated
passes through the condenser and is absorbed in the boric acid solution.

The solution in the conical flask changes from bluish purple to bluish
green as soon as ammonia reaches it. One minute after the color change,
lower the conical flask so that the condenser tip is 10 mm above the liquid
surface. Wash the condenser tip with a little distilled water and continue
distillation for another minute. Then move the burner away from the
steam generator A. Bring the conical flask to the microburet and titrate
the ammonia with standardized 0.01 $\underline{N}$ HCl until the blue color disappears.
(See Note 2.)

E. Calculation

$$\%N = \frac{(V_S - V_B) \times N_{HCl} \times 14.01 \times 100}{\text{Sample weight in mg}}$$

where V_S = ml of HCl used by sample
$\quad\quad V_B$ = ml of HCl used by blank
$\quad\quad N_{HCl}$ = normality of HCl

F. Notes

Note 1: Add mercuric oxide to the digestion mixture for N-heterocyclics.
In that case, the $NaOH/Na_2S_2O_3$ solution should be used in the distillation
step in order to decompose the mercury ammonium complex ion.

Note 2: Standardized sulfuric acid or potassium biniodate solution may be
used in place of 0.01 $\underline{N}$ HCl. If preferred, the titration may be continued
until a faint pink tinge appears; 0.02 ml is then subtracted from the buret
reading. There is no danger of missing the endpoint, because after the
pink color appears, the color intensity increases tremendously with a trace
more of the 0.01 $\underline{N}$ acid.

G. Comments

(1) Determination of nitrogen by the Kjeldahl principle is performed routinely in many laboratories, especially for biological materials. The
success of the experiment is dependent on the digestion procedure which
has been under continuous scrutiny. For instance, Concon and Soltess
[113] have recently reported two rapid digestion methods for the analysis
of fish meal, beef, keratin, blood, and urine, as well as cereals. In one
method, the ground sample (corresponding to 0.5-2.5 mg of nitrogen) is
mixed with 2.3 g of K_2SO_4/HgO (95:2) and 2.3 ml of H_2SO_4. A few drops
of 50% dodecanoic acid in ethanol (to prevent foaming) and 3 ml of 30%
H_2O_2 are added and the mixture is heated at 350°C. After 10-15 min, the
solution is cooled, 10 ml of H_2O and a few drops of octanol are added,
and ammonia is then determined. In another method, which produces less
SO_2, 1 ml of 30% H_2O_2 is added to a hot mixture of the sample, catalyst,
and H_2SO_4. Heating is continued for 5 min after the mixture is clear and
the determination of ammonia is carried out without the addition of octanol.
Hadorn and Obrist [114] have undertaken a systematic study of various
catalysts for the Kjeldahl digestion process with respect to food samples,
particularly egg products. Kramme and co-workers [115] have proposed a
comprehensive method for aqueous dispersible samples; materials with a
wide spectrum of refractability to digestion and containing as much as 10%
carbohydrate in the sample are successfully digested.
    (2) Using the catalyst mixture which contains $CuSO_4/K_2SO_4$ (3:1),
HgO, and selenized Hengar granules, excellent results are obtained for
monomeric and polymeric amines, amides, imides, imines, nitriles, iso-
cyanates, cyanates, and saturated N-heterocyclics (e.g., piperazine). In
trace analysis where as much as 500 mg of sample is heated for as long as
6 h, no loss of nitrogen occurs.

(3) While quaternary ammonium compounds are considered to belong to the aminoid type, they are not always amenable to Kjeldahl determination. It has been found that the Kjeldahl values are always too low for tetraethylammonium borohydrides, $(C_2H_5)_4N \cdot B_xH_y$, either by the regular procedure or with the modified procedure including prereduction. The only way to determine these compounds is to pass them through an anion-exchange resin (e.g., IRA-400-Mallinckrodt) and titrate the eluate with standard acid.

$$(C_2H_5)_4N^+B_xH_y^- \xrightarrow{\quad OH^- \quad} (C_2H_5)_4N^+OH^-$$

$$(C_2H_5)_4N^+OH^- + HCl \longrightarrow (C_2H_5)_4N^+Cl^- + H_2O$$

The procedure is as follows. Fill a column with the anion-exchange resin. Add 1 $\underline{N}$ NaOH to convert the resin to its OH form. Wash with distilled water until the eluate is neutral. Now place a solution of the sample on the column and rinse several times with water (total volume, 50 ml); collect the eluate in a conical flask containing 2% boric acid and the mixed indicator. Titrate with standardized 0.01 $\underline{N}$ HCl.

(4) Pyridine cannot be determined by the Kjeldahl method even with prereduction or treatment with mercuric oxide. Pyridine in aqueous solutions is determined by steam-distillation into acetic anhydride and non-aqueous titration with standardized perchloric acid in the following manner. The aqueous sample containing 5-10 mg of pyridine is steam-distilled after being made strongly alkaline with 15 ml of 30% NaOH. The distillation rate is approximately 7 ml/min, and the distillation time is about 1 min. The receiver contains acetic acid and acetic anhydride. After distillation, excess acetic anhydride is added (total volume, 150 ml), and the solution is boiled to convert acetic anhydride into acetic acid. Upon cooling, the solution is titrated with 0.01 $\underline{N}$ HClO$_4$.

(5) The micro Kjeldahl apparatus is also suitable for the determination of inorganic nitrates by the following procedure. Weigh the sample into the micro Kjeldahl flask and dissolve it in about 1 ml of water. Add a few alundum boiling chips and some chromium powder. Then add 1 ml of concentrated hydrochloric acid. When the reaction subsides, boil for a few minutes but not to dryness. Then add the catalyst and 1 ml of concentrated H_2SO_4 and proceed as in the regular micro Kjeldahl experiment.

(6) It should be mentioned that the reduction procedure for nitrates cannot be applied to the determination of inorganic nitrites. The latter will be lost in the prereduction step. Inorganic nitrites, however, can be analyzed by means of the micro Kjeldahl apparatus after being oxidized to nitrates in the following manner. Dissolve the inorganic nitrite in the micro Kjeldahl flask with 1 ml of water. Add $Ca(OCl)_2$ and boil, whereby nitrate is formed:

$$NO_2^- + OCl^- \longrightarrow NO_3^- + Cl^-$$

The resulting nitrate is then determined as described in the previous paragraph.

(7) See Appendix A for macro and semimicro Kjeldahl procedures recommended by US Pharmacopeia.

3.4 EXPERIMENTAL: DETERMINATION OF NITROGEN BY THE MICRO DUMAS METHOD

A. Principle

The organic material is mixed with copper oxide and heated at high temperature in an atmosphere of carbon dioxide. Part of the nitrogen in the sample is evolved as nitrogen gas and the remainder is converted to nitrogen oxides. The latter are reduced to elementary nitrogen by hot metallic copper. The gas stream is conducted through a solution of potassium hydroxide, which absorbs the carbon dioxide and other acidic gases, leaving the pure nitrogen to be measured in the microazotometer.

B. Apparatus

Combustion Train

The micro Dumas combustion train may be assembled in various ways. Figure 3.4 shows the American Society for Testing and Materials standard form [116], arranged for left-handed operation. It consists of the gasometer assembly A which is connected to a source of carbon dioxide, the Z-shaped connecting tube B, combustion tube C which is heated by two electric furnaces, three-way stopcock D, and microazotometer E which is connected to the leveling bulb F. The gasometer assembly A, incorporated for the purpose of measuring the volume of carbon dioxide fed into the combustion tube, can be advantageously eliminated. If the source of carbon dioxide is not pure, however, the gasometer should be used with a correction factor proportional to the volume of gas delivered into the combustion tube.

Carbon Dioxide Generator

The modified Poth generator, illustrated in Fig. 3.5, is recommended for routine analysis. The apparatus consists of two connecting bulbs in an atmosphere of carbon dioxide under a pressure about 20 mmHg higher than atmospheric. A concentrated solution of potassium bicarbonate is placed

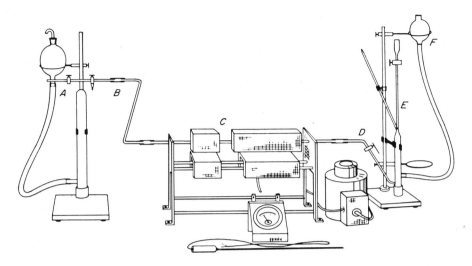

FIG. 3.4. Assembly for nitrogen determination by the micro Dumas method. (From Ref. 24, STANDARD METHODS OF CHEMICAL ANALYSIS, Vol. 2A, 6th Ed., edited by Frank J. Welcher, © 1963 by Litton Educational Publishing, Inc., reprinted by permission of Van Nostrand Reinhold Company.)

in the upper bulb, while the lower bulb holds dilute sulfuric acid. After the apparatus has been properly filled [117], the system is permanently free from air and will last for several hundred microdeterminations.

If Dry Ice (solid CO_2), free from air, is available, a simple carbon dioxide generator can be prepared from a thermos bottle fitted with a rubber stopper and gas pressure relief valve [118] (Fig. 3.6). The thermos bottle is filled with crushed Dry Ice and the stopper with the relief valve is put in place. Pressure will be built up and it is quickly released by opening the stopcock. This process is repeated several times to expel all the air in the bottle, and the generator is then ready for use. One filling will supply enough carbon dioxide for about 10 microdeterminations.

Microazotometer and Connecting Stopcock

The microazotometer has a capacity of 1.5 ml and is graduated in 0.01 ml divisions. The volume of nitrogen collected is estimated to 0.001 ml with the aid of a magnifying glass. Small tapered grooves are cut in the plug of the stopcock D in order to facilitate fine adjustment of the gas flow.

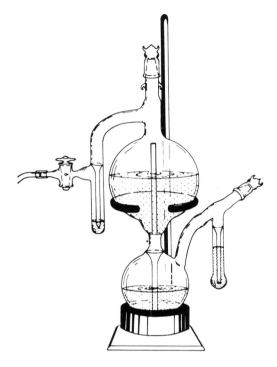

FIG. 3.5. Modified Poth carbon dioxide generator. (From Ref. 24, STANDARD METHODS OF CHEMICAL ANALYSIS, Vol. 2A, 6th Ed., edited by Frank J. Welcher, © 1963 by Litton Educational Publishing, Inc., reprinted by permission of Van Nostrand Reinhold Company.)

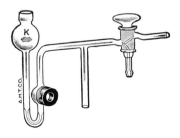

FIG. 3.6. Gas pressure relief valve. (Courtesy A. H. Thomas Co.)

Heating Devices

If all heating is done by electricity, both furnaces should be of the type that can be opened up and retracted away from the combustion tube. The two furnaces should be connected to separate variable resistors so that the long furnace is maintained at 700-800°C and the short furnace can be regulated from 250 to 900°C.

C. Reagents

Copper oxide, wire form.
Copper oxide, fine powder.
Cooper gauze, 100 mesh.
Potassium hydroxide solution (1:1, wt/wt).

D. Procedure

Filling the Microazotometer

Pour dry mercury into the microazotometer E (Fig. 3.4) so that it reaches a level about 3 mm above the side arm joined to the stopcock. Add a pinch of mercuric oxide (made by heating mercuric chloride in a small test tube) and a 10-mm headless iron nail. The former provides a rough mercury surface and prevents accumulation of gas bubbles at the bottom, while the latter can be brought up to the top (with the aid of a magnet) to break up any foam on the surface of the potassium hydroxide solution. Fill the leveling bulb F with potassium hydroxide solution until the liquid level extends about 10 mm inside the cup on top of the microazotometer. One such filling lasts for at least 50 determinations.

Packing the Combustion Tube

Pack the permanent filling of the combustion tube (Fig. 3.7) as follows. Push a wad of silver thread to the tip of the combustion tube to serve as a plug. Holding the combustion tube in vertical position, pour copper oxide wires through the funnel so that this section extends 50 mm beyond the long furnace and also 50 mm within it. Tap the tube carefully to form a compact section. Push in a little long-fiber asbestos to serve as a partition. Then introduce the roll of copper gauze which is wrapped around a heavy copper wire 50 mm long and fits the combustion tube snugly. Add another asbestos partition, followed by another compact section of copper oxide wires. (This section should reach the edge of the long burner.) Push in some asbestos and press it tightly against the copper oxide. (See Sec. 3.4F, Note 1.) This permanent filling need not be changed for at least 100 microdeterminations.

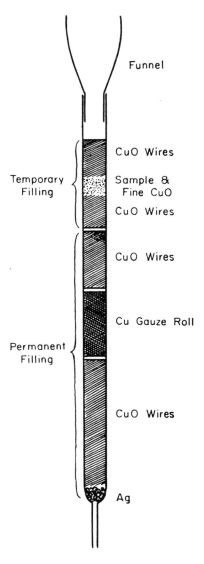

Funnel

CuO Wires

Temporary
Filling

Sample &
Fine CuO

CuO Wires

CuO Wires

Cu Gauze Roll

Permanent
Filling

CuO Wires

Ag

FIG. 3.7. Packing the micro Dumas combustion tube for a solid
sample. (From Ref. 24, STANDARD METHODS OF CHEMICAL ANALY-
SIS, Vol. 2A, 6th Ed., edited by Frank J. Welcher, © 1963 by Litton
Educational Publishing, Inc., reprinted by permission of Van Nostrand
Reinhold Company.)

The combustion tube tip is connected to the microazotometer by means of a 30-mm long paraffin-impregnated rubber tubing. The wide end of the combustion tube is fitted with a one-hole rubber stopper through which the tapered tip of the Z-tube B protrudes to deliver carbon dioxide.

Analysis of Solid Sample

Using the microweighing tube (Fig. 1.7) weigh accurately 2-8 mg of the sample into a mixing tube (Fig. 3.8) which is fitted with either a good cork stopper or ground-glass cap. Disconnect the combustion tube from the microcombustion train (Fig. 3.4). Remove the temporary filling left from the previous analysis. Insert the funnel (Fig. 3.7), which should have a smooth constriction so that the charge can be poured down readily. Introduce copper oxide wires to form a section of about 50 mm, followed by 10 mm of fine copper oxide. Now add a 10 mm layer of fine copper oxide into the mixing tube containing the sample. Replace the stopper and shake the contents to mix the sample with copper oxide. Tap the tube gently to bring down the powder which might have stuck to the stopper. Then remove the stopper and pour the mixture through the funnel into the combustion tube. Introduce some fine copper oxide to the empty mixing tube, stopper, shake, and pour the contents into the combustion tube. Repeat this once more. Finally pour in copper oxide wires to form the last section of about 50 mm.

Clean the neck of the combustion tube with a wad of cotton. Lubricate the rubber stopper for connection with Z-tube B with a trace of glycerol and insert into the combustion tube. Lubricate the paraffin-impregnated rubber tubing and join the combustion tube tip with the microazotometer

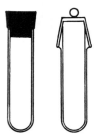

FIG. 3.8. Mixing tubes. (left) Cork stopper. (right) Glass cap. (From Ref. 24, STANDARD METHODS OF CHEMICAL ANALYSIS, Vol. 2A, 6th Ed., edited by Frank J. Welcher, © 1963 by Litton Edicational Publishing, Inc., reprinted by permission of Van Nostrand Reinhold Company.)

stopcock D, glass-to-glass. Then connect the combustion tube to the CO_2 source by inserting the tapered tip of the Z-tube B through the rubber stopper hole until it protrudes about 2 mm.

With the leveling bulb F in the lower ring and the stopcocks of the microazotometer opened, pass a rapid current of carbon dioxide through the combustion tube. After 3 min, close the stopcock D and then open it slightly to let the gas bubbles enter the azotometer slowly. If the bubbles disappear on reaching the top of the potassium hydroxide solution, the system is completely free of air. Now turn off the CO_2 source, close stopcock D and fill the microazotometer to the cup on top by raising the leveling bulb F. Replace F in the lower ring. Slowly open stopcock D, whereupon bubbles enter the azotometer but should be completely absorbed by the potassium hydroxide solution.

Heat the permanent filling, by means of the long furnace, to 700–800°C. Place the short furnace near the neck of the combustion tube in a position about 20 mm from the sample; gradually raise the temperature to 700°C. Then slowly advance the short furnace toward the sample (and the long furnace). Expansions of the gas volume in the combustion tube due to heating and oxidation of the organic sample will cause a steady stream of gas bubbles to enter the azotometer. When the short furnace meets the long furnace and the gas bubbles stop, close stopcock D, turn on the CO_2 source, and very carefully reopen stopcock D so that bubbles enter the azotometer at the rate of 2 bubbles per second. Nitrogen produced from the sample is carried by CO_2 into the azotometer and rises to the top. Now move the short furnace back to near the neck and advance it once more toward the long furnace in the course of 5 min. Then turn off both furnaces.

When all bubbles entering the azotometer again completely disappear on rising to the graduated portion, close stopcock D and place leveling bulb F in a position such that the meniscus of the liquid in the bulb is at the same level as that in the azotometer. After 5 min, read the volume of nitrogen to 0.001 ml, and record the temperature and atmospheric pressure. If a little liquid adheres to the tip of the azotometer below the cup, push it into the cup by holding the leveling bulb F above the level of the cup and carefully opening the stopcock. If a foam is formed on the surface of the potassium hydroxide solution, break it up by guiding the iron nail with the aid of a magnet to the foam and move the nail up and down. Read the nitrogen volume again after such adjustments.

Analysis of Semisolid or Oil Samples

Weigh a semisolid or oil sample of low vapor pressure in a porcelain microboat (see Sec. 1.6B) shortly before analysis and keep it in the microdesiccator. Pour into the combustion tube, after removing the temporary filling, a 50 mm section of copper oxide wires followed by 20 mm of fine copper oxide. Lay the combustion tube in a horizontal position, place the microboat containing the sample at the neck, and carefully push it toward the fine copper oxide. Then hold the combustion tube vertically and pour in fine copper oxide to cover the microboat, followed by 50 mm of copper oxide wires. Assemble the combustion train and carry out the analysis as described above. (Note 2.)

Analysis of Volatile Liquid

Weigh the volatile liquid in a double-chamber capillary (Fig. 1.8). Prepare a roll of copper oxide gauze about 50 mm long by winding copper gauze around a rod of 2-mm diameter and heating it over a strong flame. Remove the temporary filling from the previous run, introduce 50 mm of copper oxide, and place the combustion tube horizontally. Insert about 30 mm of the cold copper oxide roll into the neck of the combustion tube; cut the tip of capillary and place both in the 2-mm channel. Push the roll inside the combustion tube. Now hold the combustion tube vertically and cover the roll with fine copper oxide, followed by 50 mm of copper oxide wires. Clean the neck of the combustion tube and proceed with the experiment according to the directions given for solid samples.

E. Calculation

Obtain the corrected volume of nitrogen (corr. vol. N_2) by subtracting a blank value (see Note 3).

Corr. vol. N_2 = azotometer reading - blank

$$\%N = \frac{273}{\text{temperature}} \times \frac{\text{pressure}}{760} \times \frac{\text{corr. vol. } N_2 \times 1.2505}{\text{sample weight in mg}} \times 100$$

F. Notes

Note 1: The two sections of copper oxide wires in the permanent filling may be replaced by copper oxide gauze rolls. The latter are prepared by oxidizing rolls of 100 mesh copper gauze of suitable length over a strong flame.

Note 2: Some microanalysts also use the microboat to hold solid samples for micro Dumas determinations. The mixing tube technique, however, provides intimate contact of the organic substance with copper oxide. It gives better results except for compounds which are decomposed on shaking with copper oxide, liberating nitrogen. In the latter case (also for explosive materials), use a long boat and spread out the sample along the boat. Cover the sample with fine copper oxide before placing the boat in the neck of the combustion tube.

Note 3: The best way of obtaining the corrected volume of nitrogen is to calibrate the azotometer against pure nitrogen compounds. Use appropriate sample sizes to produce volumes between 0.3 and 1.2 ml in order to check the rectilinearity of the calibration graph. Pregl [10] proposed to deduct 2% from the azotometer reading; Niederl and Niederl [119] recommended a 1% deduction together with a blank correction; Steyermark [120] used a 2% correction.

3.5 EXPERIMENTAL: DETERMINATION OF NITROGEN USING THE COLEMAN NITROGEN ANALYZER MODEL 29

A. Principle

The organic sample is combusted in an atmosphere of CO_2 in the presence of copper oxide in a manner similar to the classical micro Dumas procedure. The combustion products are swept with CO_2 over metallic copper where the oxides of nitrogen are reduced to nitrogen and the resultant gaseous mixture is scrubbed thoroughly in an absorption chamber containing 50% KOH. All the CO_2 and other acidic gases which may be present (such as halogens and halogen acids and oxides of sulfur) are removed by the 50% KOH and the only remaining gas, nitrogen, is collected and measured in a $5000-\mu l$ stainless steel syringe linked to a digital counter by a precision micrometer screw. The operator is required only to weigh the sample and insert it into the combustion tube and, after the combustion, to adjust and read the counter indicating the nitrogen volume. All the conventional stopcock and furnace manipulations are performed automatically. The instrument is calibrated by running a pure standard compound such as acetanilide or barbituric acid in a similar manner. A determination may be completed in approximately 10 min.

B. Apparatus

Coleman Nitrogen Analyzer Model 29 (see Fig. 3.9).
Porcelain boats (for solids).
Combustion tube (quartz), 12-mm o.d. x 8-mm i.d. x 346-mm ℓ.
Post-heater tube (reduction tube), (Vycor), 12-mm o.d. x 8-mm i.d. x 205-mm ℓ.
Quartz capillary tubes (for liquids) 1.0-mm i.d. x 25-mm ℓ.

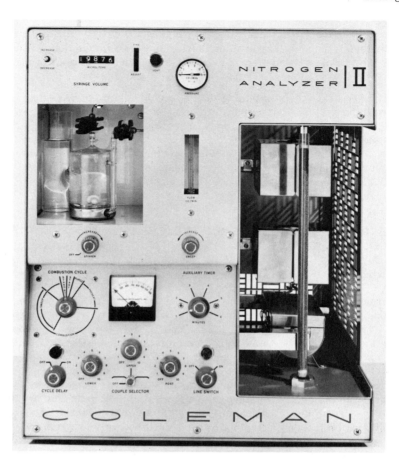

FIG. 3.9A. Coleman Nitrogen Analyzer Model 29. (Courtesy Coleman Instruments.)

C. Reagents

In combustion tube:

Copper oxide (Cuprox)
Glass wool

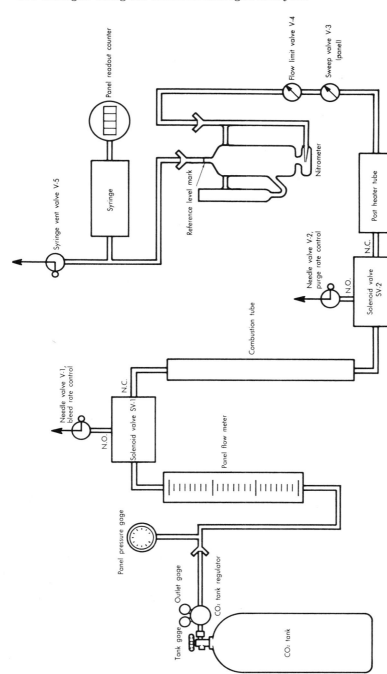

FIG. 3.9B.   Schematic flow diagram of Coleman Nitrogen Analyzer.   (Courtesy Coleman Instruments.)

In post heater tube (reduction tube):

> Copper (Cuprin).
> Copper oxide (Cuprox).
> Silver vanadate.
> Glass wool.

In nitrometer:

> 50% KOH (Causticon).
> Vanadium pentoxide (for refractory materials).
> Reagent grade (bone dry) CO_2: The CO_2 tank should be conditioned
> by packing overnight in Dry Ice, to a point just below the valve
> stem. The next day remove the tank from Dry Ice and open the
> tank valve. Allow the valve to remain open until the tank pressure
> nears atmospheric pressure and the tank contents escape slowly
> (not longer than 15 s). Allow the tank to stand for 15 min, then
> repeat the preceding step. Do this several times. The tank should
> now be sufficiently purged of contaminants to give a blank value of
> 10 μl or less. If the blank is greater than 10 μl, repeat the recti-
> fication procedure.

D.   Procedure

Fill the combustion and post-heater tubes as described in Fig. 3.10.   Fill
the nitrometer by delivering mercury into the capillary side arm until the
level reaches the inlet to the KOH chamber.   Then carefully pour 50% KOH
into the nitrometer reservoir until the liquid level just reaches the calibration mark on the central capillary.   Install the packed combustion and post-
heater tubes and filled nitrometer.   Adjust the upper and lower furnace
temperatures to approximately 850°C and the post-heater furnace to 500-
550°C.   Adjust the CO_2 pressure to 4-5 lb/in.^{2}, the flow rate to 30 ml/
min, and the meniscus of the KOH to the calibration mark on the central
capillary of the nitrometer.   With the cycle delay in the off position, the
timer on 1, the line switch on, and the stirrer speed previously adjusted
to 5-8 revolutions per second, the instrument is ready for operation.

Several blank determinations are first run.   With only an empty boat
in the combustion tube, move the combustion cycle control from standby to
start.   This starts the automatic program from purge to preheat to first
and final combustions and finally to sweep.   At the completion of the cycle,
rotate the readout counter to return the KOH to the calibration line.   The
difference between the final and initial counter readings is the magnitude of
the blank in microliters.   The main criteria for the blanks is that they are
consistent.   However, the lower the magnitude of the blank the better.   A

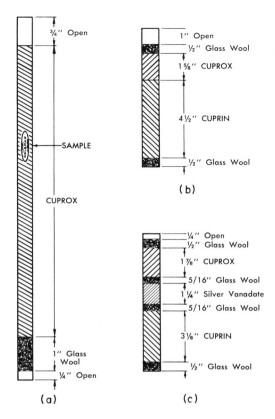

FIG. 3.10. Tube packings for Coleman Nitrogen Analyzer. (a) Combustion tube. (b) Post-heater tube. (c) Post-heater tube (sulfur and halogens). (Courtesy Coleman Instruments.)

consistent blank of 10 μl or less is quite acceptable. As soon as an acceptable blank has been established, the nitrogen factor is determined. The nitrogen factor is the microgram amount of nitrogen equivalent to each microliter of nitrogen collected. The factor is determined by running a pure sample of NBS-141 acetanilide or other suitable standard through the cycle under the same conditions that the unknown samples will be run. For best results it is advisable to use a standard which has a nitrogen content approximating that of the materials to be determined. It has been found that measurement of temperature and barometric pressure is completely unnecessary provided that the nitrogen factor is determined twice a day or

every 4 h. The elimination of all temperature and barometric pressure
readings greatly simplifies the subsequent calculations. The sample size
to be taken should contain between 0.5 and 0.8 mg of N. If less than 100
μl of N_2 is collected the determination should be repeated with at least
double the sample size. If more than 2000 μl is collected, repeat the procedure using a smaller sample. Solids are weighed in porcelain boats,
and liquids in quartz capillary tubes.

E. Calculation

As mentioned in Sec. 3.5D, elimination of temperature and barometric
pressure measurements greatly simplifies the calculations. The equation
for the calculation of %N in a sample is as follows:

$$\%N = \frac{N_f \times V_c \times 100}{W}$$

where N_f = Nitrogen factor (μg N/μl)
$\quad\quad\quad V_c$ = volume in microliters observed by subtracting the initial
$\quad\quad\quad\quad\quad$ counter reading from the final counter reading and then deduct-
$\quad\quad\quad\quad\quad$ ing the blank
$\quad\quad\quad W$ = weight of sample in micrograms

The nitrogen factor N_f is obtained by running a pure sample of known
nitrogen content through the cycle as one would run an unknown. It represents the microgram amount of nitrogen equivalent to each microliter of
nitrogen collected. The calculation of N_f is illustrated in the following
example.

A pure sample of NBS-141 acetanilide weighing 8.074 mg is combusted
in the Nitrogen Analyzer and 751 μl of N_2 are collected. The blank which
has been previously determined is 10 μl.

$$N_f = \frac{\mu g \text{ of N present in standard}}{\mu l \text{ of } N_2 \text{ collected (corrected for blank)}}$$

$$= \frac{8074 \times 0.1036 \text{ (theoretical N content of acetanilide)}}{751 - 10}$$

$$= \frac{836.5}{741}$$

$$= 1.129$$

F. Maintenance

Change the combustion tube and KOH solution every two days. Change the post-heater tube every week (normal operation).

G. Comments

(1) Refractory materials are covered with approximately twice the sample weight of vanadium pentoxide before placing in the combustion tube.

(2) For explosive samples (e.g., diazonium salts), use minimum sample weight and provide more free space (less CuO) above the sample. This provides room for rapid expansion when an explosive sample detonates.

(3) At the completion of each determination, after the combustion tube is removed, invert the neoprene washers of both the upper and lower supports. Then replace the combustion tube filled with next sample to be run. The pressure exerted by the combustion tube against the neoprene washers tends to crease them and eventually will cause leaks. Inverting the washers each time the combustion tube is removed prolongs their effective usefulness.

(4) The use of two packed combustion tubes which are used alternately is also recommended. When one is removed it is placed on a rack to cool. The other tube, which has already been filled with sample, is now ready to be installed into the apparatus. After a few minutes, the tube which has now cooled on the rack can be filled with sample. On completion of the cycle, it will be ready for insertion into the apparatus.

(5) Nonvolatile stable samples can be weighed in advance and lined up for future insertion into the combustion tube. This technique, coupled with the use of two combustion tubes, leads to a very efficient operation.

(6) At the end of the day's operation it is recommended that the spring clamps be removed from the nitrometer and that both gas tubing connections also be removed. This is an extra precaution taken to prevent backup of mercury or KOH into the valve or syringe, respectively.

(7) Knight and Inouye [121] have reported that cinchophen derivatives can be analyzed on the Coleman Nitrogen Analyzer with the addition of potassium perchlorate.

(8) Tetraethylammonium borohydrides always give high values, probably due to the formation of B_2H_6 or C_2H_6.

(9) Sels and Demoen [122] have described a modified combustion and reduction system and an improved endpoint detection method, thus achieving a 2σ confidence limit of 0.1% N. These same workers [123] have also modified the instrument in order to perform the gravimetric microdetermination of carbon and hydrogen.

(10) Inorganic nitrates can be analyzed in the Coleman Nitrogen Analyzer, but not in the Perkin-Elmer C, H, N Analyzer (Sec. 2.3).

(11) Larina and Gel'man [124] have studied the analysis of difficultly combustible substances in the Coleman Nitrogen Analyzer and recommended the following catalysts: CuO for amines that contain P, S, or Si (but not F); NiO plus MgO for fluorides, nitriles, heterocyclic nitrogen compounds, and nitro derivatives; and NiO plus PbO for polymers with carborane chains and for thermally modified nitrogenous polymers.

REFERENCES

1. N. D. Cheronis and T. S. Ma, Organic Functional Group Analysis, Wiley, New York, 1964, p. 222.
1a. Encyclopedia Britainica, Encyclopedia Britainica Inc., Chicago, 1973, Vol. 7, p. 750.
2. J. Kjeldahl, Z. Anal. Chem. 22:366 (1883).
3. H. ter Meulen, Roc. Trav. Chim. Pays-Bas 43:463 (1924).
4. T. S. Ma and G. Zuazaga, Ind. Eng. Chem., Anal. Ed. 14:280 (1942).
5. W. Horwitz (ed.), Official Methods of Analysis of the AOAC, 12th Ed., Association of Official Analytical Chemists, Washington, D.C., 1975, p. 927.
6. I. K. Phelps, J. Offic. Agr. Chem. 3:306 (1920).
7. R. L. Shirley and W. W. Becker, Ind. Eng. Chem., Anal. Ed. 17:437 (1945).
8. G. R. Lake, P. McCutchan, R. van Meter, and J. C. Neel, Anal. Chem. 23:1634 (1951).
9. T. S. Ma, R. E. Lang, and J. D. McKinley, Jr., Mikrochim. Acta 1957:368.
10. F. Pregl, Die Quantitative Organische Mikroanalyse, 3rd Ed., Springer, Berlin, 1930, p. 90.
11. E. F. Shelberg, Anal. Chem. 23:1492 (1951).
12. G. M. Gustin, Microchem. J. 1:75 (1957); 4:43 (1960).
12a. A. A. Abramyan, R. A. Megroyan, and M. A. Kazaryan, Arm. Khim. Zh. 27:706 (1974).
13. H. Wilfarth, Chem. Zontr. 56:17 (1885).
14. J. W. Gunning, Z. Anal. Chem. 28:188 (1889).
14a. J. Horacek and Z. Sir, Collect. Czech. Chem. Commun. 40:1143 (1975).
15. T. S. Ma and E. E. Jaffe, unpublished work.
16. B. M. Margosches, W. Kirsten, and E. Scheinost, Berichte 56:1943 (1923).
17. P. R. W. Baker, Analyst 80:481 (1955).

18.   F. Zinneke, Angew. Chem. 64:220 (1952).
18a. Y. Tanabe, Japan Analyst 21:747 (1972).
19.   T. S. Ma and A. T. Spencer, unpublished work.
20.   T. S. Ma, A. T. Spencer, and B. Arnowich, unpublished work.
21.   G. M. Gustin and C. L. Ogg, in I. M. Kolthoff and P. J. Elving
      (eds.), Treatise on Analytical Chemistry, Part II, Vol. 11, Wiley,
      New York, 1965, p. 405.
22.   W. J. Kirsten, in C. L. Wilson and D. W. Wilson (eds.), Compre-
      hensive Analytical Chemistry, Vol. I-B, Elsevier, Amsterdam,
      1960, p. 467.
23.   R. B. Bradstreet, The Kjeldahl Method for Organic Nitrogen,
      Academic, New York, 1965.
23a. A. Fleck, CRC Crit. Rev. Anal. Chem. 4:141 (1974).
24.   T. S. Ma, in F. J. Welcher (ed.), Standard Methods of Chemical
      Analysis, 6th Ed., Vol. 2, Van Nostrand, Princeton, 1963, p. 377.
25.   F. Canal and A. Alemanni, Attualita Lab. 5:120 (1959).
26.   G. Kainz and L. Hainberger, Z. Anal. Chem. 166:27 (1959).
27.   M. Vecera and L. Synek, Collect. Czech. Chem. Commun. 24:3402
      (1959).
28.   W. J. Kirsten, Z. Anal. Chem. 174:282 (1960).
29.   L. Dorfman, R. Oeckinghaus, F. Anderson, and G. I. Robertson,
      Anal. Chem. 34:678 (1962).
30.   M. Vecera, Talanta 8:446 (1961).
31.   S. Mizukami and K. Miyahara, Mikrochim. Acta 1962:717.
32.   G. Kakabadse and B. Manohin, Analyst 86:512 (1961).
33.   A. A. Abramyan, A. A. Kocharyan, and A. G. Karapetyen, Izv.
      Akad. Nauk Arm. SSR, Khim. Nauk 15:226 (1962).
33a. R. Fiedler, G. Proksch, and A. Koepf, Anal. Chim. Acta 63:435
      (1973).
34.   E. J. Browne and J. B. Polya, Anal. Chem. 34:298 (1962).
35.   M. F. Buckles, C. A. Rush, and J. M. Corliss, Microchem. J.,
      Symp. Ser. 2:535 (1962).
36.   M. H. Wang and M. L. Jen, Acta Chim. Sinica 29:301 (1963).
37.   C. W. Liu and P. J. Chou, Chem. Bull. (Peking) 1964:186.
38.   H. Truknovsky, Z. Anal. Chem. 198:331 (1963).
39.   M. N. Chumachenko, Izv. Akad. Nauk SSR, Ser. Khim. 1963:1893.
40.   M. N. Chumachenko and I. E. Pakhomva, Izv. Akad. Nauk SSSR,
      Sor. Khim. 1963:2090.
41.   T. Mitsui, Microchem. J. 7:277 (1963).
42.   A. Campiglio, Farmaco, Ed. Sci. 19:1021 (1964).
43.   M. N. Chumachenko and I. E. Pakhomva, Izv. Akad. Nauk SSSR,
      Ser. Khim. 1964:1561.
44.   Y. H. Sha and T. Y. Wang, Acta Chim. Sinica 30:513 (1964).

45. M. F. Buckles, Microchem. J. 9:449 (1965).

46. T. A. Mikhailova and N. V. Khromov-Borisov, Zh. Anal. Khim. 20:359 (1965).

47. A. A. Abramyan and L. E. Pogosyan, Izv. Akad. Nauk Arm. SSSR, Khim. Nauk 19:188 (1966).

48. C. S. Yeh, Microchem. J. 11:229 (1966).

49. H. Swift, Microchem. J. 11:193 (1966).

50. P. B. Olson and R. T. Knafla, Microchem. J. 13:362 (1968).

51. W. Merz, Z. Anal. Chem. 237:272 (1968).

52. V. S. Bostoganashvili and D. G. Turabelidze, Tr. Inst. Farmakohkhim., Akad. Nauk Gruz. SSR 1:36 (1967).

53. W. Dindorf and H. Hoecker, Anal. Chem. 250:29 (1970).

53a. R. Küber, W. Padowetz, and J. Pavel, Mikrochim. Acta 1972:591.

54. K. I. Nakamura, K. Ono, and K. Kawada, Microchem. J. 15:364 (1970).

55. W. Merz, Glas.-Instrum.-Tech. Fachz. Lab. 14:617 (1970).

56. J. Saran, P. N. Khanna, and S. Banerji, Mikrochim. Acta 1972:252.

56a. H. Trutnovsky, Microchem. J. 19:347 (1974).

57. M. Vecera, Collect. Czech. Chem. Commun. 26:2308 (1961).

57a. D. Fraisse and N. Schmitt, Microchem. J. 22:109 (1977).

58. M. Simek and K. Tesarik, Collect. Czech. Chem. Commun. 26:1337 (1961).

58a. I. A. Sketova, A. P. Znamenskaya, and V. A. Akinin, Tr. Gos. Nauchno-Issled Proektn. Inst. 20:75 (1973).

59. M. L. Parsons, S. N. Pennington, and J. M. Walker, Anal. Chem. 35:842 (1963).

59a. F. A. Mamina, V. M. Gorchakova, and A. I. Gerasimova, Dokl. Neftekhim. Sekt. Bashkir. Resp. Pravl 1973:140.

60. H. Hochenberg and J. Gutberlet, J. Brennst. Chemie 44:235 (1963).

61. J. Mlodecka, Chem. Anal. (Warsaw) 9:535 (1964).

62. J. M. Gautheret, S. Kubiak, and A. Nicco, Chim. Anal. 52:283 (1970).

63. W. Thürauf and H. Assenmacher, Z. Anal. Chem. 245:26 (1969).

63a. Y. Baba, Microchem. J. 21:315 (1976).

64. R. J. Barsdate and R. C. Dugdale, Anal. Biochem. 13:1 (1965).

64a. S. P. Pavlou, G. E. Friederich, and J. J. MacIsaac, Anal. Biochem. 61:16 (1974).

65. R. Fiedler and G. Proksch, Anal. Chim. Acta 60:277 (1972).

65a. C. P. Lloyd-Jones, J. S. Adam, G. A. Hudd, and D. G. Hill-Cottingham, Analyst 102:473 (1977).

66. Standard Oil Co., British Patent 1,149,832 (1967).

66a. T. A. Scott and R. P. Humpherson, Lab. Pract. 23:703 (1974).

67. T. S. Ma, in F. J. Welcher (ed.), Standard Methods of Chemical Analysis, 6th Ed., Vol. 2, Van Nostrand, Princeton, 1963, p. 383.

67a. T. Batey, M. S. Cresser, and I. R. Willet, Anal. Chim. Acta 69:484 (1974).
 68. H. Hubsch and K. Nehring, Z. Anal. Chem. 173:278 (1960).
68a. T. Chiba and Y. Takata, J. Chem. Soc. Japan 1975:472.
 69. V. Fojtova and J. Purs, Chem. Listy 55:201 (1961).
 70. O. I. Milner and R. J. Zahner, Anal. Chem. 32:294 (1960).
70a. J. G. M. M. Smeenk, Anal. Chem. 46:302 (1974).
 71. P. R. W. Baker, Talanta 8:57 (1961).
71a. P. R. Rexroad and R. D. Cathey, J. Ass. Offic. Anal. Chem. 59:1213 (1976).
 72. J. Albert, Microchem. J., Symp. Ser. 2:527 (1962).
 73. A. P. Terent'ev and B. M. Luskina, Zh. Anal. Khim. 17:227 (1962).
 74. M. Ashraf, M. K. Bhatty, and R. A. Shah, Anal. Chim. Acta 25:448 (1961).
 75. M. H. Hashimi, E. Ali, and M. Umar, Anal. Chem. 34:988 (1962).
 76. W. Kuhnert and H. J. Bauer, Pharm. Zentralh. Deut. 102:431 (1963).
 77. A. P. Terent'ev and B. M. Luskina, Tr. Komis. Anal. Khim. Akad. Nauk SSSR 13:20 (1963).
77a. W. Glowa, J. Ass. Offic. Anal. Chem. 57:1228 (1974).
 78. S. K. Mookherjea, Indian J. Chem. 2:507 (1964).
 79. D. M. Ekpete and A. H. Cornfield, Analyst 80:670 (1964).
 80. A. Gertner and V. Grdinic, Acta Pharm. Jugoslav. 15:200 (1965).
 81. J. F. Marten and E. W. Catanzaro, Analyst 91:42 (1966).
 82. W. B. T. Schultz, Z. Lebensm.-Unters. Forsch. 134:353 (1967).
82a. H. W. Schafer and N. F. Olson, Anal. Chem. 47:505 (1975).
 83. M. Dunke, Faserforsch. Text. Tech. 18:123 (1967).
83a. A. Montag, Gordian 74:203 (1974).
 84. W. C. Urban, Anal. Chem. 43:800 (1971).
 85. M. Ashraf, M. A. Siddiqui, and M. K. Bhatty, Talanta 15:559 (1968).
 86. P. N. Fedoseev and V. D. Osadchii, Izv. Vyssh. Ucheb. Zaved. Tekhnol. Legk. Prom. 1969:57.
 87. E. N. Zelenina and F. M. Shemyakin, Zh. Vses. Khim. Obshchest. 13:464 (1968).
 88. M. Kasagi and M. Ito, Japan Analyst 9:105 (1960).
88a. J. R. Quinn, J. G. Boisvert, and I. Wood, Anal. Biochem. 58:609 (1974).
 89. H. Kala, Pharmazie 18:29 (1963).
89a. M. P. Strukova and G. I. Veslova, Zh. Anal. Khim. 28:1026 (1973).
 90. A. I. Lebedeva and I. V. Novozhilova, Zh. Anal. Khim. 22:582 (1967).
90a. W. J. Blaedel and R. H. Laessig, in C. N. Reilley and F. W. McLafferty (eds.), Advances in Analytical Chemistry and Instrumentation, Vol. 5, Wiley, New York, 1966, p. 69.

91. C. W. Gehrke, F. E. Kaiser, and J. P. Ussary, J. Ass. Offic.
     Anal. Chem. 51:200 (1968).
91a. C. W. Gehrke, L. L. Wall, and J. S. Absheer, J. Ass. Offic.
     Anal. Chem. 56:1096 (1973).
92. J. Davidson, J. Mathieson, and A. W. Boyne, Analyst 95:181 (1970).
92a. J. A. Bietz, Anal. Chem. 46:1617 (1974).
93. A. Cedergreen and G. Johansson, Sci. Tools 16:19 (1969).
93a. C. A. Bostrom, A. Cedergreen, G. Johansson, and I. Pettersson,
     Talanta 21:1123 (1974).
94. F. E. Crane and E. A. Smith, Anal. Chim. Acta 31:258 (1964).
95. J. Seto, Japan Analyst 9:669 (1960).
96. R. A. Shah and A. A. Quadri, Talanta 10:1083 (1963).
96a. P. N. Pedoseev, Izv. Vyssh. Ucheb. Zaved., Khim. Khim.
     Tekhnol. 15:1885 (1972).
97. L. I. Glebko, J. I. Ulkina, and V. E. Vas'kovskii, Anal. Biochem.
     20:16 (1967).
97a. O. Tsuji, M. Masugi, and Y. Kosai, Japan Analyst 22:1363 (1973).
98. E. E. Sirotkina and R. G. Toropova, Izv. Tomsk. Politech. Inst.
     126:87 (1964).
99. M. N. Chumachenko and R. A. Mukhamedshina, Izv. Akad. Nauk
     SSSR, Ser. Khim. 1965:1262.
99a. K. Ubik, Microchem. J. 18:29 (1973).
100. G. Kainz and F. Kasler, Z. Anal. Chem. 168:425 (1959).
101. G. Pippel and S. Romer, Chem. Tech. (Berlin) 15:173 (1963).
101a. N. E. Gel'man, N. I. Larina, and I. S. Chekasheva, Zh. Anal.
     Khim. 37:612 (1972).
102. G. V. Reardon, Lab. Pract. 18:651 (1969).
103. N. T. Faithfull, Lab. Pract. 18:1302 (1969); 20:41 (1971).
104. W. Thürauf and H. Assenmacher, Z. Anal. Chem. 250:111 (1970).
104a. H. Kreutzer, Glas.-Instrum.-Tech. Fachz. Lab. 18:353 (1974).
105. M. E. Ebeling, J. Ass. Offic. Anal. Chem. 50:38 (1967); 51:766
     (1968).
105a. T. L. Lunder, Lab. Pract. 23:170 (1974).
106. C. E. Childs and E. B. Henner, Microchem. J. 15:590 (1970).
107. J. M. Stitcher, C. R. Joliff, and R. M. Hill, Clin. Chem. 15:258
     (1969).
108. G. F. Morris, R. B. Carson, and W. T. Jopkiewicz, J. Ass.
     Offic. Anal. Chem. 52:943 (1969).
109. G. F. Morris, R. B. Carson, D. A. Shearer, and W. T.
     Jopkiewicz, J. Ass. Offic. Anal. Chem. 51:216 (1968).
110. A. J. Smith, G. Myers, Jr., and W. C. Shaner, Jr., Mikrochim.
     Acta 1972:217.
111. M. A. Warwick, Microchem. J. 17:160 (1972).

111a.   B. S. Carpenter and P. D. LaFleur, Anal. Chem. 46:1112 (1974).

112.   F. G. Hamlyn and J. K. R. Gasser, Chem. Ind. (London) 1970:1142.

112a.   C. V. Lundeen, A. S. Viscomi, and F. H. Field, Anal. Chem. 45:1288 (1973).

113.   J. M. Concon and D. Soltess, Anal. Biochem. 53:35 (1973).

114.   H. Hadorn and C. Obrist, Deut. Lebensmitt. Rundsch. 69:109 (1973).

115.   D. G. Kramme, R. H. Griffin, C. G. Hartford, and J. A. Corrado, Anal. Chem. 45:405 (1973).

116.   American Society for Testing and Materials, Specification E-1148-59 T.

117.   E. Poth, Ind. Eng. Chem., Anal. Ed. 3:202 (1931).

118.   A. Steyermark, H. K. Alber, V. A. Aluise, E. W. D. Huffman, J. A. Kuck, J. J. Moran, and C. O. Willits, Anal. Chem. 21:1560 (1949).

119.   J. B. Niederl and V. Niederl, Micromethods of Quantitative Organic Analysis, 2nd Ed., Wiley, New York, 1942.

120.   A. Steyermark, Quantitative Organic Microanalysis, 2nd Ed., Academic, New York, 1961.

121.   D. M. Knight and K. Inouye, Microchem. J. 13:87 (1968).

122.   F. Sels and P. Demoen, Microchem. J. 18:107 (1973).

123.   F. Sels and P. Demoen, Mikrochim. Acta 1970:48.

124.   N. I. Larina and N. E. Gel'man, Zh. Anal. Khim. 31:189 (1976).

4

Oxygen

4.1 GENERAL CONSIDERATIONS

A. Historical Background

Next to carbon and hydrogen, oxygen is the element which is most commonly found in organic materials. As a matter of fact, during the early period of the development of organic chemistry, practically all substances investigated were oxygen-containing compounds. These were carboxylic acids and derivatives thereof, which the chemists isolated from natural products and painstakingly elucidated their compositions by chemical analysis [1]. As is now well known, most of such compounds contain only carbon, hydrogen, and oxygen; hence it would be a simple task to ascertain their compositions if these three elements could be quantitatively analyzed. Surprisingly, while carbon and hydrogen analysis was accomplished in the 1830s, reliable methods for the determination of oxygen did not come forth until more than a century later.

Perusal of the chemical literature prior to 1950 will reveal that no oxygen analysis was reported for new organic compounds containing oxygen. Whereas Pregl together with his many co-workers methodically studied the methods for organic microanalysis [2], he never tackled the problem of oxygen determination, in spite of the fact that he was drawn to quantitative organic analysis through his interest in C, H, O compounds, namely, bile acids. During the 1920s, ter Meulen [3, 4] made many attempts in Delft, The Netherlands, to determine oxygen by catalytic hydrogenation of the organic sample to produce water. Subsequently, this method was intensively studied in Pittsburgh, Pennsylvania, by Kirner [5] who gave up because he could not obtain accurate results for sugars and other compounds with high oxygen contents [6]. In 1939 Schütze [7] published a semimicro method to determine oxygen by pyrolysis of the organic substance in a large excess of carbon to obtain carbon monoxide. Zimmermann [8] adapted this technique to the milligram scale, and the validity of the micro method was confirmed by Unterzaucher [9]. Afterwards other investigators [10, 11] also reported successful determinations using similar apparatus. Thus a method for the direct determination of oxygen was finally accepted in the organic analytical laboratory during the 1950s.

It is worthy of note that oxygen was the very last element in the periodic table for which an accurate direct method of determination became available. Previously, the organic chemist had to determine oxygen in his sample by "difference," i.e., by subtracting from 100% the sum of the percentages of carbon, hydrogen, and other elements analyzed. The inadequacy of such an approach is obvious: (1) The absolute percentage error of oxygen content would be the sum total of the errors of all the other elements, and (2) an element which was actually present in the organic sample could be overlooked. As an illustration, the case of the elucidation of the structure of penicillin during World War II may be cited. First, the presence of sulfur was not expected and the missing percentages after C, H, and N values had been subtracted were ascribed to oxygen [12]; secondly, since the accuracy of C, H, N and S determinations was ±0.3% absolute, the uncertainty of the oxygen percentage was in the region of 1%; thirdly, for compounds like penicillin, having five elements and a molecular weight of over 300, the accuracy of the percent hydrogen determined could not be easily assessed. Hence it was difficult to ascertain the atomic ratios in the molecule unless all elements were analyzed and their percentages added up to 100%. The significance of direct oxygen determination, if it could be accomplished at that time, is apparent.

B. Decomposition Techniques

Pyrolytic Reduction with Carbon

The basic principle of this decomposition method is to convert the oxygen in the organic material into carbon monoxide quantitatively. When an organic substance is heated at high temperature in an inert atmosphere in a large excess of carbon, three compounds of oxygen can be formed, namely, carbon monoxide, carbon dioxide, and water vapor:

$$\text{Organic O} + C \xrightarrow{\Delta} CO + CO_2 + H_2O_{(g)} \tag{1}$$

Based on the water gas reaction,

$$H_2O_{(g)} + C \rightleftharpoons CO + H_2 \tag{2}$$

and the reduction of carbon dioxide by means of carbon,

$$CO_2 + C \rightleftharpoons 2CO \tag{3}$$

it has been established that the equilibria expressed by equations (2) and (3), respectively, proceed quantitatively to the right at 1120°C. Hence this temperature was prescribed in the early procedures [7-11] for oxygen analysis. Later Oita and Conway [13] reported that the combustion temperature could be lowered to 900°C by incorporating platinum as a catalyst, which was quickly confirmed by Oliver [14]. Stefanac and co-workers have found that the pyrolytic reduction can be carried out in an empty quartz tube by flash combustion; the temperature required is 1120°C by covering the sample in a platinum boat with carbon [15] but lowered to 900°C by using platinized carbon [16]. Pella and Andreoni [16a] have performed the hydrolysis using 40% nickel-coated carbon in a large quartz capsule.

Catalytic Hydrogenation

The aim of catalytic hydrogenation for direct oxygen determination is to obtain water from the organically bound oxygen, thus

$$C, H, O \text{ compound} + H_2 \xrightarrow{\text{catalyst}} H_2O + CH_4 \tag{4}$$

This method was investigated by ter Meulen [3] but has not been pursued by his successors in Delft [17]. Other workers [18-21], however, have revived it from time to time. Various catalysts have been proposed, including platinum, nickel, and nickel chromite. Mlinko [19] has described a procedure which involves conducting the sample and pyrolytic products over palladium, then nickel, and finally copper to yield water. Terent'ev and co-workers [20] have employed a mixture of nitrogen and hydrogen, with nickel or nickel-ThO$_2$ on pumice as catalyst. For sulfur-containing compounds [21], the sample is heated strongly for 10 min in a silica tube and sulfur is absorbed in a 15-cm layer of copper at 350°C, which precedes the hydrogen catalyst also at 350°C.

Other Methods

Sheft and Katz [22] have described an apparatus to decompose the organic compound in a mixture of bromine trifluoride and antimony pentafluoride at 500°C in a vacuum line; oxygen is liberated quantitatively in the reaction. Lee and Meyer [23] have heated the organic substance with carbon and strontium oxide in a nickel bomb to convert organically-bound oxygen into carbonate. Applications of these two techniques have not been further investigated.

Because the conventional method cannot be used to determine oxygen in fluorinated materials (see Sec. 4.1D), the isotopic dilution technique has been advocated. Kirshenbaum and co-workers [24] pyrolyze the sample in the presence of gaseous oxygen-18 in a platinum tube attached to a vacuum manifold; the liberated carbon dioxide is collected for mass spectral measurement of its ^{18}O distribution, from which the oxygen content of the unknown material is calculated. Olson and Kulver [25] have found that this technique usually gives low results with polyfluorinated compounds, apparently because of incomplete isotopic exchange. They have described an improved method, previously proposed by Boos and co-workers [26], which involves the use of ^{18}O-enriched succinic acid as tracer. The unknown material is pyrolyzed together with ^{18}O-containing succinic acid in a sealed, evacuated copper tube. After equilibration at 850°C, the tube is punctured and the pyrolysis gases are admitted to a mass spectrometer. Comparison of the m/e 46/44 ratio ($C^{16}O^{18}O/C^{16}O^{16}O$) with the ratio obtained from similar pyrolysis of the ^{18}O-compound only, provides a means of calculating the oxygen content of the unknown.

Several nuclear reactions can be utilized to determine oxygen in organic materials [27-32]. For instance, Keller and Muenzel [33] have analyzed cellulose by irradiating the sample for 60 s in a current of 0.5-2 μA using the reaction $^{16}O(n, 2n)^{15}O$. Deuterons of energy 50 MeV are produced in a cyclotron and subsequently decelerated by a graphite disk; the neutrons produced are allowed to impinge on the target material. Rafaeloff [34] has employed the reaction $^{16}O(t, n)^{18}F$ to determine oxygen in the range of 25 to 250 mg in organic materials. Lithium compounds are used as a tritium source:

$$^{6}_{3}Li + ^{1}_{0}n \longrightarrow ^{4}_{2}He + ^{3}_{1}H \qquad (5)$$

$$^{16}_{8}O + ^{3}_{1}H \longrightarrow ^{18}_{9}F + ^{1}_{0}n \qquad (6)$$

Very fine powders of the organic sample and lithium fluoride are mixed and sealed with liquid paraffin (to exclude moisture) and irradiated in a neutron flux of approximately 4×10^{11} neutrons per cm^2 per s for 15 min, and the ^{18}F produced is measured by γ, γ-coincidence counting.

C. Measurement of End Products

As discussed in Sec. 4.1B, the chemical reactions used in the determination of oxygen lead to the formation of carbon monoxide, carbon dioxide, water, or gaseous oxygen. The last-mentioned product is only obtained in one procedure [22] and is measured manometrically. Water is usually determined gravimetrically by weighing the absorption tube (see Sec. 2.6). Mlinko [19], however, has recommended a gasometric finish by converting the water to nitrogen gas in the following manner:

$$H_2O + Ba(OCN)_2 \longrightarrow NH_3 \xrightarrow{\text{CuO, } 520°C} N_2 \tag{7}$$

The nitrogen is then collected in the microazotometer (see Sec. 3.4) and measured.

A wide range of methods have been proposed for determination of the reaction products obtained in the carbon reduction of oxygenated compounds. The majority of these methods involve the conversion of the carbon monoxide into carbon dioxide, by conducting the combustion gases either through copper oxide [35, 36] at high temperature,

$$CO + CuO \xrightarrow{530-560°C} CO_2 + Cu \tag{8}$$

or through iodine pentoxide [7-9] at 120°C,

$$5CO + I_2O_5 \xrightarrow{120°C} 5CO_2 + I_2 \tag{9}$$

The carbon dioxide thus formed can be determined by one of the following methods: (1) weighing the absorption tube [8, 37] packed with Ascarite (see Sec. 2.6), (2) potentiometric titration [38] with 0.01 $\underline{N}$ Ba(OH)$_2$, (3) conductimetry [39] after absorption in NaOH solution, (4) coulometric titrimetry [35], (5) thermal conductivity measurement [36], (6) absorption in 0.5 $\underline{M}$ BaCl$_2$ at pH 10 and back titration with 0.01 $\underline{M}$ NaOH, or (7) absorption in pyridine and titration with sodium methoxide (see Sec. 4.3B).

Alternatively, the iodine produced in equation (9) can be determined. For this purpose, the procedure commonly employed involves iodometric titration [9] (see Sec. 4.4). The iodine is collected in 20% sodium hydroxide solution, and bromine in acetate buffer is then added to oxidize the iodine to iodate. Excess of bromine is removed by means of formic acid; sulfuric acid and potassium iodide are added and the iodine formed is determined by titration with standardized sodium thiosulfate solution to the starch endpoint. The sequence of reactions can be depicted as follows:

$$I_2 + Br_2 \longrightarrow 2IBr \tag{10}$$

$$2IBr + 4Br_2 + 6H_2O \longrightarrow 2HIO_3 + 10HBr \tag{11}$$

$$2HIO_3 + 10HI \longrightarrow 6I_2 + 6H_2O \tag{12}$$

$$6I_2 + 12Na_2S_2O_3 \longrightarrow 12NaI + 6Na_2S_4O_6 \tag{13}$$

Calme and Keyser [42] have constructed an absorption cell whereby the iodine vapors from the I_2O_5 tube are quantitatively driven into a 15% KI solution and automatically titrated with 0.01 $\underline{N}$ $Na_2S_2O_3$ potentiometrically. Karrman and Karlsson [43] have described a procedure (see Sec. 4.5) in which the iodine formed in the I_2O_5 tube is led by a stream of nitrogen to the cathode chamber of an electrolysis cell. The iodine is reduced at controlled potential at a rotating platinum electrode; the amount of electricity consumed is determined by an electronic integrator and read from a digital voltmeter. Bartos and Pesez [44] have proposed a method in which the iodine from the I_2O_5 tube is absorbed and reduced to iodide in a solution of hydrazine sulfate, followed by potentiometric titration with 0.01 N $AgNO_3$. Hozumi and Tamura [45,46] have described an automated technique of optical integration to determine iodine in the vapor phase. The iodine is led into a gas cell kept at 120°C and the light absorption at 530 nm through the gas cell is recorded by means of a photoelectric system. Optical integration is then attained as the maximum deflection of the recorder pen during the time when the total iodine produced from the organic sample is temporarily retained in the gas cell.

Carbon monoxide can be determined by thermal conductivity measurement [47-50], after being separated from nitrogen and other gases on a column packed with molecular sieve 5A. Without separation from any other components in the combustion gases, carbon monoxide can be quantitatively measured by means of a nondispersive infrared gas-analyzer. According to Thürauf and Assenmacher [51], the organic sample required is 2-50 mg, but the sensitivity of the instrument can be increased so that the determination can also be carried out on quantities below 1 mg.

D. Difficulties

We shall confine our discussion of the difficulties in oxygen analysis to methods based on carbon reduction only, since these methods are known to be in current practice (see Sec. 4.2A). It is obvious that the first requirement for successful direct determination of oxygen in organic materials demands the complete recovery of oxygen as carbon monoxide. Referring to equations (2) and (3), since carbon monoxide is not instantaneously removed from the combustion train, there is always the danger of reversing these reactions. It should be recognized firstly that equation (2) is endothermic, and at 500°C carbon dioxide is obtained instead of carbon monoxide [52], and secondly that increase of pressure in the system drives the equilibria to the left. Recently Kirsten [52a] found that the addition of chlorohydrocarbon to the carrier gas facilitates the quantitative conversion of oxygen to carbon monoxide.

Without modifications (see Sec. 4.2B), the methods put forward by Zimmermann [8] and Unterzaucher [9] are not suitable for organic samples which contain boron, fluorine, phosphorus, or certain metallic elements. Apparently, any element which forms an oxide not reducible by carbon will vitiate the oxygen results.

The presence of fluorine in the organic sample leads to high oxygen values due to the reaction between hydrogen fluoride and the quartz combustion tube:

$$SiO_2 + 4HF \longrightarrow SiF_4 + 2H_2O \tag{14}$$

On the other hand, when magnesium nitride is incorporated into the combustion tube packing as a means of trapping the fluorine [53], low results are obtained because magnesium fluoride reacts with water in the combustion gases to form magnesium oxide [54].

Metallic copper in the combustion tube serves to retain chlorine and sulfur. It has been reported, however, that the presence of chlorine in the organic material tends to give high blanks [49], and that the difficulties of the Unterzaucher method [9] are accentuated for samples containing high percentages of sulfur [55]. According to Brancone [56], high oxygen values are obtained from chloro compounds when a new quartz combustion tube is used with the Perkin-Elmer Elemental Analyzer (see Sec. 4.6).

When the iodometric finish is employed (see Sec. 4.1C), difficulty arises due to the action of the pyrolytic hydrogen on iodine pentoxide to produce iodine:

$$I_2O_5 + 5H_2 \longrightarrow I_2 + 5H_2O \tag{15}$$

This error is particularly noticeable when the sample contains high percentages of hydrogen [55], such as in the analysis of trace oxygen in petroleum products.

4.2 CURRENT PRACTICE

A. Apparatus

Assembly for Oxygen Analysis Only

The assemblies for oxygen analysis described in the laboratory manuals [57-59] on organic microanalysis published in the 1960s are modifications of the apparatus proposed by Zimmermann [8] and Unterzaucher [9]. The example shown in Fig. 4.1 is the assembly recommended in the latest edition of the Association of Official Analytical Chemists Official Methods of

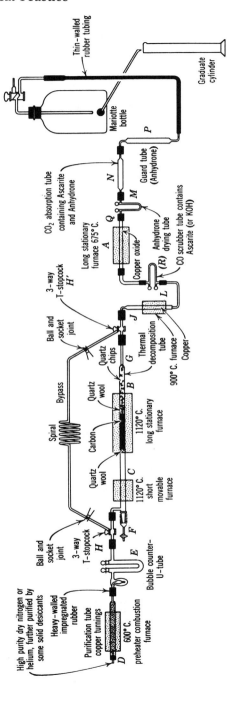

FIG. 4.1. Diagram of apparatus for microdetermination of oxygen with gravimetric finish. (Courtesy J. Wiley & Sons, Inc.)

Analysis [60]. However, there is no standard apparatus available commercially, probably because each microanalytical laboratory has its own
preference in the arrangement of the various components. The assembly
[61] employed in the Koninklijke/Shell Laboratory, Amsterdam, is shown
in Fig. 4.2; the detailed description of the quartz combustion tube in given
in Fig. 4.3. The setup used in Sandoz Pharmaceuticals, Hanover, New
Jersey, is described in Sec. 4.4. Fraisse and Andrianjafintrimo [61a]
have described an automated apparatus which uses a vertical reactor and
titrimetric finish by protonometric coulometry.

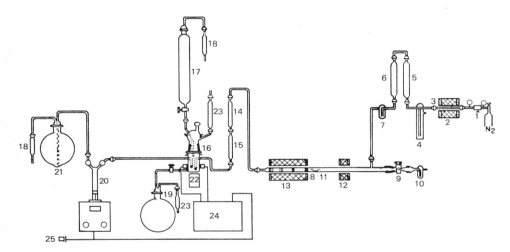

FIG. 4.2. Schematic view of the assembly for oxygen analysis, after
Gouverneur and Bruijn. 1, reducing valve; 2, furnace, 300°C; 3, purification tube filled with copper; 4, pressure regulator filled with mercury;
5, purification tube filled with BTS catalyst; 6, absorption tube filled with
soda asbestos and magnesium perchlorate; 7, bubble counter, with white
oil; 8, reaction tube; 9, wide-bore stopcock; 10, cap with tell-tale, filled
with white oil; 11, combustion boat; 12, sample heater, 1000°C; 13, furnace, 950°C; 14, absorption tube filled with soda asbestos (25 mm) and
magnesium perchlorate (25 mm); 15, absorption tube filled with "Schütze
contact" (60 mm) and copper (10 mm); 16, titration vessel; 17, absorbing
liquid container; 18, guard tube filled with soda asbestos; 19, effluent
receptacle; 20, micro piston buret; 21, supply bottle for titrant; 22, magnetic stirrer; 23, guard tube filled with active carbon; 24, photocell relay;
25, manual buret switch. (From Ref. 61, courtesy Talanta.)

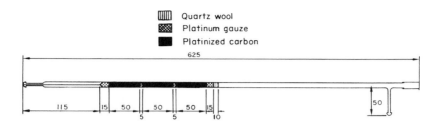

FIG. 4.3. The combustion tube for oxygen analysis, after Gouverneur and Bruijn. All dimensions in millimeters. (From Ref. 61, courtesy Talanta.)

Automated Elemental Analyzers for Oxygen Analysis

After the successful development of automated apparatus for the determination of carbon, hydrogen, and nitrogen (see Chaps. 2 and 3), it was natural to adapt the instrumentation to the determination of oxygen. Only slight changes in the combustion section of the setup are required, and the carbon monoxide obtained can be measured directly or indirectly using the same sensing device as for the C, H, N analyzer.

At the present time, there are available three commercial instruments for the automatic determination of oxygen. In the Carlo Erba [62] Elemental Analyzer Model 1104 (see Sec. 2.5) there is a built-in pyrolysis reactor and chromatographic column packed with molecular sieve 5A for the separation of carbon monoxide. The pyrolysis reactor (quartz combustion tube) filling and temperature distribution while in operation are shown in Fig. 4.4, as reported by Pella and Colombo [49]. In the Perkin-Elmer apparatus [63], the C, H, N analyzer Model 240 is modified for oxygen analysis as follows. The combustion and reduction tubes are removed and replaced with (1) a pyrolysis tube containing platinized carbon and metallic copper, and (2) an oxidation tube containing copper oxide to convert the carbon monoxide produced from the organic sample to carbon dioxide (see detailed instruction in Sec. 4.6). In the Heraeus Oxygen Automat [64], the oxygen in the sample is converted on a carbon catalyst to carbon monoxide which, together with the other combustion gases (N_2, H_2, CH_4, etc.), is carried by helium through the first catharometer cell; the carbon monoxide is subsequently separated from the other gases in a molecular sieve and is measured as a separate fraction upon passing through the second catharometer. The schematic diagram is shown in Fig. 4.5.

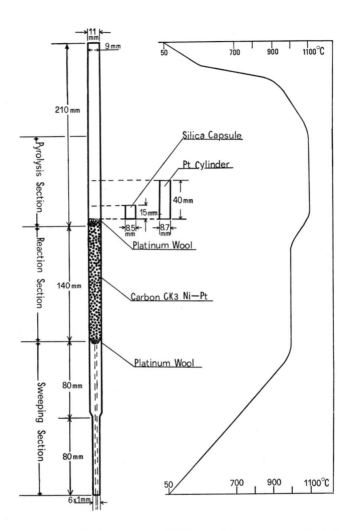

FIG. 4.4. Pyrolysis reactor: filling and temperature distribution for oxygen analysis in the Carlo Erba apparatus. (From Ref. 49, courtesy Anal. Chem.)

Dugan and Aluise [65, 66] have designed a dual-channel elemental analyzer that can determine C, H, N, and S in one channel with provisions to switch instantaneously to the other channel for oxygen analysis (see Figs. 10. 14 and 10. 15). The analyzer has two independent flow controls, two

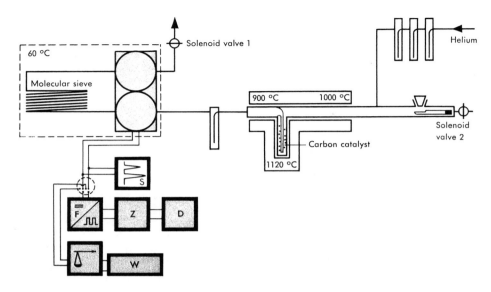

FIG. 4.5. Schematic diagram of the Heraeus Oxygen Automat: F. Voltage to frequency converter; Z. counter; D, printer; W. balance; S. recorder. (Courtesy W. C. Heraeus GmbH.)

combustion and reduction furnaces, and two trapping systems, all of which operate from one integrator-recorder-detector system. Decomposition is effected by dynamic flash pyrolysis in a quartz tube containing carbon at 1120°C in an atmosphere of helium. The carbon monoxide obtained is measured by thermal conductivity.

B. Determination Methods

As mentioned above, several automated instruments for oxygen analysis utilize thermal conductivity as the finish technique, measuring either carbon monoxide or carbon dioxide. Contrastingly, thermal conductivity measurements are not employed in the assemblies for the determination of oxygen that are modeled after Zimmermann [8] and Unterzaucher [9]. Among the variety of methods proposed (see Sec. 4.1C), the gravimetric finish is still favored by many practicing microanalysts. The operation is simple and the procedure does not require standardized solutions. As illustrated in Fig. 4.1, the Ascarite absorption tube is removed from the assembly at the end of the experiment and weighed. In order to speed up the weighing process, Trutnovsky [37] has used a device [67] by which the carbon dioxide gas is conducted from the combustion train through a Teflon tubing to the absorption tube in the microbalance case. Other workers prefer the iodometric finish (see Sec. 4.4); this method has the advantage of a sixfold increase of the original amount of iodine produced from the carbon monoxide.

Several laboratories [42,43,61,61a] have designed apparatus to perform automatic titration of the iodine or carbon dioxide formed. The method employed by Gouverneur and Bruijn [61] is depicted in Fig. 4.2. The carbon dioxide evolved by the "Schütze contact" (a material [7] consisting of iodine pentoxide and sulfuric acid on silica gel) is absorbed in a special cell containing pyridine with added thymol blue indicator and monoethanolamine. The latter compound reacts with carbon dioxide to form hydroxyethylcarbamic acid, which is subsequently titrated with 0.02 $\underline{M}$ sodium methoxide. With the photoelectric arrangement (Fig. 4.2, left-hand portion), any incoming carbon dioxide is immediately and automatically neutralized as follows. The indicator color change from blue to yellow allows sufficient light (filtered, 585 nm) to be transmitted through the cell to actuate a motor buret, which starts delivering the titrant to the cell until the color returns to blue whereupon the titrant delivery stops [68].

C. Removal of Interference

If the organic sample contains sulfur, chlorine, bromine, or iodine, incorporation of metallic copper in the combustion train (see Fig. 4.1) serves to alleviate the interference of these elements. Sulfur is retained as copper sulfide, while the halogens are absorbed by the alkali in the CO scrubber tube. Campiglio [68a] has investigated the methods to prevent the formation of COS and CS_2.

The interference due to the presence of fluorine has been circumvented in several ways. Ehrenberger and co-workers [38] perform the pyrolysis in a nickel tube and mix the sample with powdered nickel. Cruikshank and Rush [54] use a platinum combustion tube. Fadeeva [69] decomposes the organic substance in a closed system over a layer of platinum black heated at 1000-1100°C in an atmosphere of nitrogen. The resulting carbon monoxide is oxidized by copper oxide to carbon dioxide which is determined gravimetrically. Other workers [25,26] prefer to determine oxygen in fluoro compounds by the isotopic dilution technique.

Gouverneur and Bruijn [61] have reported that the interference due to phosphorus, boron, and metallic elements can be removed by mixing the organic sample with a commercial product [70], Indulin Base RM, which was suggested by Unterzaucher [71]. This reagent is a polyaminophenylphenazine compound. Upon decomposition it is supposed to leave a thin layer of finely dispersed carbon which, upon being further heated to 1000°C, liberates the oxygen from residues left in the platinum boat. According to Merz [72], the addition of a mixture of ammonium chloride, silver chloride, and hexamethylenetetramine gives the best results for the prevention of interference by metallic elements in the determination of oxygen in organic materials.

4.3 SURVEY OF RECENT DEVELOPMENTS

A. Summary of Recent Publications

The reader who is interested in the complete story of oxygen analysis is referred to the comprehensive bibliography prepared by Steyermark [112] which covers the literature up to about 1963. An excellent review on the various methods for oxygen determination was published by Davies [113] in 1969. The publications which appeared during the past decade are summarized in Table 4.1. It will be noted that most investigators employed the pyrolytic reduction technique using carbon or platinized carbon. More attention was given to titrimetric than gravimetric finish, while a variety of physicochemical measurements have been proposed.

In the early 1960s, many microanalytical laboratories added oxygen analysis to their routine operation because there were frequent requests from the organic research laboratories for complete analysis of new compounds. While such demand has slackened in recent years, the direct determination of oxygen is still a very important aspect of organic analysis. There are continuing discussions on the subject with respect to automation, attainment of accurate results, and improvement of the apparatus.

B. Comparison of Procedures

Owing to the importance of oxygen analysis and the difficulty in obtaining reliable results, a number of investigators have undertaken to compare and evaluate the various proposed methods. Several collaborative studies [88, 114, 115] were carried out under the auspices of the Association of Official Agricultural (Analytical) Chemists using C,H, C,H,O, and C,H,N,O,S compounds as test samples. Since the gravimetric procedure (see Fig. 4.1) was found by all of the collaborators to give excellent results, this method has been adopted by the Association [60]. For the evaluation of the methods of oxygen analysis in the range below 1%, the American Petroleum Institute has published a report which is discussed in Chap. 12 (see Sec. 12.3A).

TABLE 4.1. Summary of Recent Publications on the Determination of Oxygen.

Decomposition method	Product determined	Finish technique	Reference
In combustion tube with carbon or platinized carbon	$CO \rightarrow CO_2$	Titrimetric	15, 40–42, 44, 59, 61, 61a, 72–87
	$CO \rightarrow CO_2$	Gravimetric	60, 69, 88–94
	$CO \rightarrow CO_2$	Conductimetric	54, 95–97
	$CO \rightarrow CO_2$	Electrolytic	43, 98–100
	$CO \rightarrow CO_2$	Thermal conductivity	36
	$CO \rightarrow CO_2$	Manometric	101
	CO_2	Mass spectrometry	102, 103
	CO_2	Isotopic dilution	25
	CO	Gas chromatography	16a, 47, 49, 50, 66, 104–106
	CO	Infrared spectrometry	51, 107, 107a
	$CO \rightarrow I_2$	Optical integration	45, 46
In combustion tube with H_2 and catalyst	H_2O	Gravimetric	20, 21
	H_2O	Gasometric	19, 108
	CO	Thermal conductivity	108a
	$H_2O \rightarrow CO$	Radiometric	109
In quartz bomb	CO	Manometric	110
Vacuum fusion	CO	Infrared spectrometry	110a
Neutron activation		Radiometric	33, 34, 111, 111a

Belcher and co-workers [91] have investigated several modifications of the Schütze [7] procedure by analyzing the combustion gases in a mass spectrometer, and recommended the following: (1) pyrolysis over platinized carbon at 900°C, (2) both metallic copper and soda asbestos be used in the purification train, and (3) gravimetric finish after conversion of CO to CO_2 with the Schütze contact (see Sec. 4.2B) at room temperature. Pella [116] has compared helium against nitrogen or argon as carrier gas in the Unterzaucher [9] procedure and found low results owing to incomplete conversion of oxygen to carbon monoxide. The incompleteness of conversion is attributed to diffusion phenomena. Campiglio [116a] has evaluated the efficiency of different methods for eliminating interference from sulfur.

The characteristics of the different methods for oxygen analysis have been summarized by Klesment [117]. Recently, gas chromatography has been applied to the determination of oxygen. According to Karrman and Karlsson [43], those methods which use gas chromatography in the final step, either for the determination of carbon monoxide or of carbon dioxide, do not usually give such good values as Unterzaucher's original method. In order to increase the sample output while retaining the titrimetric finish, Merz [118] has designed a program for automatic combustion operation using two pyrolysis units connected to a switching device, accompanied by automatic titration of carbon dioxide with a colorimetric endpoint detection.

4.4 EXPERIMENTAL: DETERMINATION OF OXYGEN BY PYROLYTIC REDUCTION AND IODOMETRY

This section was written by William J. Bonkoski [119], Research Department, Sandoz Inc., East Hanover, New Jersey, who has been using the procedure described below for more than 5 years.

A. Principle

Oxygen in organic material is converted to carbon monoxide by catalytic pyrolysis at high temperature in the presence of platinized carbon in an atmosphere of pure nitrogen. The carbon monoxide formed is conducted to an absorption tube containing anhydroiodic acid, where carbon monoxide is oxidized to carbon dioxide with concomitant production of iodine [equation (9)]. The iodine is then determined by iodometric titration.

B. Apparatus

The schematic flow diagram of the assembly is shown in Fig. 4.6.

C. Reagents

Platinized carbon [13, 14, 91], containing 50% platinum.
Pumice chips.
Ascarite.
Sodium hydroxide solution: 25 g NaOH in 100 ml H_2O.
Sulfuric acid solution: Mix equal volumes of concentrated sulfuric acid
    and distilled water.
Copper wires, 60-80 mesh.
Phosphorus pentoxide.
Dehydrite, or anhydrous $CaCl_2$.

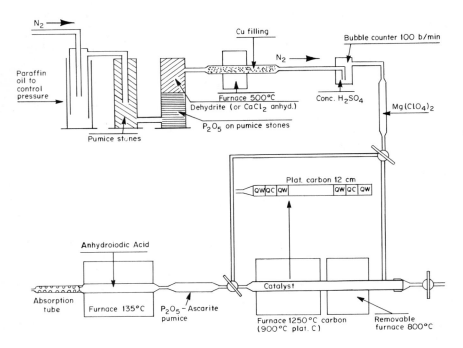

FIG. 4.6. Schematic flow diagram for the determination of oxygen by
pyrolysis.

Quartz chips.
Quartz wool.
Anhydroiodic acid.
Bromine in acetic acid solution: Add 0.4 ml of bromine to 100 ml of
    10% solution of potassium acetate in glacial acetic acid and shake
    well.
Formic acid, 10%.
Standard sodium thiosulfate solution, 0.02 N.
Potassium iodide.
Starch solution.

D. Procedure

Weigh the sample (3-6 mg) in a platinum boat (liquids in capillaries). Before inserting the substance into the combustion tube, cool the tube with
Dry Ice. Place the sample in the open end of the combustion tube, and seal
with a stopcock. Keep the long furnace at 900°C (see Sec. 4.4F).
     Sweep the combustion tube with nitrogen counterclockwise, by turning
the appropriate stopcocks, for 4 min, then clockwise for 2 min. Rinse
a Vigreaux absorption tube with 25% NaOH and place it on the end of the
anhydroiodic acid tube. Seal the sample end of the combustion tube, direct
the nitrogen flow through the combustion tube and into the absorption tube.
Remove the Dry Ice and place the 800°C furnace directly over the sample
for combustion. Set the timer for 30 min (Heraeus automated furnace).
     After 30 min, take off the Vigreaux absorption tube and rinse it tho-
roughly with 50 ml of distilled water, collecting the washings in a 200-ml
Erlenmeyer flask. Add 10 ml of bromine-acetic acid mixture, swirl, and
allow to stand for 5 min. Add about 1 ml of formic acid to the flask, swirl,
and let stand for 3 min. Add 2 ml sulfuric acid (1:1), swirl, then add 10-
15 mg of KI and titrate the liberated iodine with standardized 0.02 N
sodium thiosulfate solution to a starch endpoint.
     Run a blank each morning for the correction value and calculations.
Subtract the blank from the sample titration.

E. Calculations

$$\%O = \frac{(\text{ml } Na_2S_2O_3 - \text{ml blank}) \times 6.66 \times 100}{\text{Sample weight in mg}}$$

F. Note

The long furnace should be maintained at 1250°C when carbon only is used
as the combustion tube packing.

G. Comments

(1) Originally the combustion tube consisted of pure carbon filling and the temperature was kept at 1250°C. The heating element at that temperature would only last for 1 or 2 weeks. Using platinized carbon at 900°C, both the heating element and quartz combustion tube last at least 1 year. The only disadvantage incurred in this alteration has been to increase the combustion time from 20 to 30 min.

(2) This procedure is not suitable for organic substances containing phosphorus, fluorine, and some metals such as sodium and potassium.

(3) Honma [119a] has recommended pyrolysis in the presence of hydrogen or methane to remove oxygen-containing substances in the catalyst.

4.5 EXPERIMENTAL: AUTOMATED DETERMINATION OF OXYGEN USING COULOMETRY AT CONTROLLED POTENTIAL

The method described in this section is taken from a paper published by K. J. Karrman and Ronald Karlsson [43], Department of Analytical Chemistry, University of Lund, Lund, Sweden. The experimental procedure [43a] was verified by R. Karlsson in October 1974.

A. Principle

The organic substance is pyrolyzed in an atmosphere of nitrogen whereby the oxygen content is quantitatively converted to carbon monoxide (see Sec. 4.4). The carbon monoxide then reduces anhydroiodic acid in the absorption tube, producing an equivalent quantity of iodine. The iodine vapor formed is led by a stream of nitrogen to the cathodic chamber of an electrolysis cell, where the iodine is reduced at controlled potential at a rotating platinum electrode. The amount of electricity consumed is determined by an electronic integrator and read from a digital voltmeter.

B. Apparatus and Reagents

The first part of the assembly is similar to that proposed by Unterzaucher [71] up to and including the anhydroiodic acid tube (see Fig. 4.6). The iodine vapor which is formed here is led by the nitrogen gas stream (10 ml/min) through the electrically heated glass tube A (Fig. 4.7) down into the electrolysis cell B, also of glass, which contains 5 ml of 2 $\underline{M}$ sodium iodide and 5 ml of 2 $\underline{M}$ sodium perchlorate. To prevent the temperature

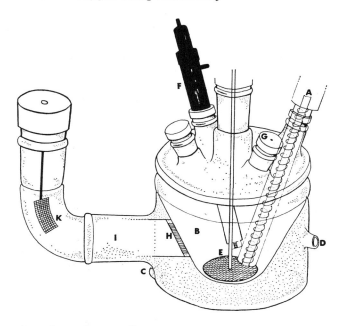

FIG. 4.7. Electrolysis cell for oxygen determination. A, double-walled electrically heated glass tube for transport of iodine vapor; B, cathode compartment with cooling jacket; C and D, inlet and outlet for cooling water; E, platinum cathode at the end of the rotating shaft; F, reference electrode; G, Teflon stopper with hole for release of nitrogen; H, Haldenwanger clay filter; I, anode compartment; K, anode, Pt-gauze. (From Ref. 43, courtesy Talanta.)

from rising, the electrolysis cell is surrounded by a glass jacket through which cooling water is led in at C and out at D. Electrode E is a rotating platinum electrode consisting of a circular platinum gauze (36 mesh), diameter about 35 mm, fixed at one end to a rotating Teflon-coated shaft. The shaft rotates at about 400 revolutions per minute. Electrode F is a calomel reference electrode (Radiometer K 401). Electrolyte can be added to, and removed from, the cell through the hole stoppered by G. The nitrogen can leave the cell via a small hole in G. A circular Haldenwanger clay filter, H, diameter about 15 mm, is fixed to the cell with special Haldenwanger cement. The anode K consists of a platinum gauze (36 mesh, 20 x 30 mm). The anode compartment contains 2 $\underline{M}$ sodium perchlorate.

The platinum electrode E is connected via a mercury contact at the upper end of the rotating shaft with the positive input of an operational amplifier in a potentiostat of exactly the same construction as described by Karlsson and Karrman [120]. The reference electrode F is connected to the negative input of the operational amplifier, and the anode K to the output. The potential of the working electrode is always 100 mV vs. the calomel electrode. The temperature of the tube A through which the iodine vapor is led from the anhydroiodic acid tube down to the working electrode is about 120°C. The quantity of electricity required to reduce the iodine is measured with an electronic integrator [120] and the value is read from a digital voltmeter (Solatron LM 1420-2).

C. Procedure

A 2-6 mg sample in a platinum boat is placed in the usual way in the pyrolysis tube (see Sec. 4.4). The cell is connected and a check made that the residual current is 15-20 μA before the movable furnace is put into position about 5 mm behind the boat. After about 3 min the current starts to increase rapidly and reaches a maximum value of 25-30 mA after 4-7 min. After 13-14 min, depending on the nature of the sample, the residual current has again reached the 15-20 μA level and the analysis is complete.

D. Calculations

The oxygen content is calculated from the value U_0 read from the digital voltmeter, using the equation

$$(I) \quad \%O = \frac{R_1 \times C \times (U_0 - U_B) \times 15 \times 32}{R_2 \times F \times nM \times 6} \times 100$$

where R_1, R_2, and C are integrator components (see Fig. 4.8 and Karlsson and Karrman [120]) and have in this case the values $R_1 = 0.5$ MΩ, $R_2 = 10$ Ω, and C = 10 μF; F = 96487 C; n = 2 (number of electrons involved); M = weight of sample (mg); U_0 = integrator reading for sample (mV); and U_B = integrator reading for blank (mV). The value of 32 is the molecular weight of oxygen.

The figures 15 and 6 come from the reaction between carbon monoxide and anhydroiodic acid. The electrical components have been accurately calibrated, and when the values are inserted in equation (I) the following simplified equation is obtained:

$$(II) \quad \%O = 0.020718 \times \frac{(U_0 - U_B)}{M}$$

The integrator readout for an analysis is 4000-5000 mV. The amount of sample is adjusted according to the oxygen content of the substance. The blank value in Equation (I), U_B, is obtained by performing the analysis in the absence of the sample and is 30 ± 3 mV. When the analysis is complete after 13-14 min (for certain substances which are more difficult to pyrolyze, a further 1-2 min will be required) the residual current is 15-20 μA. This means that the residual current contributes about 2 mV/min to the integrator readout.

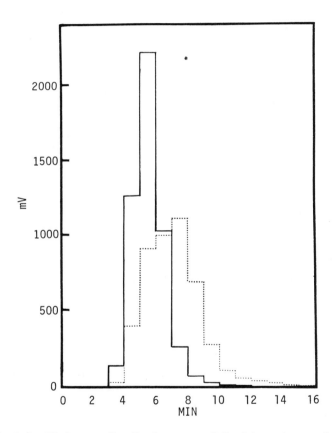

FIG. 4.8. Histogram for the increase of the integrator readout per unit time for the pyrolysis of benzoic acid (full line) and 3,5-dinitrobenzoic acid (dotted line). (From Ref. 43, courtesy Talanta.)

E. Comments

The analysis can be carried out either by placing the movable furnace just behind the boat with the sample and then letting the furnace move over the boat, or by placing the furnace directly over the boat. With the latter procedure, the pyrolysis starts earlier and the total analysis time is reduced by 1-2 min. With the apparatus described here, the pyrolysis process can be followed continuously. Figure 4. 8 shows two diagrams obtained by measuring the increase of the integrator readout each whole minute, the time being counted from the moment the movable furnace was placed just behind the boat. It may be seen that benzoic acid is pyrolyzed considerably faster than 3, 5-dinitrobenzoic acid.

4.6 EXPERIMENTAL: DETERMINATION OF OXYGEN USING A
      MODIFIED PERKIN-ELMER ELEMENTAL ANALYZER

The method described in this section is based on the report of Robert Culmo [36], Perkin-Elmer Corporation, Norwalk, Connecticut. It has been checked by Ernest Ross and Louis Brancone [121], Lederle Laboratories, American Cyanamid Company, Pearl River, New York.

A. Principle

The organic substance is pyrolyzed with platinized carbon at 950°C in a helium atmosphere. The carbon monoxide formed then passes through an oxidation tube containing copper oxide at 670°C, whereby carbon dioxide is obtained which is measured as described in Sec. 2.3.

B. Apparatus

The Perkin-Elmer Elemental Analyzer Model 240 (Perkin-Elmer Corp., Norwalk, Connecticut) is modified as follows. The combustion and reduction tubes are removed and replaced with (1) a pyrolysis tube containing platinized carbon and free copper and (2) an oxidation tube containing free copper oxide and free copper (see Fig. 4. 9a, b). A U-tube (acid gas scrubber, Fig. 4. 9c) containing Colorcarb (indicating Ascarite) along with its fitting replaces the C, H, N fitting to connect the pyrolysis and oxidation tubes as shown in Fig. 4. 10. The operating temperatures of the furnaces are the same as in the C, H, N system presently used (950-1000°C for the pyrolysis tube; 670-700°C for the oxidation tube).

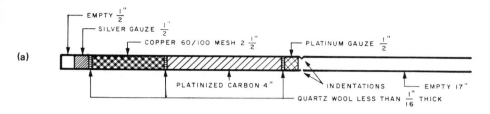

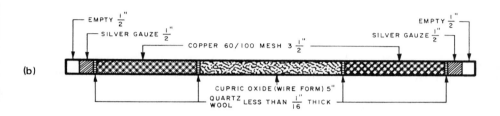

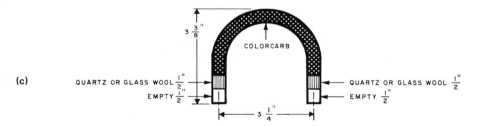

FIG. 4.9.   Tube packings for oxygen determination.   (a) Oxygen pyrolysis tube.   (b) Oxidation tube.   (c) Acid-gas scrubber.   (From Ref. 36, courtesy Mikrochim. Acta.)

C.   Reagents

   Silver gauze.
   Platinum gauze (80 mesh).
   Quartz wool.
   Copper oxide.

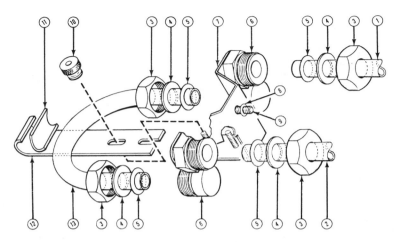

FIG. 4.10. The positions of the pyrolysis tube, U-tube, and oxidation tube. 1, Oxidation tube; 2, pyrolysis tube; 3, glass tube nut; 4, O-ring washer (brass); 5, O-ring, 7/16-in. i.d.; 6, oxygen analysis connector; 7, tube retainer bracket; 8, O-ring, 5/32-in. i.d.; 9, #8-32 1-in. long binding head screw; 10, #6-32 knurled nut; 11, 7/16-in. thick neoprene sponge; 12, tube retainer; 13, acid-gas scrubber U-tube. (From Ref. 36, courtesy Mikrochim. Acta.)

Copper.
Platinized carbon, 1/16-in. pellets, crushed carefully and broken into
    small granules which pass a 30 mesh but are retained on an 80 mesh
    sieve.
Colorcarb.

D. Procedure

A 0.1- to 3-mg sample is accurately weighed in a platinum boat; the boat is then placed in a quartz ladle and the ladle inserted into the cool zone of the pyrolysis tube. The program is started (the oxygen supply being turned off). When the inject panel lamp lights, the combust button is pressed (this bypasses the inject sequence which is used for sample insertion into the combustion zone during C, H, N analysis only). About 20 s after the high heat furnace goes on (position 4 1/4 on the program wheel) and a purging flow of helium starts, the quartz ladle containing the organic compound is introduced into the hot pyrolysis zone. The pyrolyzed gases are purged through the reaction system and finally into the mixing volume, which has been previously adjusted to a 3.5 min filling time by adjusting the pressure regulator. From this point the analysis is fully automatic and, with the exception of possible output suppression of the carbon bridge, requires no more attention from the operator.

Known compounds are first analyzed in order to determine the conversion of the detector voltage to known microgram amounts of oxygen and this is then used as a sensitivity factor for the unknown samples.

REFERENCES

1. T. S. Ma, in S. Patai (ed.), The Chemistry of Carboxylic Acids and Esters, Wiley, New York, 1969, p. 871.
2. F. Pregl, Die Quantitative Organische Mikroanalyse, 4th Ed. (revised by H. Roth), Springer, Berlin, 1935.
3. H. ter Meulen, Rec. Trav. Chim. Pays-Bas 41:112, 509 (1922); 43:899 (1924); 53:118 (1934).
4. H. ter Meulen and J. Heslinga, Nieuve Methoden voor Elementair-analyse, Meinema, Delft, 1925.
5. W. R. Kirner, Ind. Eng. Chem., Anal. Ed. 6:358 (1934); 7:363 (1935); 8:57 (1936); 9:535 (1937).
6. W. R. Kirner, personal communication, 1938.
7. M. Schutze, Z. Anal. Chem. 118:241 (1939); Naturwissenschaften 27:822 (1939).
8. W. Zimmermann, Z. Anal. Chem. 118:258 (1939).
9. J. Unterzaucher, Berichte 73:391 (1940).
10. V. A. Aluise, R. T. Hall, F. C. Staats, and W. W. Becker, Anal. Chem. 19:347 (1947).
11. W. T. Chambers, reported at the Rubber Technology Conference, London, June 1948.
12. J. F. Alicino, personal communication, 1951.
13. I. J. Oita and H. S. Conway, Anal. Chem. 26:600 (1954).
14. F. H. Oliver, Analyst 80:593 (1955).
15. Z. Stefanac, Z. Sliepecevic, and Z. Rakovic-Tresic, Microchem. J. 15:218 (1970).
16. Z. Stefanac and Z. Sliepecevic, reported at the 6th International Symposium on Microtechniques, Graz, 1970.
16a. E. Pella and R. Andreoni, Mikrochim. Acta 1976(II):175.
17. C. J. van Nieuwenberg and J. W. L. van Ligten, Quantitative Chemical Micro-Analysis, Elsevier, Amsterdam, 1963.
18. R. N. Smith, J. Duffield, R. A. Pierotti, and J. Mooi, Anal. Chem. 28:1161 (1956).

19. S. Mlinko, Mikrochim. Acta 1961:833.
20. A. P. Terent'ev, M. A. Volodina, and A. Besada, Vestr. Mosk. Gos. Univ., Ser. Khim. 1969:111.
21. A. P. Terent'ev, M. A. Volodina, and A. Besada, Zh. Anal. Khim. 24:1276 (1969).
22. I. Sheft and J. J. Katz, Anal. Chem. 29:1322 (1957).
23. T. S. Lee and R. Meyer, Anal. Chim. Acta 13:340 (1955).
24. A. D. Kirschenbaum, A. G. Streng, and A. V. Grosse, Anal. Chem. 24:1361 (1952).
25. P. B. Olson and S. Kulver, Mikrochim. Acta 1970:403.
26. R. N. Boos, A. Sohe, and N. R. Trenner, reported at the American Society for Testing and Materials committee meeting on Mass Spectrometry, New Orleans, 1962.
27. R. G. Osmond and A. A. Smales, Anal. Chem. Acta 10:117 (1954).
28. J. L. Brownlee, Microchem. J., Symp. Ser. 2:885 (1962).
29. H. Sinn and D. Aumann, Makromol. Chem. 57:105 (1962).
30. D. J. Veal and C. F. Cook, Anal. Chem. 34:178 (1962).
31. E. L. Steele and W. W. Meinke, Anal. Chem. 34:185 (1962).
32. R. A. Stallwood, W. E. Mott, and D. T. Fanale, Anal. Chem. 35:6 (1963).
33. K. A. Keller and H. Muenzel, Radiochem. Acta 12:51 (1969).
34. R. Rafaeloff, Radiochem. Radioanal. Lett. 1:199 (1969).
35. D. Fraisse and R. Levy, Bull. Soc. Chim. Fr. 1968:445.
36. R. Culmo, Mikrochim. Acta 1968:811.
37. H. Trutnovsky, reported at the 6th International Symposium on Microtechniques, Graz, 1970.
38. F. Ehrenberger, S. Gorbach, and W. Mann, Z. Anal. Chem. 198:242 (1963).
39. F. Salzer, Mikrochim. Acta 1965:835.
40. F. G. Römer, G. W. S. van Osch, W. J. Buis, and B. Griepink, Mikrochim. Acta 1972:674.
41. I. Monar, Mikrochim. Acta 1965:209.
42. P. Calme and M. Keyser, Mikrochim. Acta 1969:1248.
43. K. J. Karrman and R. Karlsson, Talanta 19:67 (1972).
43a. R. Karlsson, private communication, October 1974.
44. H. Bartos and M. Pesez, Bull. Soc. Chim. Fr. 1971:345.
45. K. Hozumi, Microchem. J. 12:218 (1967).
46. K. Hozumi and H. Tamura, Japan Analyst 17:622 (1968).
47. R. N. Boos, Microchem. J. 8:389 (1964).
48. R. Belcher, G. Dryhurst, and A. M. G. Macdonald, Anal. Lett. 1:807 (1968).
49. E. Pella and B. Colombo, Anal. Chem. 44:1563 (1972).
50. F. Ehrenberger and O. Weber, Mikrochim. Acta 1967:513.

51. W. Thürauf and H. Assenmacher, Z. Anal. Chem. 245:26 (1969).
52. W. W. Wendlandt, in McGraw-Hill Encyclopedia of Science and Tech-
     nology, Vol. 2, McGraw-Hill, New York, 1960, p. 463.
52a. W. J. Kirsten, Microchem. J. 22:60 (1977).
53. L. Mazor, Mikrochim. Acta 1956:1757.
54. S. S. Cruikshank and C. A. Rush, Microchem. J., Symp. Ser. 2:467
     (1962).
55. J. P. Dixon, Anal. Chim. Acta 19:141 (1958).
56. L. Brancone, personal communication, 1973.
57. A. Steyermark, Quantitative Organic Microanalysis, 2nd Ed.,
     Academic, New York, 1961, p. 380.
58. G. Ingram, Methods of Organic Elemental Microanalysis, Reinhold,
     New York, 1962, p. 88.
59. T. S. Ma, in F. J. Welcher (ed.), Standard Methods of Chemical
     Analysis, 6th Ed., Vol. 2, Van Nostrand, Princeton, 1963, p. 387.
60. W. Horwitz (ed.), Official Methods of Analysis of AOAC, 12th Ed.,
     Association of Official Analytical Chemists, Washington, D.C.,
     1975, p. 928.
61. P. Gouverneur and A. C. Bruijn, Talanta 16:827 (1969).
61a. D. Fraisse and R. Andrianjafintrimo, Microchem. J. 21:178 (1976).
62. Carlo Erba Elemental Analyzer Model 1104 (pamphlet DT CHN b-E),
     Carlo Erba Scientific Division, Rodano, Italy, 1972.
63. Perkin-Elmer Corp., Norwalk, Conn.
64. Heraeus Automatic CHN-O Ultra Analyzer (pamphlet EW-D3.1),
     W. C. Heraeus, Hanau, Germany, 1968.
65. G. Dugan and V. A. Aluise, Hercules Chemists 60:7 (1970).
66. G. Dugan and V. A. Aluise, Anal. Chem. 41:495 (1969).
67. H. Trutnovsky, Z. Anal. Chem. 232:116 (1967).
68. O. I. Snock and P. Gouverneur, Anal. Chim. Acta 39:463 (1967).
68a. A. Campiglio, Mikrochim. Acta 1972:631; 1973:169; Farmaco 26:333,
     349 (1971).
69. V. P. Fadeeva, Izv. Sib. Otd. Akad. Nauk SSSR, Ser. Khim. Nauk
     1969:154.
70. Indulin Base RM, Catalog No. RZ 157, W. C. Heraeus, Hanau,
     Germany, 1968.
71. J. Unterzaucher, Analyst 77:584 (1952).
72. W. Merz, Anal. Chim. Acta 50:305 (1970).
73. W. J. Kirsten, Microchem. J. 4:501 (1960).
74. N. E. Gel'man, W. Y. Wang, and I. I. Bryushkova, Zavod. Lab.
     27:24 (1961).
75. S. Mizukami and E. Hirai, Mikrochim. Acta 1960:188.
76. F. Ehrenberger, Mikrochim. Acta 1961:590.
77. M. Kapron and M. Brandt, Anal. Chem. 33:1762 (1961).

78. M. Pesez and H. Partos, Bull. Soc. Chim. Fr. 1961:1191.

79. V. S. Pansare and V. N. Mulay, Mikrochim. Acta 1961:606.

80. G. Kainz and F. Scheidl, Mikrochim. Acta 1964:539.

81. A. Campiglio, Farmaco, Ed. Sci. 19:385 (1964).

82. M. Vecera, J. Lakomy, and L. Lehar, Talanta 10:801 (1963).

83. E. A. Bondarevskaya and M. O. Korshun, Zh. Anal. Khim. 18:644 (1963).

84. Committee on Microchemical Apparatus of the American Chemical Society, Microchem. J. 8:424 (1964).

85. Z. Sliepecevic and Z. Stefanac, Mikrochim. Acta 1971:362.

86. R. Belcher, G. Ingram, and J. R. Majer, Talanta 16:881 (1969).

87. W. Merz, Anal. Chim. Acta 51:523 (1970).

88. A. Steyermark, J. Ass. Offic. Agr. Chem. 46:559 (1963).

89. O. L. Kolsto and E. F. Shelberg, Microchem. J. 7:412 (1963).

90. M. Eberling and D. Marcinkus, Microchem. J. 8:213 (1964).

91. R. Belcher, D. H. Davies, and T. S. West, Talanta 12:43 (1965).

92. P. M. Mietasch and J. Horacek, Collect. Czech. Chem. Commun. 30:2889 (1965).

93. H. J. Barton and C. W. Nash, Microchem. J. 12:568 (1967).

94. S. Mesaric, Croat. Chem. Acta 42:13 (1970).

95. F. Salzer, Mikrochim. Acta 1962:835.

96. E. J. Beck and F. E. Clark, Anal. Chem. 33:1767 (1961).

97. J. Lakomy, L. Lehar, and M. Jurecek, Collect. Czech. Chem. Commun. 34:3165 (1969).

98. K. Novak and V. Slavik, Chem. Prum. 12:193 (1962).

99. A. W. Garst and T. W. McSpadden, Chem. Eng. News 41(14):65 (1963).

100. K. I. Nakamuri, M. Nishimura, and T. Mitsui, Microchem. J. 15:461 (1970).

101. B. D. Holt, Anal. Chem. 37:751 (1965).

102. R. Belcher, G. Ingram, and J. R. Majer, Mikrochim. Acta 1968:418.

103. J. W. Taylor and I. J. Chen, Anal. Chem. 42:224 (1970).

104. R. R. Suchanec, Dissertation Abstr. 22:719 (1961).

104a. K. Imaeda, K. Ohshawa, and K. Ohgi, Japan Analyst 22:1568 (1973).

105. L. V. Kuznetsova, E. N. Stolyarova, and S. L. Dobychin, Zh. Anal. Khim. 20:836 (1965).

106. M. N. Chumachenko and N. A. Khabarova, Izv. Akad. Nauk SSSR Ser. Khim. 1970:971.

107. J. A. Kuck, A. J. Andreatch, and J. P. Mohns, Anal. Chem. 39:1249 (1967).

107a. C. Bozier, M. Grall, D. Fraisse, and C. Muller, Microchem. J. 22:249 (1977).

108. G. D. Galpern, J. K. Chudakova, and M. V. Egorushkina, Zh. Anal. Khim. 19:598 (1964).

108a. K. Ubik, J. Horacek, and V. Pechanec, Collect. Czech. Chem. Commun. 37:102 (1972).

109. S. Mlinko and I. Gacs, Mikrochim. Acta 1969:504.

110. J. W. Frazer, Mikrochim. Acta 1964:679.

110a. F. Ehrenberger, Z. Anal. Chem. 267:21 (1973).

111. D. Ishii, H. Mori, and Y. Hirose, Chem. Soc. Japan, Ind. Chem. Sect. 70:112 (1967).

111a. J. J. Lauff, E. R. Champlin, and E. P. Przybylowicz, Anal. Chem. 45:52 (1973).

112. A. Steyermark, in I. M. Kolthoff and P. J. Elving (eds.), Treatise on Analytical Chemistry, Part II, Vol. 12, Wiley, New York, 1965, p. 1.

113. D. H. Davies, Talanta 16:1055 (1969).

114. A. Steyermark, J. Ass. Offic. Agr. Chem. 41:299 (1958).

115. A. Steyermark, J. Ass. Offic. Agr. Chem. 42:319 (1959).

116. E. Pella, Mikrochim. Acta 1968:13.

116a. A. Campiglio, Mikrochim. Acta 1974:317.

117. I. Klesment, Mikrochim. Acta 1969:1237.

118. W. Merz, private communication, 1971.

119. W. J. Bonkowski, private communication, May 1973 and October 1974.

119a. H. Honma, German Patent 2,427,655 (Cl. G. Oln) (1975); Japanese Patent 73-64,454.

120. R. Karlsson and K. J. Karrmann, Talanta 18:459 (1971).

121. E. Ross and L. Brancone, private communication, May 1973 and October 1974.

The Halogens

5.1 GENERAL CONSIDERATIONS

A. Basic Principles

In this chapter we discuss the methods for the determination of fluorine, chlorine, bromine, and iodine in organic materials. It will be noted that the analysis of these elements differs from the determination of carbon, hydrogen, nitrogen, and oxygen in several aspects: (1) There is a wide range of published methods for the determination of organically bound halogen. The reader is referred to the comprehensive account on fluorine analysis prepared by Ma [1] and that on chlorine, bromine, and iodine written by Olson [2]. Since the various methods may involve different principles and differ in applicability, the analyst should judge the suitability of the method chosen to analyze his sample. (2) The equipment required for halogen analysis is relatively inexpensive and the experimental operations are simple. Therefore it is sometimes more expedient for the research chemist to analyze his own halogen compounds than to send them to an analytical laboratory far away. (3) Automated procedures for the analysis of halogens in organic samples have not been fully developed. While a number of commercial instruments are available for the automatic determination of halides, no machine has been devised to perform the operation beginning with the organic material to be analyzed.

It should be mentioned that, among the halogens, fluorine stands alone in organic elemental analysis. A method which is developed for the determination of chlorine in organic compounds can be applied, as a rule, to the bromo compounds. Extension of the same method to the iodo analogs is usually feasible, bearing in mind that inorganic iodine compounds are

more stable at the higher oxidation state. Contrastingly, analytical procedures for chloro, bromo, or iodo compounds in general are not applicable
to the analysis of organic fluorine compounds without modifications. It is
interesting to note that laboratory manuals on quantitative organic analysis
published before 1950 did not describe any method for the determination of
fluorine. One reason was that fluorine could not be treated like the other
three halogens. A second reason was that there were very few organic
fluoro compounds prior to World War II. The rapid expansion of fluorine
chemistry during the past three decades has made fluorine analysis a very
important subject. However, as late as 1960, most new organic fluoro
compounds that appeared in the literature were not characterized by their
fluorine contents. This was due to the difficulty in fluorine analysis [3]
which had required many years of investigation and was perfected only
recently.

The nature of the halogen bonding in the organic molecule has important effects on the method of analysis. For instance, acyl chlorides, quaternary ammonium iodides, and N-bromosuccinimide can be treated like
inorganic halides. For this type of compound, direct determination of the
halogen content can be performed without mineralization, provided that the
method of measurement is compatible with the remaining portion of the
carbon skeleton. Some carbon-halogen bonds, while not being ionic, can
be easily cleaved by alkali in solution; other halides (e.g., allyl and benzyl
chlorides) are hydrolyzed by refluxing with alkali for 1 h. On the other
hand, polyhalogenated compounds require very drastic treatment in order
to attain quantitative conversion to elementary halogen or to halide ions.

B. Current Practice for the Mineralization of Organic Halogen-containing
   Materials

As mentioned above, the technique for the decomposition of an organic
halogen sample depends on the nature of the substance. A mild reaction
can be employed if the cleavage of the halogen bonding is readily accomplished. When the nature of the compound is not known, however, a drastic
procedure is usually selected. In case of doubt, more than one procedure
should be tried and the analytical results compared. Unlike carbon and hydrogen analysis which is conventionally performed in the milligram range,
determination of halogen in organic materials may involve mineralization
of milligram- to gram-size samples, depending on their halogen contents.

Decomposition by Closed-flask Combustion

This technique is also known as oxygen flask combustion. The sample
is wrapped inside a piece of paper, placed in a platinum basket and burned
in a stoppered flask filled with pure oxygen (see Sec. 5.3). The paper

initiates the burning and also supports combustion. The platinum serves as a catalyst to facilitate cleavage of the carbon-halogen bonds. Alicino [4] has recommended baskets made from perforated platinum sheet that can hold 1- to 100-mg samples. Replacement of platinum by other materials such as nichrome wire [4a] has been suggested. The temperature at which thermal decomposition of the organic compound takes place in the presence of platinum is about 1200°C.

Various sizes and shapes of the oxygen-filled combustion flasks have been proposed. Hempel [5] used a 10-liter bottle for macro analysis, while Mikl and Pech [6] employed a 1-liter flask for semimicro work. Adapting this technique to microanalysis using 3-7 mg of pure organic compounds, Schöniger [7] found that a flask of 300-ml capacity was most suitable. Combustion flasks of this size are presently widely used, since they are commercially available together with the ground-glass stopper sealed to a platinum-wire basket. When the sample size deviates from the milligram range, however, combustion vessels of either larger or smaller volumes should be utilized (see Chaps. 11 and 12).

For initiating the combustion, the current procedure normally consists of lighting the tongue of the wrapper over a small flame and inserting the stopper to close the flask before the fire reaches the sample. Another technique is to close the flask first and ignite the wrapper by means of an electric circuit or radiant energy [8]. There is an apparatus [9] on the market which consists of a cabinet that keeps the closed 300-ml combustion flask on an inclined platform while the ignition is started by infrared irradiation. Ober and co-workers [10] have described a turntable which guides the ignition of eight 1-liter combustion flasks by means of a lamp shining on a cotton wick. These devices are designed for the safety of the operator.

The closed-flask combustion technique is a general method for the decomposition of organic materials. While it was initially introduced in the microanalytical laboratory for the determination of halogens, the reader will find in the surveys [11-13] on this subject that the combustion flask has been used in procedures for the analysis of a large number of elements. In fact, after extensive tests, it has been established that this technique is more suitable for the determination of other elements rather than for the halogens. Difficulties have been encountered in the analysis of bromine compounds due to incomplete decomposition [2]. According to the experience of many practicing analysts, closed-flask combustion of organic fluorine compounds is not always reliable, especially for polymeric materials and perfluoro compounds [14]. It should be emphasized that the success of closed-flask combustion depends on the instantaneous and complete oxidation of the organic sample at a very high temperature in the presence of platinum (or other catalyst). Since halogenated compounds tend to exhibit high vapor pressures, a part of the sample may escape and remain outside of the platinum basket during the burning process, leading to low results. In the case of organic fluorine compounds, the difficulty is enhanced because of several factors, namely, (1) the bond energy of C-F is higher than that of carbon linked to other elements [15], (2) the volatility of fluoro compounds increases with their fluorine content [1], and (3) formation of volatile SiF_4 and BF_3 necessitates special material for the combustion flask.

Decomposition in a Metal Bomb

This is also a general mineralization technique in organic elemental analysis. The sample is mixed with suitable reagents (usually solids) in a sealed metal bomb. Destruction of the organic material is effected through fusion of the mixture by heating the bomb externally. The experimental procedure is given in Secs. 5.6 and 5.7. Commercial metal bombs [16] are constructed of nickel and are available in two sizes: 2.5-ml bombs for microanalysis and 22-ml bombs which can accommodate 0.5 g of organic sample and 15 g of solid reagents. Metal bombs of other designs have been proposed. For example, Francis [17] has described a bomb machined from stainless steel, while Bailey and Gehring [18] have used a bomb lined with platinum.

The chemical reaction that takes place in the metal bomb for releasing the organically bound halogen is either oxidation or reduction. The reagent for oxidation is sodium peroxide. Usually an additive such as sucrose or potassium nitrate is mixed with sodium peroxide to act as an "accelerator." However, for chloro, bromo, and iodo compounds, prolonged heating with sodium peroxide alone is equally effective. Under these conditions, the halogens are converted to chloride, a mixture of bromide and bromate [2], and iodate, respectively. Contrastingly, sodium peroxide fusion of fluoro compounds is incomplete without some additives. Precautions should be taken when heating the sodium peroxide bomb since the reactions inside the bomb are exothermic and a large quantity of gas is produced.

Reduction in the metal bomb is accomplished by fusion of the organic material with metallic sodium or potassium. When sodium is used, it can be cut and weighed in the open air. Potassium, being extremely sensitive to oxygen and moisture, should be handled in capillary tubes as described in Sec. 5.7. Alkali reduction converts the halogens to the lowest valence state, hence the resulting products are all in the form of halides.

Mineralization by Reduction in Solution

While nonionic organically bound halogen does not give a silver halide precipitate in aqueous medium, if the organic compound can be dissolved in a suitable nonaqueous medium and sodium is added as the reducing agent, quantitative conversion to sodium halide can be accomplished. This phenomenon was first observed by Stepanow [19] who used ethanol as the solvent.

Subsequently, liquid ammonia [20] and an ethanolamine-dioxane mixture [21] were recommended by other workers. Then sodium dispersion was found to be more effective when used in a solution containing biphenyl [22] or naphthalene [23] and glycol ethers [24]. Lithium [25] and sodium borohydride [26] have also been suggested as the reagents for dehalogenation, and a wide range of solvents [27-29] has been proposed.

It should be noted that this method of decomposition of organic materials involves very slow reactions. The procedures described in the literature are in general for macro and semimicro determinations which deal with centigram to decigram amounts of the halogen to be measured. This technique is not satisfactory on the micro scale. It is useful, however, for trace analysis of halogens in organic materials (see Chap. 12).

Decomposition by Oxidation in Combustion Tube

The Pregl [30] technique of catalytic oxidation in a combustion tube is suitable for the decomposition of chloro, bromo, and iodo compounds. In this method, the vapor of the organic substance is mixed with oxygen and slowly conducted through sections of platinum metal heated at about 900°C. The resulting gaseous products are absorbed in a solution containing sodium carbonate and sodium bisulfite, or sodium hydroxide [31], or on a silver gauze [32,33]. Since this decomposition method is more useful for sulfur analysis, the detailed procedure is given in Chap. 6.

Decomposition in an empty combustion tube with a rapid stream of oxygen has been advocated by Belcher and Ingram [34] and is favored by some Russian workers [35]. It should be noted that decomposition of fluoro compounds is not satisfactory either in the empty tube or in the presence of platinum because of an insufficiently high temperature. A device was designed by Wickbold [36] to decompose fluoro compounds by burning them in the oxyhydrogen flame. An improved apparatus has been reported by Sweetser [37] and the combustion train for microanalysis has been described by Levy [38]. Kunkel [38a] has developed a quartz flask container for solids and liquids, a device for liquified gases, and a suction burner.

The combustion tube decomposition method is recommended for low-boiling liquid samples which are not amenable to closed-flask combustion. Needless to say, for the analysis of gaseous halogen compounds, the combustion train is the method of choice.

Other Methods of Decomposition

A technique for mineralization of organic halogen compounds which was in vogue for several decades involves the use of sealed glass tubes to permit heating at high temperatures. In the Carius [39] method, the organic sample is heated with nitric acid and silver nitrate to precipitate silver chloride or bromide [40]. In the alkali fusion method, the organic compound is reduced with metallic sodium [41] or potassium [42] to form the corresponding halides. These methods have been recently displaced because they require the skill of glass blowing.

Catalytic hydrogenation of halogenated compounds to produce hydrogen halides has been proposed [43, 44], but it has not been generally accepted because of the difficulty of finding long-life catalysts.

Chloro and bromo compounds can be decomposed by digestion in sulfuric acid mixed with silver dichromate [45] or silver persulfate [46]. Chlorine and bromine, respectively, are liberated. If iodine is present, it remains in solution as iodate. Friedrich and Kottke [46a] have decomposed the sample by refluxing it in 30% KOH solution containing $K_2S_2O_8$ and reported on the limitations of this method.

C. Current Methods for Determining the Halogens

There is a large variety of methods to be chosen for the determination of the halogens after decomposition. These methods may be based on gravimetry, visual titrimetry, or electrometric measurements. Since ion-selective electrodes for all halides are now commercially available, they can be employed with advantage [47, 48]. The scale of determination for organic halogen analysis may be in the milligram, centigram, or decigram region. Attention is called to the fact that chlorine and bromine are in general recovered from the organic substance as halides whereas iodine is usually obtained in a higher oxidation state as iodate, and that methods for fluorine determination are unique among the halogens.

Determination of Chloride or Bromide

Gravimetric Methods. While the gravimetric procedure is relatively more time-consuming per determination, it is recommended for organic chlorine or bromine analysis when the determinations are made occasionally and not for a series of samples. Gravimetric determination of silver chloride or bromide can be performed in the milligram-to-decigram range with high precision using simple equipment [49]. In the case of microanalysis (see Sec. 5.9), the filter tube can collect about 100 mg of silver halide before it is cleaned. Therefore, if 5-10 mg of precipitate is produced in each determination, as many as 20 determinations can be carried out in the same filter tube for successive runs. In our experience, dry silver chloride or bromide does not lose weight for several months when kept in the filter tube in a dark cupboard, though the color of the precipitate turns dark on standing.

Visual Titrimetric Methods. The conventional methods for visual titration of chloride or bromide are based on argentimetry using a standardized solution of silver nitrate. Thus, for bromide the reaction is represented as follows:

$$AgNO_3 + Br^- \longrightarrow AgBr + NO_3^-$$

In the Volhard method, a measured excess of silver nitrate solution is added to the bromide solution, and the unconsumed silver ion is back-titrated with standardized ammonium thiocyanate solution using ferric alum as indicator. For the determination of chloride, either the silver chloride precipitate should be separated by filtration, or nitrobenzene should be added into the mixture to coagulate the particles, before back-titration is carried out.

Absorption indicators such as eosin and dichlorofluorescein have been proposed for the direct titration of halides with silver nitrate solution. It should be noted that the equivalence point is affected by the presence of neutral salts. Therefore, this method is not applicable when the organic substance is mineralized by alkali fusion.

Understandably, the above-mentioned methods are useful only at the centigram and decigram levels because of the difficulty of locating the endpoint visually for dilute solutions. Kirsten and Alperowicz [50] have reported a micromethod for titrating bromide or chloride in an acidified ethanolic solution with ethanolic silver nitrate using mercuric chloride and diphenylcarbazide as indicator. It is apparent that this method is not suitable on the macro scale because of the limited solubility of silver nitrate in ethanol.

An indirect acidimetric titration method has been developed for the determination of halide ions in neutral solutions. Known as the mercuric oxycyanide method [51], it is based on the formation of nonionizable mercuric cyanide and halide, as depicted by the reaction

$$2NaCl + 2Hg(OH)CN \longrightarrow Hg(CN)_2 + HgCl_2 + 2NaOH$$

The alkali released is titrated with standardized 0.01 $\underline{N}$ acid. Needless to say, the sample after decomposition has to be exactly neutralized prior to the addition of the mercuric oxycyanide reagent.

Electrometric Methods.

Potentiometric Titration: Potentiometric titration with silver nitrate solution is the most frequently employed technique for the determination of chloride or bromide. The silver-calomel combination electrode is generally used (see experimental procedure in Sec. 5.4), but other electrode systems can serve as well. Huffman [52] has recommended the mercury-mercurous sulfate electrode which need not be isolated from the solution by a bridge. Although the titration can be performed in aqueous solutions, a much sharper potential break at the equivalence point is obtained in ethanolic [53] or acetone [54] solutions. Recently ion-selective electrodes [48] have come into favor. Hozumi and Akimoto [55] have proposed the titration of halides using 0.005 $\underline{N}$ or 0.0005 $\underline{N}$ silver nitrate and a sodium-sensitive glass electrode as the endpoint indicator in nonaqueous medium. The potential curves show sharp inflections close to the equivalence point, and the inflection points reproduce precisely over the range of pH 2 to pH 7. Prior to titration, the liquid which is obtained by absorbing the halide in aqueous solution after closed-flask combustion is mixed with acetone to attain 90% volume of acetone concentration. Direct sunlight or other strong light sources must be screened off the titration vessel in order to prevent photochemical reduction of colloidal silver halides. Potman and Dahmen [56] have advocated the use of mercuric nitrate solution as titrant together with an ion-selective Ag_2S electrode. When Ag^+ is used as titrant, these workers have observed adsorption of halide on the silver halide precipitate after about 18 microequivalents of chloride or bromide is titrated, leading to low results. Although the mechanism of the reaction of Hg^{2+} ions with the Ag_2S electrode is not fully understood, titrations of halides using Hg^{2+} have advantages over those with Ag^+ because Hg^{2+} ions form stable and soluble complexes with halides.

Amperometric Titration: Although seldom practiced in the organic analytical laboratory, the amperometric titration of chloride or bromide with a standardized solution of silver nitrate offers a rapid and reliable method for the determination of the halogens. According to Olson [2], the most satisfactory conditions employ a pair of silver-wire electrodes polarized by the application of 0.25 V between electrodes, and it is not necessary to remove oxygen from the solution. The endpoint can be located by plotting manually [57], or by automatic recording of the titration curve, or by the dead-stop method [58].

Coulometric Titration: Chloride or bromide can be titrated by means of coulometrically generated silver ions [53,59]. The endpoint can be located amperometrically or potentiometrically. An alternative approach has been proposed by Przybylowicz and Rogers [60] who utilized amalgamated silver or gold electrodes to generate mercurous ions. The equivalence point is determined potentiometrically using a mercury-indicating electrode. The titration should be carried out in 80% methanolic solution in order to decrease the solubilities of the mercurous halides.

Determination of Iodine

If the organically bound iodine is mineralized by reduction such as sodium fusion, the resulting iodide can be determined either gravimetrically or by titration with standardized silver nitrate as described for bromide. In most cases, however, organic iodo compounds are decomposed to yield iodine or iodate. As a rule, all iodine is then oxidized to iodate by the addition of bromine in an acetate buffer [61], and the excess bromine is removed by means of formic acid [62]. The iodate is determined by the liberation of iodine upon the addition of iodide in sulfuric acid solution, followed by titration of the liberated iodine with standardized sodium thiosulfate solution. The reactions are represented by the following equations:

$$I^- + 3Br_2 + 3H_2O \xrightarrow{\text{HAc, NaAc}} HIO_3 + Br^- + 5HBr$$

$$\text{Excess } Br_2 + HCOOH \longrightarrow 2HBr + CO_2$$

$$HIO_3 + 5KI + 5H_2SO_4 \longrightarrow 3I_2 + 3H_2O + 5KHSO_4$$

$$3I_2 + 6Na_2S_2O_3 \longrightarrow 6NaI + 3Na_2S_2O_4$$

The detailed experimental procedure is given in Sec. 5.5.

Determination of Fluorine

Gravimetric Methods. Determination of fluorine in organic substances is seldom performed by gravimetry. Unlike the other halides, silver fluoride is water-soluble. Precipitation of insoluble salts of fluorine such as calcium fluoride and lead chlorofluoride [63] requires specific conditions and is practical only on the centigram-to-decigram level. It is also apparent that the presence of sulfur or phosphorus in the sample interferes with these methods.

Titrimetric Methods. Since the fluoride-selective electrode is commercially available, the convenient method to determine fluoride is to measure the potential of the fluoride solution after mineralization. The experimental procedure is given in Sec. 5.6, after sodium peroxide decomposition [14]. Satisfactory results also can be obtained after decomposition in the combustion flask [64] or sodium hydroxide fusion [65].

Titrimetric determination of fluoride based on its reaction with thorium nitrate was the preferred procedure for some time. Although it has been displaced by more convenient methods, thorium nitrate titration is still useful because it requires only simple equipment (see Sec. 5.7). It should be noted that this titrimetric technique is different from the conventional operation of visual titration. The titrant is a standardized sodium fluoride solution which is added to the blank tube and not the tube containing the sample. This is because the reaction does not follow the equation

$$Th(NO_3)_4 + 6F^- \longrightarrow (ThF_6)^{2-} + 4NO_3^-$$

stoichiometrically; other equilibria between Th^{4+} and F^- apparently also exist in the solution.

Colorimetric Methods. Colorimetric methods based on the formation of the fluoride complex with either zirconium-alizarin or zirconium-Erichrome cyanine are recommended by the American Society for Testing and Materials [66]. Bellack and Schouboe [67] have proposed the use of zirconium-SPADNS [a contraction of the name sodium-2-(p-sulfophenylazo)-1,8-dihydroxynaphthalene-3,6-disulfonate, also called 4,5-dihydroxy-3-(p-sulfophenylazo)-2,7-naphthalene disulfonic acid trisodium salt]. Unlike the other zirconium complexes which require a waiting period of 1 h for the full development of color, there is immediate formation of color when zirconyl chloride and SPADNS are added to a solution of fluoride in 0.7 $\underline{N}$ hydrochloric acid. The color is very stable and the system obeys Beer's law over a wide range. The procedure using an automated measuring device is given in Sec. 5.8. Another spectrophotometric method for the determination of fluoride is based on the blue color of cerium alizarin complexonate-fluoride complex [68,69].

Separation of Fluorosilicic Acid. It should be noted that the presence of phorphorus, arsenic, and mercury interferes with the methods that are based on the formation of fluoride complexes. In order to circumvent the interferences, the common technique involves the steam distillation of fluorosilicic acid from the fluoride solution [70,71]. The laboratory procedure is described in Sec. 5.7.

D. Interferences and Difficulties

Presence of More Than One Halogen in the Sample

Since, as seen from the previous section, the methods for determining fluoride are unique among the halides, the presence of chlorine, bromine, and iodine in the organic sample will not affect its fluorine analysis. Contrastingly, the presence of fluorine may interfere with the determination of the other halogens. For instance, fluoride interferes with the mercurimetric determination of chloride. This interference can be circumvented by adding thorium nitrate to the solution [72].

It is apparent that silver nitrate will precipitate all three halides when the organic material contains chlorine, bromine, and iodine, and is mineralized by the reductive technique. However, if oxidative decomposition is employed, iodine will be recovered as iodate which can be determined independently. When two of these three elements are present together, they can be determined separately by potentiometric titrimetry (see Sec. 5.4). On the other hand, when chloride, bromide, and iodide are in the same solution, potentiometric titrations usually give low chlorine and high bromine values though the iodine results are acceptable [73]. Fortunately, a method has been developed for the simultaneous determination of chlorine, bromine, and iodine in one sample. This is discussed in Chap. 10.

Presence of Other Elements

The methods for the determination of fluorine are affected by the presence of other elements like arsenic, mercury, phosphorus, and sulfur. Kirsten and Shah [73a] have proposed hydrogenation to eliminate the interference of phosphorus and sulfur.

For the analysis of organic substances whose ingredients are not known, it is advisable to separate the fluorine by steam distillation of fluorosilicic acid (see Sec. 5.7) prior to the finishing measurement. Several investigators have proposed the removal of phosphate by precipitation with silver [74], zinc [75], or cadmium [76] ions, but the distillation technique is more reliable. When boron and silicon are present, BF_3 and SiF_4 may be produced in the mineralization step, and the escape of these gases should be prevented.

Precipitation of silver halides is interfered by the presence of sulfide. This is also true in the titration of halides with Ag^+ or Hg^{2+} ions. On the other hand, if the sulfur in the organic sample is converted to sulfate during mineralization, the determinations will not be adversely affected. Mercurimetric titration of chloride or bromide is interfered by the presence of phosphate; this difficulty can be overcome by adding La(III) ions, as reported by Binkowski [76a].

Polyhalogenated Compounds

Polyhalogenated compounds are usually more difficult to decompose than the monohalo compounds. Alkali fusion or sodium peroxide fusion in the metal bomb is recommended for the mineralization of such substances. When the decomposition is performed in the combustion tube in an oxygen stream, the sample should be combusted gradually [77]. If the closedflask combustion technique is employed, supplementary oxidizing agents like potassium chlorate [78] or potassium nitrate [79] should be incorporated in the wrapper.

Gases, Low-boiling Liquids, and Low-melting Solids

It is not uncommon to encounter low-boiling liquids and gaseous samples in the analysis of organic materials for halogens. In the case of fluorine compounds, the volatility of the molecule increases with its fluorine content [1]. Since most of the mineralization procedures are described for solids, modified techniques should be used when dealing with

liquids and gases. The two-chamber weighing capillary (see Sec. 1.6) is
employed for handling liquid samples in the combustion tube, while the
one-chamber capillary is used in the metal bomb. When the decomposition
is performed in the closed flask, the liquid sample is weighed in a polyethylene ampoule [80], or in a glass capillary which is then wrapped in
adhesive tape [81]. For low-melting solids [82], it is recommended to
wrap the sample in about 200 mg of surgical cotton. The cotton mass
forms a wick for even combustion and clings tightly to the platinum holder
because of the tangled fibers.

Quantitative analysis of gaseous samples is always a difficult problem.
Chambers et al. [24] have devised a special apparatus for weighing and
transferring gases which are mineralized by shaking with a biphenyl-sodium-
dimethoxymethane complex to yield sodium halides. Potman and Dahmen
[56] have constructed a combustion apparatus (see Sec. 5.2B) for analyzing
volatile chloro and bromo compounds which are introduced into the combustion tube by means of a syringe.

5.2 REVIEW OF RECENT DEVELOPMENTS

A. Methods of Decomposition

As can be inferred by the number of publications during the past decade
(see Table 5.1), many investigators have reported on the use of the closed-
flask combustion technique for halogen analysis. Various designs of the
combustion flask have been proposed, among which is the apparatus of Nuti
and Ferrarini [202] shown in Fig. 5.1. Incorporating a drain tube facilitates the recovery of the absorption liquid after combustion and also the
rinsing of the walls. This modification, however, makes the flask more
fragile. Awad and co-workers [103] have studied the conditions for successful closed-flask combustion and stated that aliphatic compounds must
be mixed with glucose or benzoic acid, whereas the absorption medium
after the destruction of aromatic compounds must contain alkaline H_2O_2
or acidified $NaNO_2$ solution. While a number of investigators have claimed
satisfactory results of fluorine analysis using the combustion flask, mineralization in the metal bomb appears to be more reliable (see Table 5.2).

Decomposition in the combustion tube has recently regained attention
since this technique is (1) suitable for handling liquids and vapors, (2)
amenable to automation, and (3) rapid for series determinations. For
example, Debal and Levy [135] have described a method to mineralize
chlorine compounds to produce HCl in a stream of moist oxygen in the
empty tube and complete one determination within 13 min. A vertical
combustion tube has been proposed by Floret [135a]. Krijgsman and co-
workers [132] have constructed a semiautomatic apparatus to decompose

TABLE 5.1. Summary of Recent Literature on the Determination of
Chlorine, Bromine, and Iodine.

Decomposition method	Product determined	Finish technique	Reference
Closed flask	Halide	Titrimetric	83–111a
	Halide	Spectrophotometric	112–116
	Halide	Coulometric	53, 101, 117
	Halide	Conductometric	118
	Halide	Polarographic	119
	Halide	Atomic absorption	120
	I_2	Colorimetric	115, 121
	Cl^-	Ion chromatographic	121a
Combustion tube	Halide	Gravimetric	77, 122–132
	Halide	Titrimetric	38a, 56, 133, 133a
	Halide	Coulometric	134–135
	I_2	Gravimetric	136
	I_2	Colorimetric	137
Sealed tube	Halide	Titrimetric	126, 138, 139
Alkali or Na_2O_2 fusion	Halide	Titrimetric	140–150
Na or Na_2O_2 fusion	AgX	Atomic absorption	120a, 150a
Na-biphenyl	Halide	Titrimetric	24, 151–153
Na-NH_3	Halide	Coulometric	148
$NaBH_4$	Halide	Titrimetric	154
KOH-C_2H_5OH	Halide	Nephelometric	155
KOH-$(CH_3)_2SO$	Halide	Titrimetric	156
KOH-$K_2S_2O_8$	Halide	Titrimetric	46a
MgS fusion	Halogen	Titrimetric	157

TABLE 5.1.   (Continued)

Decomposition method	Product determined	Finish technique	Reference
Hydrogenation in NH_3	NH_4X	Gravimetric	124
Chlorination (for I)	ICl_3	Titrimetric	158
Oxyhydrogen flame	I_2	Titrimetric	146, 159, 160
	I_2	Photometric	137
Flame emission (for Cl)	CuCl	Spectrometry	161
Photochemical	Halogen		162
Neutron activation (for Cl)	Cl	Radiometric	163
Gold electrode spark (for Br)	Br	Mass spectrometry	164

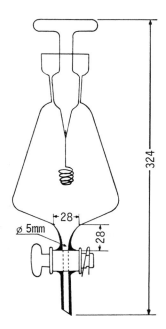

FIG. 5.1.   Combustion flask of Nuti and Ferrarini.   (From Ref. 202, courtesy Il Farmaco.)

TABLE 5.2. Summary of Recent Literature on the Determination of Fluorine.

Decomposition method	Product determined	Finish technique	Reference
Metal bomb	F^-	Colorimetric	17, 80, 165–167
	F^-	Titrimetric	168–170
	F^-	F^--selective electrode	14
	F^-	Gravimetric	171
Closed flask	F^-	Titrimetric	74, 76, 78, 172–182a
	F^-	F^--selective electrode	183–185
	F^-	Indirect polarography	185a
	F^-	Colorimetric	186, 187
	F^-	Spectrophotometric	75, 188–189a
Sealed tube	F^-	Titrimetric	190
Hydrogenation	F^-	Spectrophotometric	73a
Empty tube	SiF_4	Colorimetric	191
	NH_4F	Titrimetric	192, 193
Oxyhydrogen flame	F^-	Titrimetric	194–195
Na–biphenyl	F^-	Titrimetric	196, 197
C_2H_5ONa	F^-	Titrimetric	198
NiO–MgO	F^-	Spectrophotometric	199
Flame emission	CaF_2	Spectrometry	200
Neutron activation	^{18}F	Radiometric	201
Gold electrode spark	F	Mass spectrometry	164

0.1-10 mg of material in oxygen at 1000°C over palladium, quartz, and platinum, followed by titration of the halide solution with silver nitrate. One analysis takes 3-6 min only.

Table 5.1 also shows the variety of decomposition methods which can be used in the analysis of halogens in organic materials. Sometimes a method may appear to be simple but is difficult to adopt for routine work. For instance, Egli [154] has found that organically bound iodine, bromine, and chlorine can be dehalogenated by means of sodium borohydride in solution by stirring or boiling for 15 min in the presence of palladium, but the catalyst is easily poisoned and the halo compound may be lost due to volatility. Volodina and co-workers [154a] have described a technique for mineralization in a high-frequency discharge in hydrogen and nitrogen-hydrogen atmospheres to form ammonium halides.

B. Methods of Finish

It is understandable that titrimetry is the basis of most procedures reported for the finishing steps in halogen analysis (see Tables 5.1 and 5.2). Determination of the halides by titration is precise and rapid. Commercial instruments are available for automated operation of the buret and the recording of results. The ion-selective electrodes have been utilized for the determination of fluoride [14], chloride, bromide [56], and iodide [203] with distinct advantages. The apparatus described by Potman and Dahmen [56] is of special interest since it is constructed for the analysis of volatile samples. The schematic diagram is shown in Fig. 5.2. The sample is injected with a syringe through the septum 1. The injection port 2 (110-mm ℓ, 10.5-mm o.d.) which is surrounded by a heating tape contains an inner tube (73-mm ℓ, 6.5-mm o.d.) filled with quartz wool. The quartz

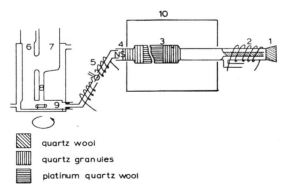

FIG. 5.2. Apparatus of Potman and Dahmen for the determination of chlorine and bromine in volatile samples. (From Ref. 56, courtesy Mikrochim. Acta.)

combustion tube is composed of section 3 (145-mm ℓ, 22-mm o.d.) and section 4 (80-mm ℓ, 10.5-mm o.d.), the latter being fitted with a 7/16-in. ground-glass joint NS. The combustion tube is connected by means of a 2-mm capillary, surrounded by a heating tape 5, to a cylindrical water-cooled absorption vessel through a three-way stopcock (2-mm bore). The double-junction reference electrode and the ion-selective electrode are placed in 6 (80-mm ℓ, 12-mm o.d.) and 7 (85-mm ℓ, 28-mm o.d.), respectively. Tube 8 (38-mm ℓ, 8-mm o.d.) is located at the center of chamber 9 (40-mm o.d., 13-mm height). The end of the 2-mm capillary, surrounded by a short piece of plastic tubing, is connected to the absorption vessel. The center of the quartz combustion tube is packed with a 20-mm length of platinum-quartz wool, whereas quartz granules are placed at the two ends. The combustion tube is maintained at 1000°C in the furnace 10. The titrant (0.1 $\underline{N}$ Hg^{2+} solution) is added from a 600-μl buret through a plastic capillary, the top of which must be placed in tube 8 well below the ion-selective electrode crystal surface. The millivolt readings are made on a pH meter, range 140 mV, connected to a recorder.

Many physicochemical methods have been proposed for the finish. Typical examples are tabulated in Tables 5.1 and 5.2. An automatic microanalyzer for chlorine, bromine, and iodine described by Hozumi and Tamura [137] is based on the optical integration method. The organic material is decomposed in a stream of oxygen in a combustion tube to yield free halogens which are further converted to iodine over silver iodide granules. The liberated iodine is conducted into a long gas cell heated at 130°C in a dark chamber where the optical integration of total iodine is carried out by light absorption at 525 nm. The maximum deflection of a recorder pen which responds to the output signal from the photoelectric system is interpreted as the quantity of the halogen in the sample by means of a calibration line previously established with several combustions of standard halogen compounds. Atomic absorption spectroscopy has been applied to halogen analysis. After flask combustion [120] and sodium [120a] or peroxide [150a] fusion, the resulting halides are converted to silver halides and the silver contents are then determined.

C. Comparison of Methods

A critical evaluation of the methods for the determination of chlorine, bromine, and iodine in organic compounds was undertaken by Debal and Levy [146]. The use of the oxyhydrogen flame was not recommended, but both empty-tube and closed-flask combustion techniques were suitable, with the former preferred for iodine-containing compounds when nitrogen, sulfur, phosphorus, or fluorine was also present in the material. Sodium peroxide fusion in a microbomb or mineralization with concentrated sulfuric acid containing a mixture of silver chromate and sodium dechromate were satisfactory.

A collaborative study on the closed-flask combustion and titrimetric finishes has been published [99]. For the determination of bromine, argentimetric titration gives better results than does iodimetric titration. For the determination of iodine, the use of hydrazine sulfate solution or hydrogen peroxide as absorbent permits greater precision.

The micro Carius sealed-tube method was considered the most reliable for several decades in organic halogen analysis. It is of interest to note that Steyermark and co-workers [126] have reported the case of a chloro compound which yielded theoretical results by the Pregl combustion-tube or the closed-flask decomposition technique but low results by the sealed-tube method. This points to the need for analyzing the sample by more than one method when the analytical data are in doubt.

The closed-flask and oxyhydrogen-flame techniques for the decomposition of fluoro compounds were compared by Levy and Debal [194]. The oxyhydrogen-flame method was shown to be universal and more precise, while the closed-flask combustion method was simpler and less expensive. The use of a polyethylene combustion flask has been advocated [64,204]. According to our experience, which has been confirmed by analysts in industrial laboratories, the closed-flask technique is not always reliable for the complete combustion of fluoro compounds, especially in the case of polymeric materials and mixtures of unknown composition.

The Commission on Microchemical Techniques of the International Union of Pure and Applied Chemistry (IUPAC) has published a list of 14 halogenated compounds recommended as test substances [205]. These compounds can serve as standard samples in the evaluation of new procedures as well as accepted methods.

Collaborative studies on the determination of chlorine, bromine, or iodine by closed-flask combustion and mercurimetric titrimetry (see Sec. 5.5) have been recently published [209,213]. It was reported that all collaborators obtained acceptable accuracy and precision. Denney and Smith [214] have undertaken a statistical comparison of the argentimetric and mercurimetric methods of finish in the closed-flask determination of bromine. While no statistical difference was found to exist, the mercurimetric method is recommended for use in laboratories that lack the facilities required for the potentiometric procedure.

5.3 EXPERIMENTAL: DECOMPOSITION BY CLOSED-FLASK
    COMBUSTION

A. Principle

The organic material is burned inside a closed flask in an atmosphere of pure oxygen, whereby halogens are liberated and absorbed, as halides, in a suitable reagent solution.

B. Apparatus

Combustion flask: Figure 5.3 shows a commercial 300-ml combustion
flask with a well; the stopper is sealed to a platinum-wire basket (sample
holder).   Figure 5.4 illustrates a home-made combustion flask assembly
which consists of a conical flask and a ground-glass stopper to which a
heavy platinum wire is sealed.   The platinum wire is either welded to a
roll of platinum sheet as shown, or is curled up to form a spiral.   Figure
5.5 is a commercial apparatus which includes an infrared igniter; the combustion flask is fitted with a ball-joint stopper and detachable sample
holder.

C. Reagents

> Oxygen cylinder.   Lecture demonstration bottle is suitable.
> Absorption solutions (for chlorine or bromine):
>> Sodium hydroxide, 0.5 $\underline{N}$.
>> Hydrogen peroxide, 30%.

D. Procedure

Preparation of the Sample

Cut out a piece of cigarette paper 20 mm square, with a tongue 20 mm
long (see Fig. 5.6).   Place the paper on a watch glass and weigh.   Add
3-10 mg of the solid sample and reweigh.   (For liquid samples see Sec.
5.3E, Note 1).   Fold the paper containing the sample and insert it into the
platinum basket.

Preparation of the Combustion Flask

Introduce 10 ml of halogen-free water, 5 ml of 0.5 $\underline{N}$ NaOH, and 10
drops of 30% H_2O_2 into the combustion flask.   Using a long glass tubing,
pass a rapid current of oxygen into the flask for about 3 min to displace all
the air.

Combustion and Absorption

Wet the neck of the combustion flask with water.   Holding the flask by
one hand (Note 2), use the other hand to bring the stopper carrying the paper
and sample over a small flame (Note 3).   Ignite the tongue and immediately
insert the stopper to close the flask; combustion of the sample takes place
instantaneously.   Meanwhile the flask should remain closed.   After combustion, shake the contents of the flask to bring all the halogens into the
alkali solution.   Determine the halide by a suitable method (e.g., see
Sec. 5.4).

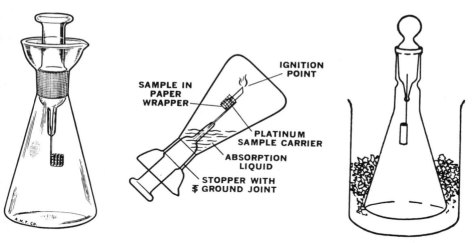

FIG. 5.3.   Combustion flask with a well. (Courtesy A. H. Thomas Co.)

FIG. 5.4.   Combustion flask made from conical flask and glass stopper. (Courtesy Van Nostrand Reinhold Co.)

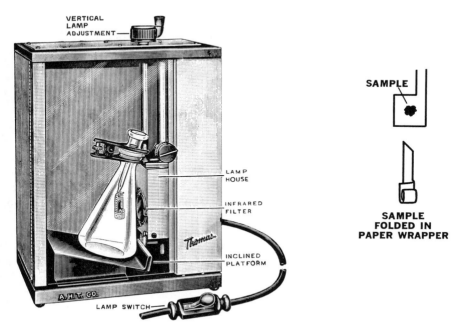

FIG. 5.5.   Commercial combustion flask and infrared igniter. (Courtesy A. H. Thomas Co.)

FIG. 5.6.   Paper holder for sample in closed-flask combustion. (Courtesy A. H. Thomas Co.)

E. Notes

Note 1: Organic liquids with low vapor pressure can be weighed in gelatin
or aluminum capsules. Low-boiling samples can be weighed in thin-walled
capillaries (see Sec. 1.6). After weighing, the capillary tip is broken
off, and the body together with the tip is quickly wrapped in the paper and
inserted into the platinum basket. Nara and Hasegawa [206] have recently
described containers made from cellulose tapes.

Note 2: It is advisable to wear a pair of gloves.

Note 3: When the infrared igniter (Fig. 5.5) is employed, the flask is
clamped and placed on an inclined platform. After the sample, wrapped
in paper, has been aligned with the beam of the infrared lamp, push-button
firing is done from outside the cabinet.

F. Comments

For compounds known to be refractory and for mixtures, it is recommended
that one checks the closed-flask combustion results against a different
method.

5.4 EXPERIMENTAL: POTENTIOMETRIC DETERMINATION OF
      CHLORINE OR BROMINE

A. Principle

After conversion of the halogen in an organic compound to its ionic form by
closed-flask combustion (see Sec. 5.3), Parr-bomb fusion (see Sec. 5.6),
or other decomposition procedure, the resultant halide is determined by
potentiometric titration with 0.02 $\underline{N}$ silver nitrate. The electrode potential
developed at the start of the titration due to excess halide slowly changes
as the halide is removed from solution, precipitating an insoluble silver
salt. When all the free halide is depleted, and excess silver ion appears,
the potential changes drastically. This is known as the equivalence point.

B. Equipment

   Metrohm Potentiograph Model 336-A (see Fig. 5.7). Figure 5.8 shows
      a new model.
   Single titrating stand (20-ml buret) Model E 436E (Brinkmann Instru-
      ments, Inc., Westbury, New York).
   Combined silver-calomel electrode with KNO_3 reference solution,
      Model PA 246.
   150-ml beakers.

FIG. 5.7. **Metrohm** Potentiograph and titration stand. (Courtesy Brinkmann Instruments.)

C. Reagents

    0.02 $\underline{N}$ AgNO$_3$: Dissolve 3.40 g of AgNO$_3$ in 1 liter of deionized water. Standardize against a known amount of pure NaCl.
    Concentrated HNO$_3$.
    Methyl red indicator, 1% in ethanol.

D. Procedure

Before starting the titration, fill the buret with titrant (AgNO$_3$) in the following manner. Turn on the main switch and allow the apparatus to warm up for 15 min. With all ground-glass connections between the glass cylinder, supply bottle, and buret in place, turn the lever on the top part of the piston buret to disengage the buret drive. [When this lever has been actuated, the piston buret can be operated by turning the horizontal drum by hand (clockwise). In this way the buret is filled to the tip. If any air bubbles are introduced during the filling, they are removed by moving the piston up and down several times with the three-way stopcock in the corresponding position.] When the buret filling is complete, reengage the buret drive by turning the lever once again, and set the three-way stopcock in the proper delivery position.

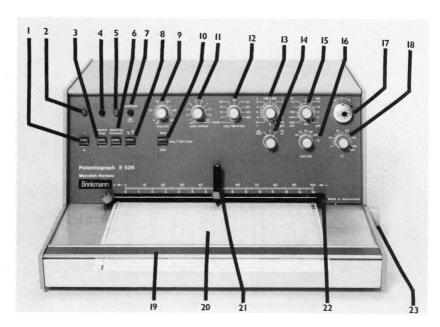

FIG. 5.8. Controls of Metrohm Potentiograph. 1, On/off switch; 2, on/off indicator light; 3, record/stop switch, for titrating; 4, record/ stop indicator light; 5, standby/measure indicator light; 6, standby/ measure switch; 7, overload indicator light; 8, lower/raise switch, for fiber tip pen; 9, potentiometer, for stopping the titration at a preset interval; 10, switch, for selecting chart paper length; 11, potentiometer, for reducing the titration speed as the equivalence point is approached; 12, titration speed control switch; 13, zero suppression control, in intervals of 100 mV or 1 pH; 14, selector switch, for pH, mV, and differential (first derivative) recording; 15, full scale sensitivity; 16, potentiometer, for the adaptation of amplifier gain to the effective electrode slope in pH measurements; 17, potentiometer, for zero point shift and pH calibration; 18, temperature control potentiometer; 19, chart paper; 20, tear bar; 21, fiber tip pen assembly; 22, chart paper guide; 23, manual chart paper advance/rewind knob. (Courtesy Brinkmann Instruments.)

Now place the solution to be titrated (about 100 ml) on the titration stand. Position the electrode and the buret tip approximately 1 in. from the bottom of the beaker and, with the addition of a 1-in. stirring bar, stir the solution at the maximum speed. Add a few drops of methyl red indicator and concentrated HNO_3 dropwise until the indicator changes from yellow to red. The potentiograph settings are as follows (see Fig. 5.7):

Range switch 3: 1000.
Compensating potential 6: 750 (for chlorine or bromine).
Titration speed 1: 4.

After the appropriate settings are made, set the pen on the extreme left line by use of the zero adjustment 4 and move the chart paper until the "0 ml" line is exactly on a level with the tip of the pen. (Using the 20-ml buret, each line represents 0.1 ml of titrant.) Now start the titration by turning the function switch 5 to the left. After the equivalence point is passed, turn off the function switch, and the buret may be refilled as previously described.

Determine the endpoint of the titration graphically by drawing the best straight line through the region of maximum $\Delta mV/vol$. The midpoint of the line between the points where the drawn line separates from the curve is the equivalence point.

E. Calculation

The halogen content (%X) is calculated from

$$\%X = \frac{N \times vol \times E_X \times 100}{W}$$

where N = normality of titrant ($AgNO_3$)
vol = volume of titrant
E_X = equivalent weight of halogen (Cl = 35.45, Br = 79.91)
W = weight of sample in milligrams

F. Comments

(1) In the closed-flask combustion procedure for samples containing chlorine or bromine, the normal $NaOH/H_2O_2$ mixture (see Sec. 5.3) is used as the absorption medium.

(2) The Parr-bomb fusion method is recommended for chloro and bromo compounds which do not burn readily. Iodine-containing compounds cannot be determined by this procedure.

(3) Scheidl and Toome [207] have recently described an automated potentiostatic procedure for determining halogens. The schematic diagram is shown in Fig. 5.9. A combustion furnace, a buret, and a titrator are modified and assembled to an apparatus. A solid-state programmer using a monostable multivibrator circuit to generate various time delays completes a halogen determination in less than 4 min.

(4) Rich et al. [207a] have employed ion chromatography to analyze chlorine and bromine.

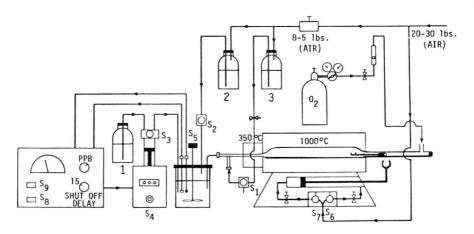

FIG. 5.9. Automated potentiostatic apparatus. Switches and solenoids are operated by the programmer (S_1 to S_9). 1, Silver nitrate (0.01 $\underline{N}$); 2, flush solution; 3, absorption solution. (From Ref. 207, courtesy Microchem. J.)

5.5 EXPERIMENTAL: DETERMINATION OF IODINE BY TITRIMETRY

A. Principle

The organic material is decomposed by the closed-flask combustion method (see Sec. 5.3), employing hydrazine sulfate as absorbent to keep all iodine in the form of iodide ions. Subsequently, the iodide ions are titrated with 0.01 $\underline{N}$ mercuric nitrate using diphenylcarbazone as the indicator. The procedure given below was described by Lalancette and co-workers [208]; it has been adopted as a standard method [209] by the Association of Official Analytical Chemists. For the determination of iodine by iodometric titration, see Sec. 5.5G (2).

B. Apparatus

Combustion flask, 500-ml capacity.

C. Reagents

Ethanol, 95%.
Diphenylcarbazone, 1.5%, in ethanol.
Bromophenol blue, 0.05%, in ethanol.
Hydrogen peroxide, 30%.
KOH, 0.05 $\underline{N}$.
HNO_3, 0.05 $\underline{N}$.
Saturated solution of hydrazine sulfate in H_2O.
Standard mercuric nitrate solution, 0.01 $\underline{N}$: Dissolve 1.7 g of
   $Hg(NO_3)_2 \cdot H_2O$ in 500 ml of water that contains 2 ml of concentrated
   HNO_3. Then add water to make up to 1 liter. Adjust the pH to 1.7
   (pH meter) with concentrated HNO_3 by adding one drop at a time.
   Weigh between 4 and 6 mg of KCl and place in a 250-ml Erlenmeyer
   flask. Add 20 ml of water and 80 ml of ethanol. Introduce a 1-in.
   magnetic stirring bar and place the flask on a magnetic stirrer.
   With the solution stirring at moderate speed, add five drops of the
   bromophenol blue solution, and adjust the pH by adding 0.5 $\underline{N}$ HNO_3
   to the yellow endpoint, and then add three drops in excess. Add
   five drops of diphenylcarbazone as indicator. Titrate (same buret
   as for Sec. 5.5D) with mercuric nitrate solution taking as endpoint
   the change in color from faint yellow to orchid pink. Run a blank
   determination on the reagents. Subtract the volume (ml) of mercuric nitrate used in this blank from that used in KCl standardization before calculating normality (N).

$$N = \frac{\text{mg KCl}}{74.555 \text{ x ml } Hg(NO_3)_2}$$

D. Procedure

Weigh the sample (containing 6-9 mg of iodine) and fold it in the paper carrier (Fig. 5.6). Insert the paper carrier in the platinum holder attached
to the stopper of the 500-ml combustion flask. Add 2.0 ml of 0.5 $\underline{N}$ KOH,
four drops of saturated hydrazine sulfate solution, and 10 ml of H_2O to the
flask. Flush the flask with a rapid stream of oxygen for at least 3 min.
Add one drop of a long-chain alcohol (e.g., dodecanol) to the paper carrier
in the platinum basket (not on the "tail") just prior to combustion. Combust the sample as described in Sec. 5.3.
      After combustion is complete, shake the stoppered flask for 10 min
or until all visible cloudiness has disappeared. After shaking has been
completed, allow the stoppered flask to stand at room temperature for 5 min.

Add approximately 3 ml of H_2O at the funnel portion of the stoppered flask as a water seal and stopper wash. Remove the stopper from the flask and rinse the stopper, platinum holder, and flask walls with 15 ml of H_2O. Add eight drops of 30% H_2O_2 to the flask and boil until the small bubbles no longer come off (about 10 min). Caution: Do not allow the contents of the flask to go to dryness--add water if necessary.

Cool the flask under the tap (Sec. 5.5F, Note 1). Add enough H_2O to bring the volume to approximately 75 ml, rinsing down the flask walls; then add 150 ml of ethanol. Add the stirring bar and place on a magnetic stirrer, stirring at moderate speed. Add 15 drops of the bromophenol blue solution. Neutralize the alkali with 0.5 $\underline{N}$ HNO_3 to the yellow endpoint and then add three drops of HNO_3 in excess. Add eight drops of diphenyl-carbazone solution as indicator. Titrate the sample with mercuric nitrate standard solution (Note 2), taking as endpoint the change in color from faint yellow to orchid pink. Determine the blank by repeating the determination in the absence of a sample and subtract this amount from the volume used for the sample (Note 3).

E. Calculation

$$\%I = \frac{ml \times normality\ of\ Hg(NO_3)_2 \times 126.91 \times 100}{mg\ sample}$$

F. Notes

Note 1: Do not let the flask sit for more than 5 min without proceeding.

Note 2: The buret should be graduated in 0.01 ml divisions and should allow one to estimate the third decimal place. The tip of the buret should be fine enough so that one drop is approximately 0.015 ml.

Note 3: A typical blank is about 0.15 ml of $Hg(NO_3)_2$ standard solution.

G. Comments

(1) The above titrimetric procedure also can be used for the determination of chlorine or bromine [210] after closed-flask combustion, employing $NaOH/H_2O_2$ as absorbent.

(2) In order to take advantage of the sixfold amplification for the determination of iodine (see Sec. 5.1C), iodometric titration is recommended.

After closed-flask combustion, 10 ml of bromine-acetic acid mixture (prepared by adding 4 ml of bromine and 10 g of potassium acetate to 100 ml of glacial acetic acid) is introduced. Then 5 ml of 1 $\underline{M}$ H_2SO_4 and a few crystals of KI are added. The flask is again stoppered and the contents are mixed by swirling. After 5 min, the iodine liberated is titrated with 0.02 $\underline{N}$ $Na_2S_2O_3$. Starch indicator is added when the yellow color nearly disappears.

5.6 EXPERIMENTAL: DETERMINATION OF FLUORINE USING AN ION-SPECIFIC ELECTRODE

A. Principle

Organic fluoro compounds are decomposed by fusion in a metal bomb with sodium peroxide, potassium nitrate, and sucrose. The fluoride ions thus produced can be conveniently determined by measurement of the potential of the reaction mixture by means of a fluoride-specific electrode. The presence of phosphorus or boron in the sample does not interfere with the determination [14].

B. Apparatus

Parr microbomb, 2.5-ml capacity (Fig. 5.10), with lead gaskets
    (Parr Instrument Co., Moline, Illinois).
Nickel beaker (of copper, nickel plated) with spout, 250-ml capacity
    (Macalaster Bicknell Co., New Haven, Connecticut).
Orion Fluoride Electrode, Model 94-09.
Calomel reference electrode, filled with saturated KCl.
Beckman Research pH Meter, Model 1019.
Nalgene (polypropylene) flasks, 250-ml.
Magnetic stirrer.

C. Reagents

Fusion mixture: Mix two parts of potassium nitrate with one part of
    sucrose and store in a glass-stoppered bottle. Use a scoop which
    holds about 100 mg of this mixture. Use a separate scoop (delivering about 1.5 g) for sodium peroxide, kept in another container.
Standard sodium fluoride solution, containing 2 μg of F per ml, prepared by dissolving 4.421 mg of NaF in 1000 ml of water.

FIG. 5.10.   Parr microbomb and stand.   (Courtesy Parr Instrument Co.)

Total ionic strength buffer (abbreviated TISAB) prepared as follows:
   Add 57 ml of glacial acetic acid, 58 g of NaCl, and 0.3 g of sodium
   citrate to 500 ml of distilled water and mix thoroughly.   Then titrate the solution to pH 5.0 with 5.0 $\underline{M}$ NaOH.   After cooling to room
   temperature, dilute to 1 liter.

D.   Procedure

Combustion in Microbomb

Weigh accurately up to 25 mg of the sample, containing between 0.5
and 5.0 mg of F, into the microbomb.   Cover the sample with one scoop-
full (100 mg) of KNO_3-sucrose mixture and one scoop-full (1.5 g) of Na_2O_2.
Assemble the microbomb and ignite it over a Bunsen burner for 50 s.
After cooling, rinse the microbomb with distilled water, open the lid, and
place the microbomb and contents in a nickel beaker.   Add distilled water
so that the microbomb is just submerged, and boil the solution on a hot
plate until effervescence ceases.   Cool the solution to room temperature,
remove the microbomb, transfer the solution quantitatively into a 250-ml
Nalgene flask, and make up to volume.

Measurement of Potential

Before proceeding further, turn on the pH meter.   Set the sensitivity
at "low," the function switch at "standby," and the scale switch at "+mV x
100."   Immerse the electrodes in a solution containing 10-100 μg of fluoride

in a 150-ml plastic beaker and stir the solution rapidly with the magnetic stirrer. (This operation is essential to condition the electrodes before measuring the potential of the samples.)

Pipet a 5.00-ml [or smaller (Sec. 5.6F)] aliquot from the Nalgene flask into a 100-ml volumetric flask. Add 10 ml of the TISAB buffer and dilute to volume with distilled water. At the same time prepare a 10-μg F and a 100-μg F standard by taking 5.00-ml and 50.0-ml aliquots, respectively, of the standard NaF solution, 10 ml of TISAB buffer, and diluting to 100 ml.

Now turn the function switch to "read" and the sensitivity control to "high." When the reading is stable, proceed with the measurement of the samples and standards. After each measurement, turn the function switch to "standby" and the sensitivity control to "norm." Pour the total contents of the 100-ml volumetric flask into a plastic beaker and, with the contents being stirred rapidly, turn the function switch to "read" and rotate the "read" control until the null meter is balanced. Simultaneously advance the sensitivity control to the highest range. When the reading is steady, record the millivolt reading to the nearest 0.1 mV. Repeat this procedure for all samples and for the two standards.

E. Calculation

Although one can obtain the concentration of F^- in the sample solution from a calibration graph constructed from the millivolt measurements of the 10-μg and 100-μg F standards, it is much simpler and more accurate to determine the F^- concentration mathematically. Using the equation of a straight line, one can derive the following relationship between the F^- concentration in a sample and the millivolt values of the two standards.

F^- (concentration in sample)

$$= \text{antilog} \left[\frac{\text{mV reading of 10-}\mu\text{g std} - \text{mV reading of sample}}{\text{mV reading of 10-}\mu\text{g std} - \text{mV reading of 100-}\mu\text{g std}} \right]$$

Example: mV reading of 10-μg standard = 137.5
mV reading of 100-μg standard = 78.3.
mV reading of sample solution (5-ml aliquot) = 91.7.
Weight of sample taken = 9835 μg.

Then

$$F^- \text{ (concentration in sample)} = \text{antilog} \left[\frac{137.5 - 91.7}{137.5 - 78.3} \right] = 59.38 \ \mu\text{g/5 ml}$$

$$\text{Total F in sample} = \frac{59.38 \times 250}{5} \ \mu\text{g}$$

$$\%F \text{ in sample} = \frac{59.38 \times 250 \times 100}{5 \times 9835} = \frac{59.38 \times 5000}{9835} = 30.19\%.$$

F. Note

Not more than 5 ml of the sample solution should be delivered into the 100-ml volumetric flask, otherwise excessive buffer would then have to be added to neutralize the OH⁻ which interferes.

G. Comments

It should be emphasized that, in order to obtain accurate millivolt readings in the present method, the electrodes should be conditioned by immersing them into a solution containing 10-100 μg of fluoride and being rapidly stirred.

5.7 EXPERIMENTAL: DETERMINATION OF FLUORINE BY ALKALI FUSION, FLUOROSILICIC ACID DISTILLATION, AND THORIUM NITRATE TITRATION

A. Principle

Upon strong heating with alkali metal in a microbomb, organically bound fluorine is converted to alkali fluoride. The fluoride ions are then separated from the fusion mixture by steam distillation of fluorosilicic acid in a perchloric acid solution, and determined by means of thorium nitrate using the match-titration technique. This method is suitable for refractory materials containing interfering elements [70].

B. Apparatus

   Metal bomb: Use the Parr microbomb, 2.5-ml capacity (see Fig. 5.10) and copper gaskets.
   Distilling apparatus: The apparatus for steam distillation of fluoro-silicic acid is shown in Fig. 5.11. The lower section is the steam generator made from a 1-liter Florence flask. It is provided with a safety tube and a side arm that serves as the outlet which is closed by a screw clamp. The upper section (b) comprises the distilling flask and condenser. As indicated by the arrows, steam S travels along the ground-glass joint J_1 passes through two

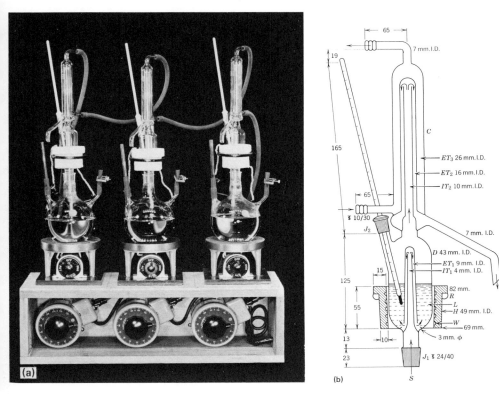

FIG. 5.11. Distilling apparatus for fluorine determination. (a) Complete unit. (b) Details of upper part. (From Ref. 70, courtesy Anal. Chem.)

concentric tubes, IT_1 and ET_1, and reaches the distilling flask D through the two openings. The vapors enter the condenser C which consists of three concentric tubes. In IT_2 and ET_2, the vapors are condensed, and in ET_3 the cooling water circulates. The ground-glass joint J_2 serves as the opening for the introduction of the alkali fluoride solution, as well as the seat of a thermometer during distillation, with the mercury bulb immersed in the liquid L. Since steam distillation from a perchloric acid solution requires a temperature of 135°C, an additional heating system is provided by the electric heating jacket H, wound by nichrome wire W, and covered with insulating cement.

Equipment for visual titration: A set of 100-ml Nessler tubes, with polyethylene caps.

C. Reagents

Metallic potassium (see Sec. 5.7E).
Perchloric acid, 70%.
Silver perchlorate solution, 25%.
Thorium nitrate standard solution, 200.00 mg $Th(NO_3)_4 \cdot 4H_2O$ in
    1000 ml of water.
Sodium fluoride standard solution, 100 μg F per ml.
Sodium alizarin sulfonate indicator, 0.01%.

D. Procedure

Preparation of the Potassium Capillary

Use the device illustrated in Fig. 5.12. Keep molten potassium under
kerosene in a test tube placed in a heating block. Weigh the glass capillary
of about 2-mm bore and join it to the assembly as shown. Apply suction to
draw potassium into the capillary. Then disconnect the capillary. After
the potassium solidifies, reweigh the capillary to get the amount of potassium per 10-mm length of the capillary.

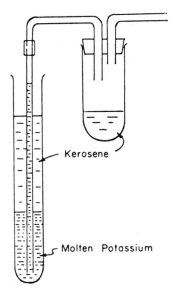

FIG. 5.12. Preparation of the potassium capillary. (Courtesy Microchem. J.)

Preparation of the Sample

Use the weighing tube (Sec. 1.6) to transfer 1-5 mg of solid sample (containing 0.1-1.0 mg F) into the microbomb. Use a glass microcup, made by sealing one end of a 4-mm tubing, to weigh a viscous liquid sample, and a 2-mm glass capillary with a tip of 15-mm length to weigh a volatile liquid.

Alkali Fusion

After introducing the sample in the microbomb, add a section of the potassium capillary containing about 50 mg of potassium. Close the bomb tightly (the copper gasket provides a resistant seal) and then heat it over a Bunsen burner for 10 min. Upon cooling, open the bomb and allow it to stand in the air for 5 min. Wash the underside of the lid with water into a 100-ml beaker. Cautiously add one to two drops of water into the bomb to destroy the last traces of residual potassium. Then add more water to dissolve the fusion mixture, and transfer the solution into the 100-ml beaker. Remove material adhering to the walls of the bomb by means of a glass rod, and rinse the bomb with a jet of water.

Steam Distillation of Fluorosilicic Acid From Perchloric Acid Solution

Transfer quantitatively the contents of the 100-ml beaker into the distilling apparatus through a funnel with a ground-glass joint to fit opening J_2 (Fig. 5.11b). Rinse the beaker with 20 ml of 70% $HClO_4$. Add 1 ml of 25% $AgClO_4$ and 10 glass beads into the still. Insert the thermometer and start the distillation. Regulate the opening of the outlet tube of the steam generator by means of the screw clamps so that pressure corresponding to a 25-mm column of water in the safety tube is maintained inside the flask. When the temperature of the fluoride solution reaches 130°C as indicated by the thermometer, close the screw clamp to force the steam into the distilling flask. Meanwhile adjust the variable resistance of the hot plate so that the temperature of the liquid L is kept at 135 ± 2°C. Collect 250 ml of distillate in a polystyrene container in about 45 min.

Determination of Fluoride by Thorium Nitrate Titration

Measure a suitable aliquot (not more than 75 ml) of the distillate into a polyethylene beaker. Adjust the pH to 3.0 ± 0.05 with drops of dilute HCl or NaOH. Transfer the solution quantitatively into a Nessler tube. Add 2 ml of 0.01% sodium alizarin sulfonate and make up to 100 ml. Prepare a blank in another Nessler tube containing fluorine-free water adjusted to the same pH and 2 ml of the indicator. Both solutions are of the same green color. Introduce a measured volume of thorium nitrate standard solution from a microburet into the Nessler tube containing the sample until a pronounced pink color is obtained. Add exactly the same volume of the thorium nitrate solution to the blank. Since the latter contains no fluoride, its pink color is much darker. Now back-titrate the blank with the sodium fluoride standard solution until the color matches the sample tube under the fluorescent lamp. Add the sodium fluoride solution in small increments and mix the contents by inverting the capped Nessler tube after each addition. The amount of fluoride required for the bleaching of the lake in the blank tube corresponds to the concentration of fluoride in the sample tube. Calculate the total amount of fluoride in the original sample by proportion.

E. Note

For the analysis of monofluoro compounds, metallic sodium can be used in place of potassium. Since sodium is less hazardous than potassium, pieces of sodium can be cut and added into the bomb directly.

5.8 EXPERIMENTAL: DETERMINATION OF FLUORINE BY AUTOMATED SPECTROPHOTOMETRY USING THE ZIRCONIUM-SPADNS REAGENT

A. Principle

After the decomposition of the organic material, the absorbent solution containing fluoride ions is diluted to a suitable volume. Zirconium-SPADNS (see Sec. 5.1C) reagent is added and the color is measured at 588 nm in the automated multichannel machine. The following procedure has been described by Ryland and Pickhardt [211] of DuPont Co. Experimental Station, Wilmington, Delaware; the experimental detail was checked by W. Pickhardt and N. Brown [211a] in 1975.

B. Apparatus

  Combustion train.
  Technicon AutoAnalyzer, from Technicon Industrial Systems, Tarrytown, New York.

C. Reagents

  Zirconyl chloride octahydrate.
  SPADNS, under the name 4,5-dihydroxy-3-(p-sulfophenylazo)-2,7-naphthalene disulfonic acid trisodium salt (Eastman Organic Chemicals, Rochester, New York).

Concentrated hydrochloric acid.

Sodium fluoride, CP, dry at 110°C.

Color reagent solution: Prepare this by mixing thoroughly equal volumes of reagents A and B. Reagent A: Dissolve 958 mg of SPADNS in distilled water and dilute to 500 ml with distilled water. Reagent B: Dissolve 133 mg of zirconyl chloride octahydrate in about 25 ml of distilled water, add 350 ml of concentrated HCl, and dilute to 500 ml with distilled water.

Reference solution. Mix equal volumes of reagent A and an HCl solution containing 350 ml concentrated HCl per 500 ml aqueous solution.

Sodium fluoride stock solution (0.5 mg of F per ml): Dissolve 1.1053g of dry sodium fluoride in distilled water and dilute to 1000 ml with distilled water.

Fluoride standard solutions: Prepare fluoride standard solutions by making appropriate dilutions of the stock solution with distilled water to produce the desired concentrations.

D. Procedure

Decompose the organic fluoro compound in the combustion tube using the oxyhydrogen flame (see Sec. 5.1B). Collect the HF in 1 $\underline{N}$ NaOH. Dilute the absorbent solution containing fluoride ions to a large volume so that its F content is between 1.0 and 10.0 parts per million (ppm).

Refer to the Technicon AutoAnalyzer manual. Prepare the appropriate manifold for the 0-1 ppm F range or the 0-10 ppm F range. Select the appropriate filter, aperture, and flow cell. Use 588 nm as the analytical wavelength. With distilled water flowing in the sample line, and the reference solution in the color reagent line, adjust the colorimeter to give 100% transmission on the recorder. Then remove the reference solution and replace it with the color reagent solution. Run standards and samples, using distilled water for the zero fluoride standard. The flow diagram for the 0-10 ppm F range is shown in Fig. 5.13. (Note 1.) For other AutoAnalyzer methods see Note 2.

E. Notes

Note 1: For ranges below the 1 ppm F level, use the following ml/min flow rates: sample, 3.90; air, 0.80; reagent, 0.80.

Note 2: At present there are three AutoAnalyzer methods for determining fluoride, including a colorimetric method and two electrochemical methods [211b].

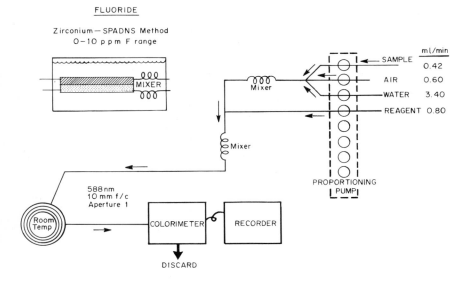

FIG. 5.13.   Flow diagram for fluorine determination using the Auto-
Analyzer.   (Courtesy Technicon Industrial Systems.)

5.9 EXPERIMENTAL: GRAVIMETRIC DETERMINATION OF CHLORINE OR BROMINE

A. Principle

Occasionally, analysis of organic chloro and bromo compounds can be carried out advantageously by using the gravimetric finish which does not require a standardized silver nitrate solution and the electrometric titration
assembly.   After decomposition of the organic material, the halide solution
obtained is treated with silver nitrate in nitric acid.   The silver halide
precipitate is then collected on a sintered glass filter, dried, and weighed.
The procedure described below illustrates a general microtechnique for
the quantitative collection of milligram amounts of precipitate by gravity
filtration.

B. Equipment

Large test tube, 30-mm diameter, 120-mm length.
Filtration assembly (Fig. 5.14) comprising the sintered glass filter
    tube, siphon tube, and filter flasks.
Drying device, such as the simple arrangement [212] shown in Fig. 5.15.

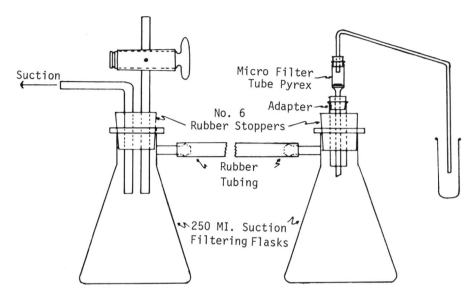

FIG. 5.14. Sintered glass filter tube and filtration assembly. (From
Ref. 122, STANDARD METHODS OF CHEMICAL ANALYSIS, Vol. 2A, 6th
Ed., edited by Frank J. Welcher,   1963 by Litton Educational Publishing,
Inc., reprinted by permission of Van Nostrand Reinhold Company.)

C. Reagents

    Concentrated nitric acid.
    Silver nitrate solution, 5%.
    Ethanol, 95%.

D. Procedure

    Decomposition

        See Sec. 5.3 for decomposition by closed-flask combustion (see also
    Sec. 5.9F, Note 1).

    Precipitation of AgCl or AgBr

        Transfer quantitatively the contents of the combustion flask containing
    sodium chloride or bromide into the large test tube which is placed in a
    600-ml beaker that serves as the water bath. Add 2 ml of concentrated

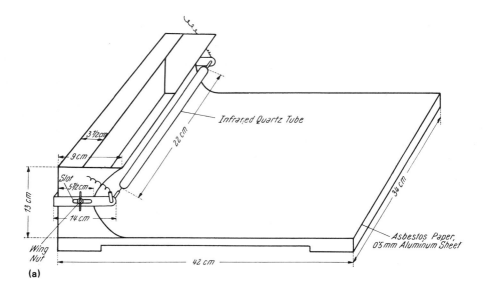

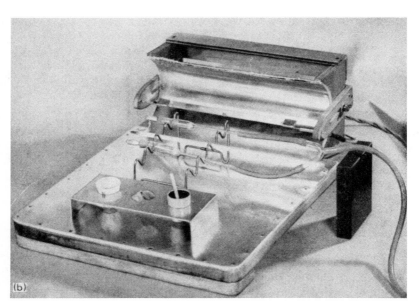

FIG. 5.15.   Infrared drying device for micro apparatus.   (a) Construction of drying device.   (b) Device in use.   (From Ref. 212, courtesy Mikrochim. Acta.)

HNO_3 to acidify the solution, followed by 2 ml of 5% $AgNO_3$. Precipitation of silver halide is indicated by a cloudy appearance. Cover the test tube with a small beaker, heat the water bath just to boiling, and then let the test tube cool in the water bath in the dark until the silver halide coagulates and settles at the bottom, leaving a clear supernatant liquid.

Filtration and Drying of Silver Halide

Prepare the sintered glass filter tube by placing a piece of filter paper on the sintered glass bed, and washing with water and ethanol. Dry at 105°C and weigh (Note 2). Assemble the filtration apparatus as shown in Fig. 5.14. Immerse the siphon tube in the clear liquid in the large test tube until its lower end is approximately 10 mm above the silver halide precipitate at the bottom of the test tube. Carefully add a layer of ethanol from the wash bottle without disturbing the silver halide. Apply a slight suction to draw the supernatant liquid through the sintered glass filter. After the aqueous solution has been replaced by ethanol (while the precipitate is covered by some liquid), gradually raise the test tube so that the siphon touches the precipitate and conducts the latter into the sintered glass filter. Then rinse the test tube with water and ethanol successively. Remove the siphon, insert an air-inlet tube packed with cotton into the mouth of the sintered glass filter, and detach the latter from the filtration assembly. Dry the filter and contents at 105°C while a slow current of air is conducted through the tube (see Fig. 5.15). Reweigh after cooling to room temperature.

E.   Calculation

$$\% \text{ Halogen} = \frac{\text{mg precipitate x factor x 100}}{\text{mg sample}}$$

Factor is 0.2474 for Cl, and 0.4255 for Br.

F.   Notes

Note 1: For the decomposition of low-boiling liquids, it is better to use the Pregl catalytic combustion technique (see Sec. 5.1B).

Note 2: It is not necessary to prepare a new filter for each determination. Successive precipitates can be collected on the sintered glass filter tube until it has received about 200 mg of silver halide. When stored in the dark, AgCl and AgBr are stable indefinitely.

REFERENCES

1. T. S. Ma, in I. M. Kolthoff and P. J. Elving (eds.), Treatise on
    Analytical Chemistry, Part II, Vol. 12, Wiley, New York, 1965,
    p. 117.
2. E. C. Olson, in I. M. Kolthoff and P. J. Elving (eds.), Treatise on
    Analytical Chemistry, Part II, Vol. 14, Wiley, New York, 1971,
    p. 1.
3. C. A. Horton, in C. N. Reilley (ed.), Advances in Analytical Chem-
    istry and Instrumentation, Vol. 1, Interscience, New York, 1960,
    p. 151.
4. J. F. Alicino, Microchem. J. 19:32 (1974).
4a. W. D. Conway, A. J. Grace, and J. E. Rogers, Anal. Biochem.
    14:491 (1966).
5. W. Hempel, Z. Angew. Chem. 5:393 (1892).
6. O. Mikl and J. Pech, Chem. Listy 46:382 (1953).
7. W. Schöniger, Mikrochim. Acta 1955:123.
8. C. L. Ogg, R. B. Kelley, and J. A. Connelly, Microchem. J.,
    Symp. Ser. 2:427 (1962).
9. A. H. Thomas Co., Philadelphia, Pa.
10. R. E. Ober, A. R. Hansen, D. Mourer, J. Bankema, and G. E.
    Gwynn, J. Appl. Radiat. Isotopes 20:703 (1969).
11. A. M. G. Macdonald, in C. N. Reilley (ed.), Advances in Analytical
    Chemistry and Instrumentation, Vol. 4, Wiley, New York, 1965,
    p. 75.
11a. W. Schöniger, Analysenvorschriften und Literaturnachweise für die
    Arbeiten mit der Kolben-Verbrennungsapparatur, Heraeus, Hanau,
    Germany, 1968.
12. J. M. Corliss, Oxygen Flask Combustion of Chemical Agents,
    Armed Services Technical Information Agency, Arlington, 1962.
13. J. E. Zarembo and I. Cohen, The Closed Flask Combustion, FMC
    Corp., Princeton, 1961, mimeographed.
14. R. C. Rittner and T. S. Ma, Mikrochim. Acta 1972:404.
15. T. L. Cottrell, The Strengths of Chemical Bonds, 2nd Ed., Butter-
    worths, London, 1958.
16. Parr Instrument Co., Moline, Ill.
17. H. J. Francis, Jr., Microchem. J. 7:150 (1963).
18. J. J. Bailey and D. G. Gehring, Anal. Chem. 33:760 (1961).
19. A. Stepanow, Berichte 39:4056 (1906).
20. T. H. Vaughn and J. A. Nieuwland, Ind. Eng. Chem., Anal. Ed.
    3:274 (1931).
21. W. H. Raucher, Ind. Eng. Chem., Anal. Ed. 9:296 (1937).
22. L. M. Liggett, Anal. Chem. 26:748 (1954).

23. F. L. Benton and W. H. Hamill, Anal. Chem. 20:269 (1948).
24. R. D. Chambers, W. K. R. Musgrave, and J. Savory, Analyst 86:356 (1961).
25. R. D. Strahm, Anal. Chem. 31:615 (1959).
26. R. A. Egli, Helv. Chim. Acta 51:2090 (1968).
27. R. R. Umhoffer, Ind. Eng. Chem., Anal. Ed. 15:383 (1943).
28. K. Sisido and H. Yagi, Anal. Chem. 20:677 (1948).
29. A. K. Ruzhentseva and V. S. Letina, Zh. Anal. Khim. 3:139 (1948).
30. F. Pregl, Die quantitative Organische Mikroanalyse, 3rd Ed., Springer, Berlin, 1930, p. 131.
31. T. Leipert and O. Watzlawek, Z. Anal. Chem. 98:113 (1934).
32. A. Lacourt, C. T. Chang, and R. Vervoort, Bull. Soc. Chim. Belges 50:67, 115 (1941); 52:175 (1943).
33. T. Mitsui and H. Sato, Mikrochim. Acta 1956:1603.
34. R. Belcher and G. Ingram, Anal. Chim. Acta 7:319 (1952).
35. J. A. Kuck (ed.), Methods in Microanalysis (translated from Russian), Gordon and Breach, New York, 1964.
36. R. Wickbold, Angew. Chem. 64:133 (1952); 66:173 (1954).
37. P. B. Sweetser, Anal. Chem. 28:1768 (1956).
38. R. Levy, in Proceedings of the International Symposium on Microchemistry (1958), Pergamon, London, 1959, p. 114.
38a. E. Kunkel, Mikrochim. Acta 1976(II):1.
39. L. Carius, Ann. 116:1 (1860).
40. A. Steyermark, Quantitative Organic Microanalysis, 2nd Ed., Academic, New York, 1961, p. 316.
41. P. J. Elving and W. B. Ligett, Ind. Eng. Chem., Anal. Ed. 14:449 (1942).
42. G. Kainz and F. Schöller, Mikrochim. Acta 1956:843.
43. H. ter Meulen, Rec. Trav. Chim. Pays-Bas 41:112 (1922); 53:118 (1934).
44. A. Lacourt, Mikrochemie 23:308 (1938).
45. M. K. Zacherl and H. G. Krainick, Mikrochemie 11:61 (1932).
46. F. Vieböck, Berichte 65:493, 586 (1932).
46a. F. Friedrich and K. Kottke, Zentbl. Pharm. Pharmakother. 115:235 (1976).
47. R. A. Durst, American Scientist 59:353 (1971).
48. R. A. Durst (ed.), Ion-Selective Electrodes, National Bureau of Standards Special Publication No. 314, U.S. Government Printing Office, Washington, D.C., 1970.
49. J. M. Corliss, C. A. Rush, S. S. Cruikshank, and E. J. W. Rhodes, Methods of Microanalysis, III. Gravimetric Determination of Sulfur, Phosphorus, Bromine, and Chlorine in Organic Compounds, U.S. Department of Commerce, Office of Technical Services, Washington, D.C., 1960.

50. W. J. Kirsten and I. Alperowicz, Mikrochim. Acta 39:234 (1952);
    Anal. Chem. 25:74 (1953).
51. R. Belcher, A. M. G. Macdonald, and A. J. Nutten, Mikrochim.
    Acta 1954:104.
52. E. W. D. Huffman, Microchem. J., Symp. Ser. 2:811 (1962).
53. E. C. Olsen and A. F. Krivis, Microchem. J. 4:181 (1960).
54. J. J. Lingane, Anal. Chem. 26:748 (1954).
55. K. Hozumi and N. Akimoto, Anal. Chem. 42:1312 (1970).
56. W. Potman and B. A. M. F. Dahmen, Mikrochim. Acta 1972:303.
57. J. J. Lingane, Electroanalytical Chemistry, 2nd Ed., Wiley, New
    York, 1958.
58. C. W. Foulk and A. T. Bawden, J. Amer. Chem. Soc. 48:2045
    (1926).
59. O. E. Sundberg, H. C. Craig, and J. S. Parsons, Anal. Chem.
    30:1842 (1958).
60. E. P. Przybylowicz and L. B. Rogers, Anal. Chem. 28:799 (1956).
61. P. L. Kirk and K. Dod, Mikrochemie 18:179 (1935).
62. F. Viebock and C. Brecher, Berichte 63:3297 (1930).
63. T. S. Ma, Microchem. J. 2:91 (1958).
64. M. B. Terry and F. Kasler, Mikrochim. Acta 1971:569.
65. R. L. Baker, Anal. Chem. 44:1326 (1972).
66. American Society for Testing and Materials, ASTM Standards, 1961,
    part 10, p. 1328.
67. E. Bellack and P. J. Schouboe, Anal. Chem. 30:2032 (1958).
68. M. A. Leonard and T. S. West, J. Chem. Soc. 1960:4477.
69. R. Belcher and T. S. West, Talanta 8:853, 863 (1961).
70. T. S. Ma and J. Gwirtsman, Anal. Chem. 29:140 (1957).
71. W. Horwitz (ed.), Official Methods of Analysis of AOAC, 12th Ed.,
    Association of Official Analytical Chemists, Washington, D.C.,
    1975, p. 926.
72. A. F. Colson, Analyst 90:35 (1965).
73. R. C. Rittner and T. S. Ma, Mikrochim. Acta 1976(I):243.
73a. W. J. Kirsten and Z. H. Shah, Anal. Chem. 47:184 (1975).
74. F. H. Oliver, Analyst 91:771 (1966).
75. N. I. Larina and N. E. Gel'man, Zh. Vses. Khim. Obshchest.
    15:231 (1970).
76. J. Horacek and V. Pechanek, Collect. Czech. Chem. Commun.
    34:323 (1969).
76a. J. Binkowski, private communication, November 1976.
77. C. Tomlinson, Talanta 9:1065 (1962).
78. F. W. Cheng, Mikrochim. Acta 1969:441.
79. I. A. Favorskaya and V. I. Lukina, Vestn. Leningrad. Univ., Ser.
    Fiz. Khim. 1961:148.

80. E. N. Tulyakov and R. A. Kalabina, Zavod. Lab. 30:1449 (1964).
81. J. M. Corliss and N. B. Scholtz, Microchem. J. 17:135 (1972).
82. L. Okolski, Chemist-Analyst 55:116 (1966).
83. T. S. Ma, in F. J. Welcher (ed.), Standard Methods of Chemical
     Analysis, 6th Ed., Vol. 2, Van Nostrand, Princeton, 1963, p. 389.
84. J. Haslam, J. B. Hamilton, and D. C. M. Squirrell, Analyst
     85:556 (1960).
85. F. W. Cheng, Microchem. J. 3:537 (1959).
86. D. G. Newman and C. Tomlinson, Mikrochim. Acta 1961:73.
87. D. C. White, Mikrochim. Acta 1961:449.
88. N. V. Sokolova, V. A. Orestova, and N. A. Nikolaeva, Zh. Anal.
     Khim. 14:472 (1959).
89. J. E. Fildes and A. M. G. Macdonald, Anal. Chim. Acta 24:121
     (1961).
90. W. A. Cook, Microchem. J. 5:67 (1961).
91. A. Nara and K. Ito, Japan Analyst 11:454 (1962).
92. L. Mazor, K. M. Papzy, and P. Klatsmanyi, Talanta 10:557 (1963).
93. C. E. Childs, E. E. Meyers, J. Cheng, and E. Laframboise,
     Microchem. J. 7:266 (1963).
94. W. R. Bontoyan, J. Ass. Offic. Agr. Chem. 47:255 (1964).
95. D. Pitre and M. Grandi, Mikrochim. Acta 1965:193.
96. M. Leclarcq and A. Mace de Lepinay, Mem. Poudres 46:129 (1964).
97. N. V. Sverkunova, V. M. Aksenenko, and R. D. Glukhovskaya,
     Izv. Tomsk. Politech. Inst. 128:160 (1964).
98. T. M. Downer, Jr., Microchem. J. 8:365 (1964).
99. A. Steyermark, J. Ass. Offic. Agr. Chem. 48:709 (1965).
100. W. I. Awad, Y. A. Gawargious, and S. S. M. Hassan, Mikrochim.
     Acta 1967:852.
101. D. F. Ketchum and H. E. P. Johnson, Microchem. J. 11:139 (1966).
102. K. Y. Kuus and L. A. Lipp, Zh. Anal. Khim. 21:1103 (1966).
103. W. I. Awad, Y. A. Gawargious, S. S. M. Hassan, and N. E. Milad,
     Anal. Chim. Acta 36:339 (1966).
104. E. Celon and S. Bresadola, Mikrochim. Acta 1969:441.
105. I. K. Chudakova and A. M. Simongauz, Zh. Anal. Khim. 22:409
     (1967).
106. D. Pitre and M. Grandi, Mikrochim. Acta 1967:347.
107. H. Wiele and E. Horak, Mikrochim. Acta 1966:1324.
108. V. I. Skorobogatova, Y. M. Faershstein, and G. A. Kravchenko,
     Zh. Anal. Khim. 23:1876 (1968).
108a. A. B. Sakla and S. A. Abu-Taleb, Talanta 20:1332 (1973).
109. E. Celon, L. Volponi, and S. Bresadola, Gazz. Chim. Ital. 99:514
     (1970).
109a. K. Narita, Japan Analyst 22:158 (1973).

110.   E. Celon and G. Marangoni, Gazz. Chim. Ital. 100:521 (1970).

110a. F. W. Cheng, Microchem. J. 21:125 (1976).

111.   Z. Pikulikova, V. Springer, and B. Kopecka, Farm. Obz. 37:503 (1968).

111a. S. S. M. Hassan, Z. Anal. Chem. 266:272 (1973).

112.   I. Lysyz, Microchem. J. 3:529 (1959).

113.   S. Mizukami, T. Ieki, and U. Kasugai, Mikrochim. Acta 1962:717.

114.   C. I. Wang, H. Tuan, and H. Y. Chi, Chem. Bull. (Peking) 1962:53.

115.   R. Willemart and J. Robin, Ann. Pharm. Fr. 21:423 (1963).

115a. V. P. Fadeeva, N. F. Zaslavskaya, and I. M. Moryakina, Zh. Anal. Khim. 29:785 (1974).

116.   H. Fukamauchi and R. Ideno, J. Pharm. Soc. Japan 787:1025 (1967).

117.   G. E. Secor and L. M. White, Microchem. J. 15:409 (1970).

118.   W. Wielopolski, J. Karjewski, and J. Swierkot, Chem. Anal. (Warsaw) 7:1139 (1962).

119.   G. M. Habashy, Y. A. Gawargious, and B. N. Faltaoos, Talanta 15:403 (1968).

120.   E. D. Truscott, Anal. Chem. 42:1657 (1970).

120a. Y. Kidani, H. Takemura, and H. Koike, Japan Analyst 22:604 (1973).

121.   N. Akimoto and K. Hozumi, Japan Analyst 18:1021 (1969).

121a. F. Smith, Jr., A. McMurtrie, and H. Galbraith, Microchem. J. 22:45 (1977).

122.   T. S. Ma, in F. J. Welcher (ed.), Standard Methods of Chemical Analysis, 6th Ed., Vol. 2, Van Nostrand, Princeton, 1963, p. 398.

123.   E. Meier, Mikrochim. Acta 1960:204.

124.   S. Mlinko, Mikrochim. Acta 1961:854.

125.   E. Pella, Mikrochim. Acta 1965:369.

126.   A. Steyermark, R. R. Kaup, and B. Blazenko, Microchem. J. 8:329 (1964).

127.   E. Kozlowski and E. Sienkowska-Zyskowska, Acad. Pol. Sci., Ser. Sci. Chim. 12:627 (1964); 13:361 (1965).

128.   M. Kan, Y. Tsukamoto, and H. Suzuki, Takeda Kenkyusho Nemko 24:187 (1965).

129.   A. Campiglio, Farmaco, Ed. Sci. 19:1033 (1964).

130.   O. Hadzija, Croat. Chem. Acta 40:247 (1968).

131.   M. A. Volodina and T. S. Gorshkova, Zh. Anal. Khim. 24:1437 (1969).

132.   W. Krijgsman, B. F. A. Griepink, J. F. Mansveld, and W. J. van Oort, Mikrochim. Acta 1970:793.

133.   G. Raspanti, Z. Anal. Chem. 225:24 (1967).

133a. A. A. Abramyan, R. A. Megroyan, and A. S. Tevosyan, Arm. Khim. Zh. 27:646 (1974).

134. H. V. Drushel, Anal. Lett. 3:353 (1970).
134a. J. Solomon and J. F. Uthe, Anal. Chim. Acta 73:149 (1974).
135. E. Debal and R. Levy, Microchem. J. 16:277 (1971).
135a. A. Floret, Bull. Soc. Chim. Fr. 1974:2350.
136. K. Hozumi and S. Kinoshita, Japan Analyst 11:113 (1962).
137. K. Hozumi and H. Tamura, Microchem. J. 14:47 (1969).
138. M. Ishidate and E. Kimura, Japan Analyst 8:739 (1959).
139. A. A. Abramyan and R. S. Sorkisyan, Izv. Akad. Nauk Arm. SSR Khim. Nauk 12:341 (1959).
140. F. Popescu, A. Carpov, and M. Dima, Stud. Cercet. Stiint. Chi. Iasi 14:203 (1963).
141. J. Jenik, M. Jurecek, and V. Patek, Collect. Czech. Chem. Commun. 24:4040 (1959).
142. L. Mazor, T. Meisel, and L. Erdy, Mikrochim. Acta 1960:417.
143. V. K. Bukina and M. Y. Moizhas, Dokl. Akad. Nauk Uzb. SSR 1959:27.
144. R. A. Shah and S. A. Jabbar, Pakistan J. Sci. Ind. Res. 5:162 (1962).
145. C. Hennart, Mikrochim. Acta 1961:543.
146. E. Debal and R. Levy, Mikrochim. Acta 1964:272.
147. M. P. Federovskaya, I. M. Khaskina, and M. N. Chumachenko, Tr. Inst. Goryuch. Iskop. Akad. Nauk SSR 21:197 (1963).
148. S. Suzuki, Japan Analyst 11:228 (1962).
149. G. F. Petukhov and T. V. Guseva, Zavod. Lab. 30:1071 (1964).
150. M. P. Strukova and I. I. Kashiricheva, Zh. Anal. Khim. 24:1244 (1969).
150a. M. A. T. M. de Almedia, S. de Moraes, and J. C. Barberio, Relat. Inst. Energia Atom. IEA-285 (1973).
151. R. D. Chambers, T. F. Holmes, and W. K. R. Musgrave, Analyst 89:369 (1964).
152. R. Liebermann, S. Hacker, and P. Hennings, Chem. Tech. (Berlin) 20:169 (1968).
153. E. P. J. Czech, J. Ass. Offic. Anal. Chem. 51:568 (1968).
154. R. A. Egli, Z. Anal. Chem. 247:39 (1969).
154a. M. A. Volodina, N. S. Moroz, M. M. Bogorodskii, A. S. Arutyunova, and S. V. Medvedev, Zh. Anal. Khim. 28:180 (1973); 29:1402 (1974).
155. A. Alon, D. Alder, F. Bernas, and N. Levite, Bull. Res. Council Israel, Sect. C 11:225 (1962).
156. J. A. Vinson and J. S. Fritz, Anal. Chem. 40:2194 (1968).
157. V. D. Osadchii and P. N. Fedoseev, Zavod. Lab. 36:104 (1970).
158. J. Körbl, D. Mansfoldeva, and E. Vanickova, Mikrochim. Acta 1963:920.

159. F. Ehrenberger, Mikrochim. Acta 1961:590.
160. F. Ehrenberger, S. Gorbach, and K. Homel, British Patent
     1,098,407 (1965).
161. M. Maruyama and S. Seno, Bull. Chem. Soc. Japan 32:486 (1959).
162. M. D. Vitalina, G. P. Shipulo, and V. A. Klimova, Mikrochim.
     Acta 1971:513.
163. Y. Kusaka and H. Tsuji, J. Radioanal. Chem. 5:359 (1970).
164. S. C. Tong, W. H. Gutenmann, L. E. St. John, Jr., and D. J.
     Lisk, Anal. Chem. 44:1069 (1972).
165. T. S. Ma, in F. J. Welcher (ed.), Standard Methods of Chemical
     Analysis, 6th Ed., Vol. 2, Van Nostrand, Princeton, 1963, p. 392.
166. C. A. Johnson and M. A. Leonard, Analyst 86:101 (1961).
167. J. Joos and R. Ruyssen, J. Pharm. Belg. 19:525 (1964).
168. A. Kondo, Japan Analyst 8:561 (1959).
169. A. I. Lebedeva, N. A. Nikolaeva, and V. A. Orestova, Zh. Anal.
     Khim. 16:469 (1961).
170. V. A. Klimova and M. D. Vitalina, Izv. Akad. Nauk SSSR, Otd.
     Khim. Nauk 1962:2245.
171. W. Selig, Z. Anal. Chem. 234:261 (1968).
172. A. Steyermark, R. R. Kaup, D. A. Petras, and E. A. Pass,
     Microchem. J. 4:55 (1960).
173. A. Konovalov, Ind. Chim. Belge 26:1257 (1961).
174. H. Trutnovsky, Mikrochim. Acta 1963:498.
175. N. H. Gel'man and L. M. Kiparenko, Zh. Anal. Khim. 20:229
     (1965).
176. T. V. Reznitskaya and V. P. Grigoryan, Zavod. Lab. 31:1329
     (1965).
177. J. Horacek and V. Pechanec, Mikrochim. Acta 1966:17.
178. D. Malysz, Acta Pol. Pharm. 25:275 (1968).
179. I. M. Maryakina, Izv. Sib. Otd. Akad. Nauk SSSR, Ser. Khim.
     Nauk 1967:125.
180. T. S. Light and R. F. Mannion, Anal. Chem. 41:107 (1969).
181. W. Selig, Z. Anal. Chem. 249:30 (1970).
181a. K. Mizuno and N. Miyaji, Japan Analyst 21:631 (1972).
182. F. W. Cheng, Mikrochim. Acta 1970:841.
182a. W. F. Heyes, Analyst 98:546 (1973).
183. J. Pavel, R. Kuebler, and H. Wagner, Microchem. J. 15:192
     (1970).
184. H. J. Francis, Jr., J. H. Deonarine, and D. D. Persing, Micro-
     chem. J. 14:580 (1969).
185. D. A. Shearer and G. F. Morris, Microchem. J. 15:199 (1970).
185a. Y. A. Gawargious, A. Besada, and B. N. Faltaoos, Anal. Chem.
     47:502 (1975).

186. H. J. Ferrari, F. C. Geronimo, and S. M. Brancone, Microchem. J. 7:617 (1961).

187. G. Bussmann and W. Hanni, Pharm. Acta Helv. 42:41 (1967).

188. E. C. Olson and S. R. Shaw, Microchem. J. 5:101 (1961).

189. W. van den Bossche, A. Haemers, and P. de Moerloose, Pharm. Weekbl. Ned. 104:445 (1969).

189a. M. Hanocq, Mikrochim. Acta 1972:707.

190. A. A. Abramyan and R. S. Sarkisyan, Izv. Akad. Nauk Arm. SSR, Khim. Nauk 14:401 (1961).

191. L. M. Dubnikov and I. A. Kalyagin, Zaved. Lab. 29:1298 (1963).

192. A. I. Filina, G. P. Shcherbachev, and V. A. Zariskii, Zh. Anal. Khim. 24:1121 (1969).

193. M. A. Volodina, T. A. Gorshkova, and A. P. Terent'ev, Zh. Anal. Khim. 24:1121 (1969).

194. R. Levy and E. Debal, Mikrochim. Acta 1962:224.

194a. B. Schreiber and R. W. Frei, Mikrochim. Acta 1975(I):219.

195. F. Ehrenberger, S. Gorbach, and K. Hommel, Z. Anal. Chem. 20:349 (1965).

196. W. K. R. Musgrave, Analyst 86:842 (1961).

197. P. P. Wheeler and M. I. Fauth, Anal. Chem. 38:1970 (1966).

198. S. Sass, N. Beitsch, and C. V. Morgan, Anal. Chem. 31:1970 (1959).

199. N. I. Larina and W. E. Gel'man, Zh. Anal. Khim. 22:582 (1967).

200. I. Schoenfeld, Israel J. Chem. 6:959 (1968).

201. F. A. M. England, J. B. Harnsby, W. T. Jones, and D. R. Terry, Anal. Chim. Acta 40:365 (1968).

202. V. Nuti and P. L. Ferrarini, Farmaco, Ed. Sci. 24:930 (1969).

203. B. Paletta and K. Panzenbeck, Clin. Chim. Acta 26:18 (1969).

204. M. Hanocq, Mikrochim. Acta 1972:705.

205. Commission on Microchemical Techniques, Section of Analytical Chemistry, IUPAC, Pure Appl. Chem. 5:759 (1962).

206. A. Nara and K. Hasegawa, Microchem. J. 19:157 (1974).

207. F. Scheidl and V. Toome, Microchem. J. 18:42 (1973).

207a. W. E. Rich, R. C. Chang, and F. C. Smith, Jr., reported at the American Chemical Society Central Regional Meeting, Charleston, October 1977.

208. R. A. Lalancette, D. M. Lukazewski, and A. Steyermark, Micro-Chem. J. 17:665 (1972).

209. A. Steyermark and R. A. Lalancette, J. Ass. Offic. Anal. Chem. 56:888 (1973).

210. A. Steyermark, R. A. Lalancette, and E. M. Contreras, J. Ass. Offic. Anal. Chem. 55:680 (1972).

211.   A. L. Ryland and W. Pickhardt, private communication, 1963; see Ref. 1, p. 159.

211a. W. Pickhardt and N. Brown, private communication, November 1974 and January 1975.

211b. M. Margoshes, Technical Director, Technicon Instruments Corp., private communication, February 1978.

212.   R. M. Maurmeyer and T. S. Ma, Mikrochim. Acta 1957:563.

213.   R. Lalancette and A. Steyermark, J. Ass. Offic. Anal. Chem. 57:26 (1974).

214.   R. C. Denney and P. A. Smith, Analyst 99:166 (1974).

Sulfur, Selenium, and Tellurium

6.1 GENERAL CONSIDERATIONS

It is logical to treat sulfur, selenium, and tellurium in one chapter since these three elements belong to the same group in the Periodic Table and possess similar chemical properties. It should be noted, however, that the methods for determining these elements in organic materials are different, although the principles utilized may be identical. For instance, organic compounds containing sulfur, selenium, or tellurium can be decomposed by wet oxidation in the micro Kjeldahl flask, but the reagents used and conditions required are not identical. The modes of finish also vary considerably.

On account of the large number of known sulfur-containing compounds and the continuing synthesis of new sulfur compounds as potential pharmaceuticals, the demand for sulfur analysis is great. In contrast, organic selenium and tellurium compounds are still rare, in spite of the promotion of these two elements in recent years [1]. Hence there are few publications on the analysis of selenium and tellurium while the literature on sulfur analysis is legion. Alicino et al. [2] have written an excellent survey of the development of methods for the determination of sulfur in organic compounds covering the literature up to 1961. It is of interest to note that although a wide range of analytical methods were proposed from time to time, few received attention. Unlike the procedures for halogens, determination of sulfur in the organic analysis laboratories seems to be restricted to only three or four methods.

Organically bound sulfur can be mineralized by reduction to the sulfide or by oxidation to the valence state of +6 (sulfate or SO_3). In certain analytical procedures, the sulfur is maintained at the oxidation state of +4 (SO_2)

to be measured. Reduction of sulfur compounds can be effected by heating
the samples with potassium [3], magnesium [4], Raney nickel [5], and
calcium hydride [6]; or by catalytic hydrogenation in the presence of platinum [7, 8]. Oxidation can be performed in several ways: (1) in nitric acid
solution in a sealed glass tube (Carius [9] method) to raise the reaction
temperature [10]; (2) in a metal bomb (Parr [11] method) by means of solid
reagents comprising sodium peroxide, sugar and potassium chlorate [12]
or sodium nitrate [13]; (3) in a specially designed bomb in the presence of
oxygen under 25-40 atmospheres pressure [14]; (4) in a combustion tube
with platinum rolls (Dennstedt [15], Pregl [16], and Grote [17] methods);
(5) in an empty tube (Belcher-Ingram [18] method) which is provided with
a double-surface chamber fitted with baffle plates; (6) in an oxyhydrogen
flame inside a combustion tube [19]; (7) in a flame inside a closed flask [20];
(8) in a crucible or microboat by fusion with sodium carbonate and potassium permanganate [21] or cobaltic oxide [22]; (9) in an open vessel by digestion with cupric nitrate and sodium chlorate (Benedict's mixture [23])
or other oxidants [24], with nitric acid and perchloric acid [25, 26], with
phosphoric acid and potassium dichromate [27], or with sodium hydroxide,
sodium nitrate, and magnesium nitrate [28].

After reductive mineralization to yield sulfide, the latter can be determined (1) gravimetrically as silver sulfide or mercuric sulfide [29], (2)
titrimetrically with standardized potassium iodate [30] or potassium ferri-
cyanide [31] solution, and (3) colorimetrically using the methylene blue
[32] or molybdenum blue [33] method. When the mineralization of the organic material converts the sulfur quantitatively to SO_3, the simplest procedure for the finish is to titrate the resulting sulfuric acid with a standardized sodium hydroxide solution (see Sec. 6.6). Numerous other procedures
have been published; some examples are presented below.

Huffman [34] suggested absorbing the SO_3 in the combustion tube by
means of a silver gauze. The Ag_2SO_4 obtained can be determined gravimetrically as such, or by other suitable methods after leaching out the
sulfate. When sulfate is produced in solution, the common gravimetric
procedure involves the precipitation of barium sulfate [35] (see Sec. 6.5)
or benzidine sulfate [36], albeit many reagents have been proposed to replace benzidine as the precipitant [37, 38]. For visual titration of sulfate
ions with barium chloride, THQ (tetrahydroquinone) [39] and potassium
rhodizonate [40] were the recommended indicators for some time in spite
of their disadvantages. Later thorin [2-(2-hydroxy-3, 6-disulfo-1-
naphthylazo)benzenearsonic acid] was shown to be superior [41], and barium perchlorate was proposed as the titrant [42, 43]. Since lead sulfate is
insoluble in nonaqueous solvents, methods have been developed to use lead
nitrate as the titrant [44] with dithizone as the indicator which turns from
green to red at the endpoint [45]. Other methods are based on the quantitative precipitation of barium sulfate with a known amount of barium ions,

followed by the titration of the excess barium complexometrically using EDTA (ethylenediamine tetraacetic acid, also called ethylenedinitrilotetra-acetic acid) as the reagent. The barium sulfate precipitate is either separated by filtration [46], or prevented from redissolving by adding ethanol to the solution to repress its solubility [47].

Many instrumental methods have been reported for sulfur determination after conversion to sulfate. Barium acetate has been recommended as the reagent for conductimetric titration [48], since acetate has a lower mobility than chloride and other inorganic anions. Sulfate can be titrated potentiometrically with barium chloride using the platinum-calomel electrode pair [49], or with lead nitrate using a lead ferri-ferrocyanide indicator electrode [50]. Lead nitrate has also been used for the amperometric determination of sulfate after the closed-flask combustion [51]. Turbiti-metric measurement of barium sulfate [52,53] and photometric titration have been advocated [54,55]. Spectrophotometric methods for the determination of sulfate are based on the measurement of the absorbance of its salt with chloranilic acid [56], benzidine [57], and 4-chloro-4'-amino-diphenyl [58], respectively, or on the release of chromate ions [59] or the thorium lake of certain dyes [60,61].

Decomposition of organoselenium and organotellurium compounds is generally carried out by wet digestion with a sulfuric-nitric acid mixture (see Secs. 6.7 and 6.8), although other mineralization procedures have been recommended [62,63]. Selenium can be determined gravimetrically by weighing the elementary selenium obtained by reduction of selenious acid with sulfur dioxide [64]. Titrimetric determination of selenium can be performed by titrating sodium hydrogen selenite with 0.01 $\underline{N}$ sodium hydroxide using methyl orange as indicator [65], or by reacting the selenious acid with iodide to yield iodine which is then titrated with sodium thiosulfate [66]. Tellurium is determined spectrophotometrically by means of potassium dichromate [67].

6.2 CURRENT PRACTICE

A. Methods of Mineralization

Notwithstanding the numerous mineralization techniques that have been proposed for the analysis of organic sulfur compounds, oxidative decomposition is the currently accepted practice. Most analysts use closed-flask combustion (see Sec. 6.4) because the apparatus is inexpensive and the operation is simple and rapid. It should be emphasized that, although the combustion takes place in a large excess of pure oxygen, the conversion of organically bound sulfur to sulfur trioxide is incomplete unless hydrogen peroxide is incorporated into the liquid absorbent. Oxidation by fusion with sodium peroxide in a metal bomb is recommended for compounds like sulfones which are difficult to decompose (see Sec. 6.5). On the other hand, the open-tube combustion method (see Sec. 6.6) is useful for the analysis of low-boiling liquids and gaseous samples; it is also amenable to automated operation of the subsequent finishing step. Sulfur in silanethiols can be titrated mercurimetrically after dissolution in isopropyl alcohol [67a].

Unlike sulfur analysis, mineralization of organoselenium and organotellurium compounds is conveniently effected by wet oxidation in sulfuric acid solution. This can be performed in test tubes or micro Kjeldahl flasks (see Secs. 6.7 and 6.8).

B.   Methods of Determination

For Sulfur Analysis

Since practicing analysts currently favor oxidative decomposition of sulfur compounds, the methods of finish in general are based on the determination of SO_3. If the organic material contains only the elements C, H, O, and S, closed-flask or open-tube combustion will lead to the formation of H_2SO_4, which is conveniently determined by titration with standard alkali (see Sec. 6.6). This method is applicable at the macro-to-micro region. When the material also contains nitrogen and halogens, the interference due to nitric and halogen acids can be circumvented by evaporating the acidic solution in a quartz dish on a steam bath [68]. The dish is then removed from the steam bath, water is added, and the sample is titrated with 0.01 $\underline{N}$ NaOH. The evaporation procedure, however, is cumbersome and hence less frequently used than the determination of SO_3 as sulfate.

Titrimetric determination of sulfate is commonly performed in aqueous-alcoholic solution using standardized barium perchlorate as the titrant and thorin as indicator [2] (see Sec. 6.4). While the endpoint is far from ideal, it is considerably better than the older method employing THQ which requires an illuminated titration stand and orange filter plate for comparison [69]. Barium chloride [2, 70], barium nitrate [71], and lead perchlorate [72] have been used in place of barium perchlorate, though there are no distinct advantages. The presence of nitrogen, chlorine, bromine, and iodine in the original sample does not affect the result. In contrast, phosphorus and fluorine interfere with the titration and should be removed. When the open-tube combustion is used, a layer of zinc oxide serves to retain the oxides of phosphorus while SO_3 passes through quantitatively [73]. When closed-flask combustion is employed, phosphate can be masked with ferric ion [70], or it may be removed by ion-exchange precipitation [74] by adding solid magnesium carbonate to the solution and filtering at 10°C. The interference due to fluoride can be eliminated by treatment with boric acid

[72, 75]. Bishara et al. [70a] have used cation-exchange resin to remove metals, and $Th(NO_3)_4$ to precipitate the phosphate.

If the organic sample is decomposed by sodium peroxide fusion, the sulfate obtained is preferably determined gravimetrically and weighed as barium sulfate (see Sec. 6.5). Since the precipitation is carried out in hydrochloric acid solution, the presence of phosphate does not interfere with the procedure. If closed-flask combustion is employed, the resulting absorbent solution should be adjusted to between 0.20 $\underline{N}$ and 0.25 $\underline{N}$ in hydrochloric acid in order to eliminate the interference of phosphate [76].

Iodimetric determination of sulfur is favored in the petroleum industry, using a commercial apparatus. Since it deals with trace amounts and low levels of sulfur in the organic material, this method is described in Chap. 12.

For Selenium and Tellurium Analysis

Since the oxidative mineralization of organo-selenium compounds produces selenious acid, selenium is usually determined by iodimetry. The selenious acid is reduced with iodide to yield iodine which is titrated with standardized sodium thiosulfate solution (see Sec. 6.7). In contrast, while organotellurium compounds are also oxidized to the +4 valence state, the Te(IV) obtained is determined by oxidation with potassium dichromate to Te(VI) and the excess dichromate is measured either titrimetrically or spectrophotometrically (see Sec. 6.8). Masson [76a] has proposed a method for the argentometric titration of selenite and tellurite, respectively, after flask combustion or H_2SO_4/HNO_3 digestion.

C. Difficulties

Whereas much improvement has taken place during the past several decades in the techniques for the analysis of sulfur in organic materials, the current methods are still beset with difficulties. The simple closed-flask combustion procedure may give low results because of incomplete decomposition of compounds like sulfones and sulfonamides. Sodium peroxide fusion leads to high concentrations of inorganic salts which may interfere with subsequent finishing procedures. This is also true with sealed-tube digestion using nitric acid or alkali metal fusion. Thus, nitrate has to be removed before the sulfate in solution can be determined by conductimetric titration [77], and interfering cations are separated by means of an ion-exchange column [78].

Extensive investigations which have been made on the visual titration of sulfate attest to the difficulties and complications of this procedure. According to Alicino and co-workers [2], barium perchlorate in dilute perchloric acid is the titrant of choice. Accurate results are obtained by

concentrating the sulfate solution to between 5 and 10 ml, adding ethyl
alcohol or isopropyl alcohol to make the solution about 80% nonaqueous,
and adjusting the pH to between 2.5 and 4.0 with 0.01 $\underline{N}$ perchloric acid.
Salts present in high concentration should be removed. Metal ions are removed with a Dowex 50 x 8 ion-exchange column. Chloride, nitrate, and
several other anions are eliminated by passing the solution through a per-
chlorated alumina column to retain sulfate. The sulfate is then eluted.
Some workers have employed a mixed indicator containing thorin and
methylene blue, but it is doubtful if this mixture offers any advantage over
thorin alone. Pietrogrande and Dalla Fini [78a] have pointed out the difficulties in applying blank corrections when fluorine or phosphorus is present
in the sample.

The drawbacks of barium sulfate as a gravimetric factor (namely, coprecipitation and extremely fine particle size) are well known. However,
attempts to find better precipitants, such as benzidine and many of its
analogs, have not yet produced a satisfactory replacement. The requirements of mixed solvent system, restricted precipitation conditions, and
drying temperature ranges make the proposed procedures unappealing.
Furthermore, these procedures are interfered by other anions, notably
phosphate. The presence of phosphorus in the organic material affects the
sulfur determination in many ways. As mentioned in Sec. 6.1, sulfur can
be reduced to hydrogen sulfide by catalytic hydrogenation, but phosphorus
will poison the catalyst.

In the determination of selenium, the main difficulty is the volatility
of selenium dioxide. During the mineralization step, rapid oxidation with
heavy charring of the sample tends to cause loss of selenium.

6.3 REVIEW OF RECENT DEVELOPMENTS

A. Techniques of Decomposition

The literature which appeared during the past decade is summarized in
Table 6.1. A wide range of decomposition techniques for the analysis of
sulfur in organic materials has been actively investigated. It is of interest
to note that the number of publications employing the combustion flask is
about equal to those employing the combustion tube with or without a platinum catalyst. Whereas the flask combustion technique is simple and
relatively inexpensive, decomposition of the organic sample in a combustion tube has the advantage of being applicable to liquids and solids, as
well as adaptable to automation and continuous serial determinations. Thus
Dokladalova and Nekovarova [114] have advocated a rapid method for the
decomposition of the sample in an empty quartz tube to produce sulfur

TABLE 6.1. Summary of Recent Literature on the Determination of Sulfur.

Decomposition method	Product determined	Finish technique	Reference
Closed flask	SO_4^{2-}	Titrimetric	45, 48, 70–72, 78–95
	SO_4^{2-}	Gravimetric	76
	H_2SO_4	Conductimetric	96
	SO_3	Gas chromatography	97
	(Ba^{2+})	Polarography	98, 98a
	(Sr)	Spectrometry	99
	SO_4^{2-}	Nephelometry	100–102
	SO_4^{2-}	Ion chromatographic	102a
Combustion tube	SO_4^{2-}	Gravimetric	73, 103–105
	SO_4^{2-}	Titrimetric	106–110
	SO_2	Spectrophotometric	111–114
	SO_2	Gas chromatography	115
	SO_2	Thermal conductivity	115a
	H_2S	Titrimetric	116–119
	H_2S	Colorimetric	120
	H_2S	Gas chromatography	121
	$(^{14}CO_2)$	Radiometric	122
Empty tube	SO_2	Colorimetric	123
	SO_2	Spectrophotometric	114, 124
	H_2SO_4	Conductimetric	125
	H_2SO_4	Coulometric	126
Sealed tube	SO_4^{2-}	Gravimetric	81
	SO_4^{2-}	Titrimetric	127, 128
	H_2S	Gas chromatography	129
Alkali fusion	S^{2-}	Titrimetric	130–135
	S^{2-}	Colorimetric	136

TABLE 6.1. (Continued)

Decomposition method	Product determined	Finish technique	Reference
Oxygen bomb	SO_4^{2-}	Gravimetric	136a
Flame	SO_3	Titrimetric	137–139
	S_2	Spectrometry	140
Wet oxidation	SO_4^{2-}	Gravimetric	141, 142
	(Ba)	Atomic absorption	143
Hydrogenation	S^{2-}	Titrimetric	117, 144
	S^{2-}	Fluorimetric	145
	H_2	Manometric	145a
Neutron activation	(^{32}P)	Radiometric	146

dioxide which is determined spectrophotometrically. The techniques of hydrogenation have been revived again. Wronski and Bald [117] have reduced the organic material with Raney nickel and then liberated hydrogen sulfide by adding hydrochloric acid, while Grünert and Tölg [145] have obtained nanogram amounts of hydrogen sulfide by hydrogenation of a 2-μg sample in the presence of a platinum catalyst (see Chap. 11). Campbell et al. [130a] have reinvestigated the potassium fusion method.

Decomposition by digestion in a test tube or micro Kjeldahl flask is the most economical technique and is recommended wherever applicable. Diuguid and Johnson [142] have shown that digestion with 70% perchloric-nitric acid mixture is suitable for the mineralization of many sulfur compounds. Similarly, organotellurium compounds are conveniently decomposed by digestion with sulfuric acid and nitric acid [67].

In contrast to the voluminous literature on sulfur analysis, there are very few recent publications on the mineralization of selenium and tellurium compounds. Srp [147] and Kuus and Lipp [148] have studied the closed-flask combustion of selenium compounds whereby the selenium is obtained in the +6 oxidation state. As new organic compounds containing selenium and tellurium are being compiled [149], interest on the analytical chemistry of these elements will be aroused. In this connection, it is worthy of note that recent publications on new organotellurium compounds reported only their carbon and hydrogen percentages and no tellurium analysis [150]. This deficiency will soon be corrected since a simple method for microdetermination of tellurium is now available (see Sec. 6.8). Masson [149a] has surveyed the methods for determining selenium or tellurium; Cole [150a] has reviewed the literature on the analysis of selenium in biological materials.

B. Modes of Finish

As can be seen in Table 6.1, determination of the final products by titrimetry predominates the research reports which appeared during the past decade, but gravimetry is still employed. Instrumental methods like spectrophotometry, polarography, thermal conductivity, and gas chromatography have been investigated.

Though barium perchlorate remains the favorite reagent for the titration of sulfate, Vickers and Wilkinson [80] have reported that this titrant is not satisfactory for compounds with a high proportion of nitrogen. Budesinsky [91] has recommended sulfonazo III as the visual indicator for titration with barium perchlorate, claiming that a sharp endpoint is obtained which is not affected by pH changes.

With the development of ion-selective electrodes, it was natural to utilize them for organic sulfur analysis. Heistand and Blake [139] have employed a lead ion-selective electrode to titrate sulfate. Slanina and coworkers [151] have titrated sulfide with a silver sulfide ion-specific electrode. Determination using a sulfate ion-selective electrode, however, has not yet been accomplished. According to Rechnitz and co-workers [152], it is relatively easy to achieve potentiometric response to the sulfate ion but extremely difficult to obtain potentiometric selectivity for sulfate. Recently these investigators utilized the crystal membrane electrode technique [153] to prepare electrodes with good response and selectivity for the sulfate ion. Optimum results were obtained with membranes consisting of 32 mole percent Ag_2S, 31 mole percent PbS, 32 mole percent $PbSO_4$, and 5 mole percent Cu_2S. The most successful membranes were formed by pressing with an applied pressure of 102,000 lb/in.^{2} for 18 h at temperatures up to 300°C. The mechanism of operation of this electrode is still obscure. It has been tested with univalent anions like perchlorate, nitrate, and chloride but has not been used to determine sulfate produced from organic materials.

Gas chromatography has been applied to organic sulfur analysis by several groups of workers. Beuerman and Meloan [115] combusted the sample in oxygen, condensed the SO_2 produced in liquid nitrogen, and later vaporized the gases into a chromatographic column, with helium as the carrier gas, and a thermistor detector. Schuessler [97] used closed-flask combustion, subsequently equilibrated the resulting gases in a gas-sample loop, and then injected them into a gas chromatograph. Okuno and co-workers [121] injected the organic sample into a stream of hydrogen flowing over a platinum catalyst, condensed the H_2S gas, and then vaporized it into the gas chromatograph. In contrast, Chumachenko and Alekseeva [129] sealed the organic material in a tube with hexadecane as a source of hydrogen and pyrolyzed it in an argon atmosphere, and then flushed the gases into a gas chromatographic column to separate H_2S which was determined conductimetrically.

The conductimetric finish has been proposed by Pell et al. [125] for automated operations. The sample is decomposed in the combustion tube at 1400°C and the gases are absorbed in a mixture of sulfuric acid and hydrogen peroxide. The sulfur content is calculated from the increase in conductivity of the sulfuric acid solution.

Bishara [98] has utilized indirect polarography to determine sulfate in the combustion flask. The solution is treated with excess of $BaCl_2$ and the unconsumed Ba^{2+} is measured polarographically in a medium 0.25 $\underline{M}$ in $CaCl_2$. A more simple and convenient procedure is to titrate the excess Ba^{2+} complexometrically [88] with 0.02 $\underline{M}$ EDTA.

Gawargious and Farag [154] have studied the amplification reaction for the iodimetric determination of sulfur. A commercial apparatus [155] for sulfur determination uses a colorimetric finish based on iodimetry. Stephen [102] has proposed 2-aminoperimidine hydrochloride while Mendes-Bezerra and Uden [100] have recommended chloroaminobiphenyl as the precipitant for the nephelometric determination of sulfate. Veillon and Park [140] have studied the determination of organic sulfur based on the blue emission from S_2 molecules in the hydrogen flame. Mlinko and Dobis [122] have described a procedure for the radiometric determination of the sulfur content of organic compounds. These modes of finish are useful for the analysis of samples which contain low percentages of sulfur (see Chap. 12).

C. Comparison of Methods

Mazor et al. [156] compared a number of methods for the microdetermination of sulfur in organic compounds. Debal and Levy [132] undertook a critical study of four methods of decomposition and various titrimetric methods for determining the resulting SO_4^{2-} or S^{2-}. They concluded that the preferred procedure is the reduction of the sulfur compound by fusion with potassium metal, followed by decomposition of the K_2S with HCl and precipitation of CdS by the evolved H_2S, and finally the determination of the CdS by iodimetry.

Jenik and Kalous [136] compared the closed-flask combustion with mineralization using magnesium metal. Steyermark and co-workers [81] reported that the closed-flask technique yielded theoretical results for sulfhydryl and sulfoamide groups while the Carius sealed-tube method gave low results. Malissa and Machherndl [87] made a systematic investigation of the closed-flask combustion method and reported no correlation for sulfur content, standard deviation, and nature of sulfur bonding.

Soep and Demoen [45] compared three methods for the titrimetric determination of sulfate after closed-flask combustion: (1) barium perchlorate as titrant and thorin as indicator; (2) barium nitrate as titrant and sodium alizarin sulfonate as indicator; (3) lead nitrate as titrant and dithizone as indicator. The last procedure was recommended as the best. Scroggins [45a] has conducted a collaborative study on the use of flask combustion in the analysis of sulfur compounds.

No report has been published on the critical evaluation of the procedures available for analysis of organoselenium compounds. The recent literature is summarized in Table 6.2, which shows that the closed-flask combustion technique is most frequently used. However, Ihn and co-workers [159] reported that the platinum basket was attacked by selenium and they recommended the use of a quartz basket with electrical ignition. After decomposition in the closed flask, Kuus and Lipp [148] found that the permanganimetric finish gave results slightly lower than those obtained by a gravimetric method. A review on the fluorometric methods for determining selenium in organic materials has been prepared by Turner [168].

TABLE 6.2. Summary of Recent Literature on the Determination of Selenium.

Decomposition method	Product determined	Finish technique	Reference
Closed flask	SeO_3^{2-}	Titrimetric	76a, 147, 148, 157-160
	Se	Gravimetric	148
Acid digestion	Se(VI), Se(IV)	Titrimetric	76a, 161, 162
	SeO_3^{2-}	Spectrophotometric	163
	Se(IV)	Fluorometric	164
Combustion tube	SeO_2	Titrimetric	165
	SeO_2	Colorimetric	166
	SeO_2	Gravimetric	167

Needless to say, fluorimetry is employed for the determination of selenium at very low levels. When the amount of selenium to be measured is in the milligram region, titrimetric procedures are preferred. Watkinson [168a] has found no significant difference between the fluorimetric and polarographic methods for the determination of selenium in blood samples.

6.4 EXPERIMENTAL: DETERMINATION OF SULFUR BY TITRATING THE SULFATE

A. Principle

The sulfur compound is oxidized by closed-flask combustion. Using aqueous hydrogen peroxide as absorbent, all sulfur is converted to sulfate. Titration of the sulfate is then carried out at pH 3 in acetone-water mixed solvent with 0.01 $\underline{M}$ barium perchlorate as titrant and dimethylsulfonazo III as indicator. The procedure given below has been recommended [169] by the Association of Official Analytical Chemists (AOAC). (See Sec. 6.4G.)

B. Equipment

Combustion flask (see Sec. 5.3).
pH meter, equipped with glass-calomel electrodes.
Magnetic stirring apparatus.
Titration lamp, stirrer-lamp, or any similar source of glare-free
    fluorescent illumination with flat white background.

C. Reagents

Acetone, reagent grade.
Potassium sulfate, ACS, powdered and dried.
Dowex 50W-X87, cation exchange resin, H^+ form.
Hydrogen peroxide, 30%, reagent grade.
Barium perchlorate standard solution, 0.01 $\underline{M}$: Dissolve 4.0 g
    $Ba(ClO_4)_2 \cdot 3H_2O$ in 1 liter of distilled water and adjust the pH
    value of the solution to 3.0 with 0.5 $\underline{N}$ HCl. Standardize by titrat-
    ing 5-7 mg of freshly dried potassium sulfate (weighed to the near-
    est 0.01 mg) using the method employed for sample titration (see
    Sec. 6.4D). Correct the titration for indicator error by making a
    blank determination and then calculate the factor (Sec. 6.4E).

Dimethylsulfonazo III indicator solution, 0.1%: Dissolve 100 mg of dimethylsulfonazo III in 30 ml of distilled water. Elute the solution through a column of Dowex 50 ion-exchange resin (pretreated with HCl). Dilute the eluent to 100 ml with distilled water.

D. Procedure

Combustion

Weigh the sample (containing 0.75-1.5 mg S) and put it in a folded paper carrier. Combust the sample using 10 ml H_2O and six drops of 30% H_2O_2 as absorbent. After combustion (see Sec. 6.4F, Note 1), shake the closed flask for 10 min and let it stand for 20 min. Open the flask and rinse the stopper and sample carrier with distilled water. Boil the solution for 10 min to remove carbon dioxide and hydrogen peroxide.

Titration

Cool to room temperature and transfer with rinsings of distilled water to a 200-ml tall-form graduated beaker. Adjust the pH to 3.0 ± 0.2 by addition of 0.5 $\underline{N}$ NH_4OH. (The total volume should be within 50 ml.) Add six drops (0.3 ml) of dimethylsulfonazo III indicator solution and 50 ml of acetone. Stirring vigorously, titrate with 0.01 $\underline{M}$ $Ba(ClO_4)_2$ to a permanent sky-blue color which persists while stirring for at least 30 s (Note 2).

E. Calculation

$$\text{Factor} = \frac{\text{mg } K_2SO_4 \times 0.1840}{\text{ml } Ba(ClO_4)_2 - \text{blank}} = \text{mg S/ml } Ba(ClO_4)_2$$

$$\%S = \frac{(\text{ml } Ba(ClO_4)_2 - \text{blank}) \times \text{factor} \times 100}{\text{mg sample}}$$

F. Notes

Note 1: If sodium or potassium is present in the sample, allow the platinum basket to fall into the absorption solution just before beginning the shaking period.

Note 2: A light blue color may appear at the start of titration but the normal purple color will return upon continuing the titration. The titration endpoint is a permanent change from mauve-purple to azure-blue.

G. Comments

An alternate AOAC method prescribes titration in ethanolic solution using
thorin-methylene blue mixed indicator. We have found that this procedure
is suitable for the analysis of organic materials containing approximately
2.0 mg of sulfur. Besides, it is better to (1) employ standardized Ba^{2+}
solution that contains 4 g of $Ba(ClO_4)_2 \cdot 3H_2O$ per liter, (2) limit the total
volume to 60 ml of ethanol, and (3) use four drops of the thorin indicator
(0.2%) together with two drops of the methylene blue indicator (0.0125%).
The endpoint is a faint pink color.

6.5 EXPERIMENTAL: DETERMINATION OF SULFUR BY GRAVIMETRY

A. Principle

The organic sample containing sulfur is combusted in a Parr microbomb
with KNO_3, sucrose, and Na_2O_2, whereupon sulfur is converted to Na_2SO_4.
The sulfate is precipitated as $BaSO_4$ by the addition of $BaCl_2$ to the aqueous
solution, and then collected in a microcrucible and dried in a muffle furnace to constant weight. From the weight of $BaSO_4$ obtained, the percent
sulfur in the sample is calculated. This method is recommended for oc-
casional analysis and for sulfur compounds (e.g., sulfones) which are
difficult to decompose.

B. Equipment

    Parr microbomb, 2.5-ml capacity (see Sec. 5.6).
    Selas crucibles, 1.2-ml capacity, No. 10 porosity (Selas Corporation
        of America, Dresher, Pennsylvania).
    Muffle furnace, 800°C.
    No. 541 Whatman filter paper.

C. Reagents

    Fusion mixture: Mix two parts of potassium nitrate with one part of
        sucrose and store in a glass-stoppered bottle. Use a scoop which
        holds about 100 mg of this mixture. Use a separate scoop (deliver-
        ing 1.5 g) for sodium peroxide, kept in another container.
    Barium chloride solution: Dissolve 10 g of $BaCl_2 \cdot 2H_2O$ in 90 ml of
        distilled water.

D. Procedure

Weigh accurately a sample (containing 3-5 mg sulfur) into the microbomb.
Cover the sample with one scoop-full (100 mg) of KNO_3-sucrose mixture
and one scoop-full (1.5 g) of Na_2O_2. Assemble the microbomb and shake
the bomb and its contents. Ignite over a microburner for 50 s. After
cooling, rinse the microbomb with distilled water, open the lid and place
the microbomb and contents into a 100-ml beaker. Add distilled water so
that the microbomb is just submerged, and boil the solution over a hot
plate until effervescence ceases. Remove the microbomb with the aid of
forceps and rinse thoroughly with distilled water. Cover the beaker with a
watch glass and boil the solution down to a volume of approximately 20 ml.
If any insoluble material is present, cool and filter contents into another
100-ml beaker using Whatman No. 541 filter paper. Add 5 ml of concentrated HCl (Sec. 6.5F), and bring the solution to a boil. Slowly add 3 ml
of $BaCl_2$ solution. Boil the mixture gently until its volume is between 15
and 20 ml. Cool and let stand overnight. Filter the precipitate into a
previously tared Selas crucible and place into the muffle furnace at 800°C
for 1 h. Remove the crucible, cool in a desiccator, and weigh. Repeat to
constant weight.

E. Calculation

$$\%S = \frac{mg\ BaSO_4 \times 0.1374 \times 100}{mg\ sample}$$

F. Note

Use 10 ml of concentrated HCl if phosphorus is present in the sample.

G. Comments

(1) The most common interference associated with the Parr bomb method
for sulfur determination is the presence of phosphorus. It has been found
that increasing the acidity of the solution by addition of excess HCl prevents the formation of the insoluble $Ba_3(PO_4)_2$.
    (2) Another less common interference is the presence of certain heavy
metals (e.g., Pb) which may form insoluble sulfates. These can be removed by means of EDTA and proper pH adjustment [170].
    (3) Liquid samples are weighed in methyl cellulose or gelatin capsules.
If gelatin capsules are used, a blank must be run.

(4) The technique of "inverted filtration" is also applicable to the gravimetric determination of sulfate [103]. The apparatus is shown in Fig. 6.1. After combustion, the sulfate solution is quantitatively transferred to a black glazed crucible which has been previously weighed with the accompanying filterstick, the latter being now joined to the rubber connection of the siphon in the filtration assembly. Then 1 ml of $BaCl_2$ solution is added, and the contents of the crucible are evaporated at 100°C to nearly dryness in order to coagulate the precipitate. Upon cooling, 5 ml of distilled water containing one drop of dilute HCl is added, and then the solution is filtered through the filterstick. The residue in the crucible is rinsed twice with distilled water. The crucible together with the filterstick is dried at 125°C and reweighed [35, 171].

FIG. 6.1. Assembly for "inverted filtration." (Courtesy A. H. Thomas Co.)

6.6 EXPERIMENTAL: DETERMINATION OF SULFUR BY ACIDIMETRY

A. Principle

When the organic material to be analyzed contains only C, H, O, and S, decomposition by closed-flask combustion or by oxidation in a combustion tube [79] will produce sulfur oxides. Absorption of the vapors in aqueous H_2O_2 converts all sulfur oxides to sulfuric acid which is conveniently determined by titration with 0.01 $\underline{N}$ NaOH.

B. Equipment

Combustion flask (see Sec. 5.3), or combustion tube and platinum contact [79] (Fig. 6.2; see Sec. 6.6F).

C. Reagents

NaOH, 0.01 $\underline{N}$.
Methyl red indicator solution, 0.1%.

FIG. 6.2. Combustion tube and platinum contact. (Courtesy A. H. Thomas Co.)

D. Procedure

After combustion and absorption in aqueous H_2O_2 (neutralized before use),
transfer the solution into a 100-ml Pyrex conical flask. Boil for 2 min
to remove the residual H_2O_2. Titrate, while the solution is still warm,
with standardized 0.01 $\underline{N}$ NaOH using 0.1% methyl red as indicator. The
endpoint is a distinct pink color which persists for 10 s.

E. Calculation

$$\%S = \frac{\text{ml NaOH x normality* x 0.1603 x 100}}{\text{mg sample}}$$

F. Note

The procedure for combustion in the combustion tube is as follows. Into a
large test tube of 30-mm o.d. and 130-mm ℓ is added 10 ml of 6% hydrogen
peroxide solution. With the tip of the combustion tube immersed in the
liquid in the test tube, gentle suction is applied to draw the absorbent up
through the glass spiral (but not beyond it). The liquid is then allowed to
drain back to the test tube. The combustion tube is now placed on the combustion stand in such a way that the glass spiral protrudes outside the combustion furnace. The large test tube is inserted to cover the tip of the
combustion tube and part of the glass spiral. Two platinum contacts are
pushed into the combustion tube until they are inside the long furnace. The
open end of the combustion tube is connected, by means of a rubber stopper
carrying a fine-tapered glass tubing, through a bubble counter to a cylinder
of oxygen. The long furnace is heated to 900°C while oxygen passes through
the combustion tube at the rate of one to two bubbles per second. The
rubber stopper is now removed to introduce the sample which is weighed
in a platinum boat or glass capillary (see Sec. 1.6). The sample is pushed
forward until it stands within 60 mm of the long furnace. The rubber
stopper is replaced and the combustion is started by placing the Bunsen
burner (or the short furnace) to the right of the sample and gradually ad-
vancing the burner toward the furnace. The organic substance is slowly
vaporized and enters the hot platinum contacts where it is catalytically
oxidized. The resultant sulfuric acid is collected on, or ahead of, the
glass spiral.

---

*This is the normality correction factor, required only when the sodium
hydroxide standard solution is not exactly 0.01 $\underline{N}$.

6.7 EXPERIMENTAL: DETERMINATION OF SELENIUM

A. Principle

The organoselenium compound is decomposed by heating in concentrated
sulfuric acid with the addition of fuming nitric acid. Selenious acid is
produced. It is then caused to react with iodide to generate iodine which
is titrated with standardized sodium thiosulfate solution. The procedure
given below is based on the method described by Gould [172].

B. Equipment

    Kjeldahl flask, 30-ml capacity, and digestion rack (see Sec. 3.3).
    Iodine flask, 125-ml capacity.

C. Reagents

    Concentrated sulfuric acid.
    Fuming nitric acid.
    Urea.
    Potassium iodide, CP.
    Sodium thiosulfate standard solution, 0.01 $\underline{N}$.
    Starch indicator solution, 2%.

D. Procedure

Accurately weigh a sample containing 0.5-2 mg of selenium into the
Kjeldahl flask. Add 3 ml of concentrated H_2SO_4 and heat over a small
flame until the reaction mixture is well charred. Then add fuming nitric
acid dropwise until the solution becomes yellow in color. Continue heating
for 15 s. If charring occurs again, repeat, adding more fuming nitric
acid. (In order to prevent loss of selenious acid by volatilization at high
temperatures, overdigestion must be avoided. and no attempt should be
made to boil off all the nitric acid.) After cooling, transfer the clear solution to the iodine flask with about 25 ml of distilled water which has been
boiled to remove dissolved oxygen. Introduce a few pieces of Dry Ice or
a current of nitrogen to displace the air in the flask. Add 3 g of urea to
destroy any nitrous acid present, 5 ml of 2% starch solution, and 1 g of
potassium iodide crystals. Swirl the reaction mixture, allow to stand for
15 s, and titrate [173] with the standardized 0.01 $\underline{N}$ $Na_2S_2O_3$.

E. Calculation

$$H_2SeO_3 + 4I^- + 4H^+ = Se + 2I_2 + 3H_2O$$

$$I_2 + 2Na_2S_2O_3 = 2NaI + Na_2S_4O_6$$

$$\%Se = \frac{ml\ 0.01\ \underline{N}\ Na_2S_2O_3 \times 0.197 \times 100}{mg\ sample}$$

F. Comments

(1) The above method is also applicable to organic materials containing chlorine or bromine. In this case, however, 1 ml of fuming nitric acid should be added to the sulfuric acid before heating, so as to prevent escape of the volatile selenium tetrahalide during digestion.

(2) Kainz [174] has employed sodium peroxide fusion for decomposition, followed by the iodometric finish.

6.8 EXPERIMENTAL: DETERMINATION OF TELLURIUM

A. Principle

Tellurium in organic materials is quantitatively converted to Te(IV) by heating with a nitric-sulfuric acid mixture in the micro Kjeldahl flask. After removal of the residual oxides of nitrogen by means of urea, the Te(IV) can be determined by oxidation to Te(VI) with a 0.02 $\underline{N}$ solution of potassium dichromate in the sulfuric acid medium. The amount of tellurium is calculated by determining the dichromate remaining in the reaction mixture, either titrimetrically or spectrophotometrically [67].

B. Equipment

Micro Kjeldahl flasks, 10-ml capacity, and digestion rack (see Sec. 3.3).
Pipets, 5-ml capacity.
For titrimetric finish: microburet, 10-ml capacity, graduated in 0.02-ml intervals.
For spectrophotometric finish: spectrophotometer, visual range.

C. Reagents

Nitric acid, specific gravity 1.43.

Sulfuric acid, specific gravity 1.84.

Potassium dichromate standard solutions, 0.0200 $\underline{N}$, and 0.0100 $\underline{N}$, prepared by dissolving 980.52 mg and 490.26 mg, respectively, of the primary standard reagent in exactly 1 liter of distilled water.

Ferrous ammonium sulfate solution, 0.02 $\underline{N}$, standardized against the above potassium dichromate standard solution.

Buffer solution: Mix equal volumes of distilled water and phosphoric acid.

Indicator solution: p-Diphenylaminesulfonic acid sodium salt, 3% aqueous.

D. Procedure

Digestion

Accurately weigh a sample containing 1-4 mg tellurium into the micro Kjeldahl flask by means of the microweighing tube (see Sec. 1.6). Add 0.5 ml of nitric acid and 1.0 ml of sulfuric acid. Heat until strong fumes of sulfur trioxide appear. Cool the flask, grease its lip with Vaseline, and transfer the contents into a 50-ml Erlenmeyer flask. Rinse the micro Kjeldahl digestion flask three times with distilled water into the 50-ml flask, care being taken to see that the volume of the solution does not exceed 10 ml. Add a microspatula-full (60-80 mg) of urea and heat to boiling. Cool, pipet exactly 5 ml of 0.0200 $\underline{N}$ potassium dichromate solution into the flask, and allow the reaction mixture to stand for not less than 30 min in order to complete the oxidation of Te(IV) to Te(VI).

Titrimetric Finish

At the end of the prescribed standing period, add exactly 5 ml of standardized 0.02 $\underline{N}$ ferrous ammonium sulfate solution and swirl the flask. Then add 2.0 ml of 1:1 phosphoric acid-water buffer and two drops (0.1 ml) of p-diphenylaminesulfonic acid indicator, and immediately back-titrate with the 0.0100 $\underline{N}$ potassium dichromate solution to a neutral gray endpoint.

Spectrophotometric Finish

After the nitric-sulfuric acid digestion, cool the solution, add 5 ml of distilled water followed by a microspatula-full of urea. Boil the solution gently and transfer it into a 50-ml volumetric flask. Introduce exactly 5 ml of 0.0200 $\underline{N}$ potassium dichromate solution and fill to the mark with distilled water. Mix thoroughly. Allow to stand for 30 min, then take aliquots for absorbance measurements at 348 nm. Prepare a calibration graph using the same procedure and 0.5-5 mg of tellurium.

E. Calculation

For titrimetric finish:

$$3Te(IV) + Cr_2O_7{}^{2-} + 14H^+ = 3Te(VI) + 2Cr^{3+} + 7H_2O$$

$$\%Te = \frac{ml\ 0.01\ \underline{N}\ K_2Cr_2O_7\ consumed\ x\ 0.638\ x\ 100}{mg\ sample}$$

REFERENCES

1. Selenium-Tellurium Development Association, Inc., Darien, Conn., Symposium on Organic Selenium and Tellurium Chemistry, Lund, Sweden, August 1975; Symposium on Selenium-Tellurium in the Environment, Notre Dame, Ind., May 1976.
2. J. F. Alicino, A. I. Cohen, and M. E. Everhard, $\underline{in}$ I. M. Kolthoff and P. J. Elving (eds.), Treatise on Analytical Chemistry, Part II, Vol. 12, Wiley, New York, 1965, p. 57.
3. G. Bussmann, Helv. Chim. Acta $\underline{33}$:1566 (1950).
4. W. Schöniger, Mikrochim. Acta $\underline{1954}$:74.
5. L. Granatelli, Anal. Chem. $\underline{31}$:434 (1959).
6. M. Schmidt and G. Talbky, Berichte $\underline{90}$:1683 (1957).
7. H. ter Meulen and J. Heslinga, Neue Methoden der Organischemischen Analyse, Akad. Verlag, Leipzig, 1927.
8. I. Irimescu and E. Chirnoga, Z. Anal. Chem. $\underline{128}$:71 (1947).
9. L. Carius, Ann. $\underline{116}$:1 (1860); Berichte $\underline{3}$:697 (1870).
10. A. Friedrich and F. Mandel, Mikrochemie $\underline{22}$:14 (1937).
11. S. W. Parr, J. Amer. Chem. Soc. $\underline{30}$:764 (1903).
12. J. F. Alicino, Ind. Eng. Chem., Anal. Ed. $\underline{13}$:506 (1941).
13. A. Elek and D. W. Hill, J. Amer. Chem. Soc. $\underline{55}$:3479 (1933).
14. R. K. Siegfriedt, J. S. Wiberley, and R. W. Moore, Anal. Chem. $\underline{23}$:1008 (1951).
15. M. Dennstedt, Berichte $\underline{30}$:1590 (1897).
16. F. Pregl, Die Quantitative Organischer Mikroanalyse, 3rd Ed., Springer, Berlin, 1930, p. 133.

17. W. Grote and H. Krekeler, Angew. Chem. 46:106 (1933).
18. R. Belcher and G. Ingram, Anal. Chim. Acta 7:319 (1952).
19. W. Kirsten, Mikrochemie 35:174 (1950).
20. O. Mikl and J. Pech, Chem. Listy 46:382 (1952); 47:904 (1953).
21. F. Feigl and R. Schorr, Z. Anal. Chem. 63:10 (1923).
22. G. H. Young, Ind. Eng. Chem., Anal. Ed. 10:686 (1938).
23. S. R. Benedict, Biol. Chem. 6:363 (1909).
24. G. Zdybek, D. S. McCann, and A. J. Boyle, Anal. Chem. 32:558 (1960).
25. G. F. Smith and A. G. Deem, Ind. Eng. Chem., Anal. Ed. 4:227 (1932).
26. R. J. Evans and J. L. St. John, Ind. Eng. Chem., Anal. Ed. 16:630 (1944).
27. T. T. Garsuch, Analyst 81:501 (1956).
28. H. Schreiber, J. Amer. Chem. Soc. 32:977 (1910).
29. J. Dick and V. Bica, Acad. Rep. Populare Romine, Baza Cercetari Stiint, Timisoara, Studii Cercetari, Stiinte Chim., Ser. I 2:103 (1955).
30. W. Zimmermann, Mikrochemie 33:122 (1946).
31. G. Charlot, Bull. Soc. Chim. Fr. 6:1447 (1939).
32. H. Stratmann, Mikrochim. Acta 1956:1031.
33. M. Vecera and A. Spevak, Collect. Czech. Chem. Commun. 21:1278 (1956).
34. E. W. D. Huffman, Ind. Eng. Chem., Anal. Ed. 12:53 (1940).
35. T. S. Ma, K. Kaimovitz, and A. A. Benedetti-Pichler, Mikrochim. Acta 1954:651.
36. A. Friedrich, Die Praxis der Quantitative Mikroanalyse, Deuticke, Leipzig, 1933, p. 103.
37. R. Belcher, M. Kapel, and A. J. Nutten, Anal. Chim. Acta 8:146 (1953).
38. R. Belcher, A. J. Nutten, E. Parry, and W. I. Stephen, Analyst 81:4 (1956).
39. C. L. Ogg, C. O. Willits, and F. J. Cooper, Anal. Chem. 20:83 (1948).
40. J. F. Alicino, Anal. Chem. 20:85 (1948).
41. J. S. Fritz and S. S. Yamamura, Anal. Chem. 27:1461 (1955).
42. S. Inglis, Mikrochim. Acta 1956:1834.
43. H. Wagner, Mikrochim. Acta 1957:19.
44. D. C. White, Mikrochim. Acta 1959:254.
45. H. Soep and P. Demoen, Microchem. J. 4:77 (1960).
45a. L. H. Scroggins, J. Assoc. Offic. Anal. Chem. 56:892 (1973).
46. O. N. Hinsvark and F. J. O'Hara, Anal. Chem. 29:1318 (1957).

47. J. R. Munger, R. W. Nippler, and R. S. Ingols, Anal. Chem.
    22:1455 (1950).
48. J. P. Dixon, Analyst 86:597 (1961).
49. B. E. Christensen, H. Wymore, and V. H. Cheldelin, Ind. Eng.
    Chem., Anal. Ed. 10:413 (1938).
50. I. I. Zhukov and T. G. Raikhinshtein, J. Gen. Chem. (USSR) 4:962
    (1934).
51. L. Gildenberg, Microchem. J. 3:167 (1959).
52. J. Haslam and D. C. M. Squirrel, J. Appl. Chem. 11:244 (1961).
53. C. J. van Nieuwenberg and B. F. E. van Bevervoode, Anal. Chim.
    Acta 19:32 (1958).
54. R. Belcher, A. D. Campbell, P. Gouverneur, and A. M. G.
    Macdonald, J. Chem. Soc. 1962:3033.
55. O. Menis, D. L. Manning, and R. G. Ball, Anal. Chem. 30:1772
    (1958).
56. I. Lysyj and J. E. Zarembo, Microchem. J. 3:173 (1959).
57. L. Anderson, Acta Chem. Scand. 7:689 (1953).
58. A. S. Jones and D. S. Letham, Chem. Ind. (London) 1954:622.
59. E. W. McChasney and W. F. Banks, Jr., Anal. Chem. 27:987 (1955).
60. J. L. Lambert, S. K. Yasuda, and M. P. Grotheer, Anal. Chem.
    27:800 (1955).
61. V. Palaty, Chem. Ind. (London) 1960:176.
62. L. T. Hallett, Ind. Eng. Chem., Anal. Ed. 14:956 (1942).
63. J. D. McCullough, T. W. Campbell, and N. J. Krilanovich, Ind.
    Eng. Chem., Anal. Ed. 18:638 (1946).
64. H. D. K. Drew and C. R. Porter, J. Chem. Soc. 11:2091 (1929).
65. F. Wrede, Z. Physiol. Chem. 109:272 (1920).
66. G. Kainz, Mikrochemie 40:332 (1953).
67. T. S. Ma and W. G. Zoellner, Mikrochim. Acta 1971:329.
67a. M. Wojnowska and W. Wojnowski, Chem. Anal. (Warsaw) 18:1117
    (1973).
68. E. L. Brewster and W. Rieman, III, Ind. Eng. Chem., Anal. Ed.
    14:820 (1942).
69. A. Steyermark, Quantitative Organic Microanalysis, 2nd Ed.,
    Academic, New York, 1961, p. 286.
70. R. B. Balodis, A. Comerford, and C. E. Childs, Microchem. J.
    12:606 (1967).
70a. S. W. Bishara, M. E. Attia, and H. N. A. Hassan, Rev. Roum.
    Chim. 19:1099 (1974).
71. K. F. Novikova and N. N. Basargin, Tr. Komis, Anal. Khim. Akad.
    Nauk SSSR 13:27 (1963).
72. W. Selig, Mikrochim. Acta 1970:168.
73. J. M. Corliss and E. J. W. Rhodes, Anal. Chem. 36:394 (1964).

74. J. S. Fritz, S. S. Yamamura, and M. J. Richard, Anal. Chem. 29:158 (1957).

75. A. M. G. Macdonald, Analyst 86:3 (1961).

76. N. Kramer, Mikrochim. Acta 1965:144.

76a. M. R. Masson, Mikrochim. Acta 1976(I):399.

77. G. R. Jamieson, J. Appl. Chem. 7:81 (1957).

78. D. C. White, Mikrochim. Acta 1960:282.

78a. A. Pietrogrande and G. Dalla Fini, Mikrochim. Acta 1974:997; 1976(II):371.

79. T. S. Ma, in F. J. Welcher (ed.), Standard Methods of Chemical Analysis, 6th Ed., Vol. 2, Van Nostrand, Princeton, 1963, p. 396.

80. C. Vickers and J. V. Wilkinson, J. Pharm. Pharmacol. 13:72 (1961).

81. A. Steyermark, E. A. Bass, C. C. Johnston, and J. C. Dell, Microchem. J. 4:55 (1960).

82. U. Bartels and H. Hoyme, Chem. Tech. (Berlin) 11:600 (1959).

83. K. F. Novikova, N. N. Basargin, and M. F. Tsyganova, Zh. Anal. Khim. 16:348 (1961).

84. W. J. Kirsten, K. A. Hansson, and S. K. Nilsson, Anal. Chim. Acta 28:101 (1963).

85. P. Fabre, Ann. Pharm. Fr. 20:563 (1962).

86. M. Ellison, Analyst 87:389 (1962).

87. H. Malissa and L. Machherndl, Mikrochim. Acta 1962:1089.

88. G. Vetter, Chem. Tech. Leipzig 15:43 (1963).

89. A. F. Colson, Analyst 88:26 (1963).

90. R. Aragones-Apodaca, Quim. Ind. (Bilbao) 10:46 (1963).

91. B. Budesinsky, Anal. Chem. 37:1159 (1965).

92. N. N. Basargin and K. F. Novikova, Zh. Anal. Khim. 21:473 (1966).

93. R. Stoffel, Z. Anal. Chem. 251:303 (1970).

94. A. Campiglio, Farmaco, Ed. Sci. 24:748 (1969).

95. C. M. Yih and D. F. Mowery, Jr., Microchem. J. 16:194 (1971).

96. A. Nara and N. Oe, Japan Analyst 13:847 (1964).

97. P. W. H. Schuessler, J. Chromatogr. Sci. 7:763 (1969).

98. S. W. Bishara, Microchem. J. 15:211 (1970).

98a. Y. A. Gawargious, A. Besada, and B. N. Faltaoos, Mikrochim. Acta 1976(I):75.

99. K. Gersonde, Anal. Biochem. 25:459 (1968).

100. A. E. Mendes-Bezerra and P. C. Uden, Analyst 94:308 (1969).

101. N. K. Loginova, V. G. Barenova, and T. P. Nesterova, Zavod. Lab. 34:1192 (1968).

102. W. I. Stephen, Anal. Chim. Acta 50:413 (1970).

102a. F. Smith, Jr., A. McMurtrie, and H. Galbraith, Microchem. J. 22:45 (1977).

103. T. S. Ma, in F. J. Welcher (ed.), Standard Methods of Chemical
      Analysis, 6th Ed., Vol. 2, Van Nostrand, Princeton, 1963, p. 397.
103a. A. B. Sakla, Anal. Chim. Acta 65:147 (1973).
104. H. Swift, Analyst 86:621 (1961).
105. K. Hozumi and H. Miura, Japan Analyst 10:640 (1962).
106. J. P. Dixon, Chem. Ind. (London) 1959:156.
107. M. Ishidate and E. Kimura, Japan Analyst 8:733 (1959).
108. B. Smith and A. Hoglund, Acta Chim. Scand. 14:1349 (1960).
109. M. Sterescu and C. Ioan, Rev. Chim. (Bucharest) 14:689 (1963).
110. B. Budesinsky and D. Vrzalova, Chemist-Analyst 55:110 (1966).
111. M. A. Volodina, M. Abdukarimova, T. A. Gorshkova, V. G.
      Borodina, and V. N. Zhardetskaya, Vest. Mosk. Gas. Univ., Ser.
      Khim. 1968:114.
112. M. A. Volodina, T. A. Gorshkova, M. Abdukarimova, V. G.
      Borodina, and L. V. Kozlovskaya, Vest. Mosk. Gas. Univ., Ser.
      Khim. 1968:110.
113. M. A. Volodina, M. Abdukarimova, and L. V. Kozlovskaya, Vest.
      Mosk. Gas. Univ., Ser. Khim. 1968:109.
114. J. Dokladalova and M. Nekovarova, Chem. Tech. (Berlin) 21:490
      (1969).
115. D. R. Beuerman and C. E. Meloan, Anal. Chem. 34:319 (1962).
115a. R. F. Culmo, Microchem. J. 17:499 (1972).
116. M. A. Volodina, M. Abdukarimova, and A. P. Terent'ev, Zh.
      Anal. Chim. 23:1420 (1968).
117. M. Wronski and E. Bald, Chem. Anal. (Warsaw) 14:173 (1969).
118. S. Covic and M. Sateva, Nafta (Zagreb) 19:351 (1968).
119. M. N. Chumachenko and N. N. Alekseeva, Izv. Akad. Nauk SSSR,
      Ser. Khim. 1968:1646.
119a. M. A. Volodina and G. A. Martynova, Zh. Anal. Khim. 27:1856
      (1972).
120. T. Takeuchi, I. Fujichima, and Y. Wakayama, Mikrochim. Acta
      1965:635.
121. I. Okuno, J. C. Morris, and W. E. Haines, Anal. Chem. 34:1427
      (1962).
122. S. Mlinko and E. Dobis, Acta Chim. Acad. Sci. Hung. 61:133 (1969).
123. J. Dokladolova, E. Korbel, and M. Vecera, Collect. Czech. Chem.
      Commun. 29:1962 (1964).
124. J. Dokladalova and S. Banas, Mikrochim. Acta 1969:741.
124a. D. Fraisse and S. Raveau, Talanta 21:633 (1974).
125. E. Pell, L. Machherndl, and H. Malissa, Mikrochim. Acta
      1963:615.
126. H. V. Drushel, Anal. Lett. 3:353 (1970).
127. L. Mazor, T. Meisel, and L. Erdey, Mikrochim. Acta 1960:412.

128. A. A. Abramyan and R. S. Sarkisyan, Izv. Akad. Nauk Arm. SSR, Khim. Nauk 16:131 (1963).

129. M. N. Chumachenko and N. N. Alekseeva, Izv. Akad. Nauk SSSR, Ser. Khim. 1969:964.

130. M. H. Hashmi, M. Elahi, and E. Ali, Analyst 88:140 (1963).

130a. A. D. Campbell, M. J. Brown, and D. J. Hannah, Anal. Chim. Acta 78:234 (1975).

131. A. F. Colson, Analyst 88:791 (1963).

132. E. Debal and R. Levy, Mikrochim. Acta 1966:202.

133. V. V. Mikhailov and T. I. Tarasenko, Zavod. Lab. 15:1380 (1967).

134. J. Binkowski and M. Wronski, Mikrochim. Acta 1971:429.

135. M. P. Strukova and A. A. Lapshova, Zh. Anal. Khim. 24:1577 (1969).

136. J. Jenik and J. Kalous, Paliva 41:329 (1961).

136a. British Standards Institution, B.S. 4454 (1969).

137. F. Barat and B. Laine, Chim. Anal. 46:83 (1964).

138. R. A. Megroyan and S. N. Tonakanyan, Izv. Akad. Nauk Arm. SSR, Khim. Nauk 18:219 (1965).

139. R. N. Heistand and C. T. Blake, Mikrochim. Acta 1972:212.

140. C. Veillon and J. Y. Park, Anal. Chim. Acta 60:293 (1972).

141. G. D. Tiwari, G. S. Johar, and S. R. Trivedi, Indian J. Appl. Chem. 32:191 (1969).

142. L. I. Diuguid and N. C. Johnson, Microchem. J. 13:616 (1968).

143. A. Wollin, At. Absorption Newslett. 9:43 (1970).

144. J. Slanina, P. Vermeer, J. Agterdenbos, and B. Griepink, Mikrochim. Acta 1973:607.

144a. M. Wronski and E. Bald, Chem. Anal. (Warsaw) 12:863 (1967).

145. A. Grünert and G. Tölg, Talanta 18:881 (1971).

145a. J. W. Frazer and R. K. Stump, Mikrochim. Acta 1967:651.

146. R. Heslop and S. K. Tay, Anal. Chim. Acta 47:183 (1969).

147. L. Srp, Chem. Prum. 17:390 (1967).

148. K. Y. Kuus and L. A. Lipp, Zh. Anal. Khim. 21:1266 (1966).

149. D. H. Reid (ed.), Organic Compounds of Sulphur, Selenium, and Tellurium, The Chemical Society, London, Vol. I, 1969; Vol. II, 1973.

149a. M. R. Masson, Mikrochim. Acta 1976(I):419.

150. J. L. Piette and M. Renson, Bull. Soc. Chim. Belges 80:521 (1971).

150a. L. E. Coles, J. Ass. Publ. Analysts 12:69 (1974).

151. J. Slanina, E. Buysman, J. Agterdenbos, and B. Griepink, Mikrochim. Acta 1971:657.

152. G. A. Rechnitz, G. H. Fricke, and M. S. Mohan, Anal. Chem. 44:1098 (1972).

153. J. W. Ross, Jr., in R. Durst (ed.), Ion-Selective Electrodes, U.S. Government Printing Office, Washington, D.C., 1969, p. 57.

154. Y. A. Gawargious and A. B. Farag, Talanta 19:64 (1972).

155. Sulfur Titrator, Leco Co., St. Joseph, Mich.

156. L. Mazor, T. Meisel, and L. Erdey, Magy. Kem. Lapja 14:494 (1959).

157. E. Meier and N. Shaltiel, Mikrochim. Acta 1960:580.

157a. E. Debal, G. Madelmont, and S. Peynot, Talanta 23:675 (1976).

158. A. S. Zabrodina and A. P. Khlystova, Vestn. Mosk. Univ., Ser. Khim. 1960:66.

159. W. Ihn, G. Hesse, and P. Newland, Mikrochim. Acta 1962:628.

160. L. Barcza, Acta Chim. Acad. Sci. Hung, 47:137 (1966).

161. H. Bieling and W. Wagenknecht, Z. Anal. Chem. 201:419 (1964).

162. A. Kotarski, Chem. Anal. (Warsaw) 10:541 (1965).

163. D. Dingwall and W. D. Williams, J. Pharmacol. 13:12 (1961).

164. J. A. Raihle, Environ. Sci. Technol. 6:621 (1972).

165. Z. Stefanac and Z. Rakovic, Mikrochim. Acta 1965:81.

166. N. Kunimine, H. Ugajin, and M. Nakamura, J. Pharm. Soc. Japan 83:59 (1963).

167. A. P. Terent'ev, M. A. Volodina, E. G. Fursova, and G. A. Martynova, Zh. Anal. Khim. 23:953 (1968).

168. G. K. Turner Associates, Fluorometry Reviews: Selenium, September 1972.

168a. J. H. Watkinson, N. Z. Med. J. 80:202 (1974).

169. L. H. Scroggins, private communication, 1974.

170. F. J. Welcher, The Analytical Uses of EDTA, Van Nostrand, Princeton, 1958, p. 321.

171. R. M. Maurmeyer and T. S. Ma, Mikrochim. Acta 1957:563.

172. E. S. Gould, Anal. Chem. 23:1502 (1951).

173. J. D. McCullough, T. W. Campbell, and N. J. Krilanovich, Ind. Eng. Chem., Anal. Ed. 18:638 (1946).

174. G. Kainz, Mikrochemie 40:332 (1953).

Phosphorus, Arsenic, Antimony, and Bismuth

7.1 GENERAL CONSIDERATIONS

In this chapter we consider the analysis of organic materials for Group V elements, with the exception of nitrogen which is treated in Chap. 3. It is convenient to discuss phosphorus, arsenic, antimony, and bismuth together since these four elements have some common analytical characteristics. For example, in the course of mineralization of the organic samples, they are usually converted to the +5 oxidation state. On the other hand, they also can be obtained as trivalent products, e.g., PH_3, AsH_3, and SbH_3.

Determination of phosphorus, arsenic, and bismuth is frequently performed on biological, agricultural, or pharmaceutical samples. Phosphorus is a constituent of many proteins and certain insecticides. Arsenicals are used as chemotherapeutic agents, insecticides, and exterminating poisons. Bismuth is employed in pharmacy and it is considered a heavy metal of significance in the analysis of food products [1].

7.2 CURRENT PRACTICE

A. Methods of Decomposition

Acid digestion in an open vessel is the method of choice for all four elements. A mixture of nitric acid and sulfuric acid is the most common reagent. Hydrogen peroxide may be added after the reaction mixture has cooled down, and then the flask is reheated. It should be noted that commercial hydrogen peroxide may contain phosphorus, as reported by Chen and co-workers [2]. A very effective oxidizing medium has been advocated by Diuguid and Johnson [3]; it consists of 3 ml of nitric acid (specific gravity 1.42) and 2 ml of perchloric acid (70-72%) for 3-5 mg of sample. For larger samples, however, preliminary digestion with nitric acid should be utilized in order to prevent violent reaction in the presence of perchloric acid. Williams [4] has employed sulfuric acid mixed with potassium and copper sulfates for the decomposition of arsenical war gases.

Sodium peroxide fusion has been recommended by Fennell and co-workers [5,6] and by Terent'ev and co-workers [7] for the decomposition of organic samples containing phosphorus, silicon, and fluorine. In contrast, as reported by Bamford [8], Parr bomb decomposition of antimony compounds was unsatisfactory.

Decomposition by means of closed-flask combustion is often beset with complications for the determination of these Group-V elements. Pfirter [8a] has discussed phosphorus compounds that are difficult to combust. Romer and Griepink [9] have reported loss of phosphorus [10, 11] unless the platinum basket is heated in the absorption solution. Arsenic and bismuth form alloys with platinum. For this reason, a quartz spiral has been recommended as a replacement for the platinum gauze [12, 13]. Celon et al. [13a] have used benzoyl peroxide as additive. However, flask combustion is not well suited for the analysis of arsenic compounds [14, 15], though it is frequently employed for phosphorus determinations.

Special methods of decomposition have been proposed for certain types of compounds. For instance, Sharma et al. [15a] have determined P(III) in tertiary phosphines or As(III) in tertiary arsines by dissolving the sample in acetic acid, adding 0.1 $\underline{N}$ KMnO$_4$ and KI, and titrating the liberated iodine; P(V) and As(V) compounds can be reduced by LiAlH$_4$ to P(III) and As(III), respectively, followed by iodometric determination.

B. Methods of Determination

    Phosphorus

    Spectrophotometric and Colorimetric Methods. Since the spectrophotometer has become standard equipment in the analytical laboratory, spectrophotometric procedures for phosphorus determination are currently the prevailing practice. This is particularly true in the analysis of biological materials for which multisamples are usually submitted. Generally speaking, the yellow color produced by vanadiphosphomolybdate complexes is recommended for determinations of amounts of phosphorus at the milligram level, while the blue color produced by the reduction of phosphomolybdate complexes (molybdenum blue methods) is favored for determinations at the microgram level.

The experimental details utilizing the yellow color [16] are given in Sec. 7.4, and the molybdenum blue procedure is given in Sec. 7.5. Mention may be made here that the yellow solution is stable and its absorbance is not easily affected by slight variation of operational conditions. In contrast, the blue solution is very sensitive to ambient changes and requires strict adherence of the specified details in order to obtain accurate results. The latter has the advantage, however, of producing an intense color with micro amounts of phosphate. As a matter of fact, when the phosphorus content of the organic sample is high, the blue solution obtained must be diluted before being transferred to the absorption cell of the spectrophotometer.

It should be noted that the solutions being measured contain more than one colored species. The situation is particularly complicated for "molybdenum blue." The exact nature of the complex ions responsible for the color has been under investigation for half a century [17, 18] and the problem still awaits elucidation. Development of the blue color by different procedures gives complexes with absorption maxima at 650, 725, and 830-900 nm, respectively, and the absorbance is also dependent on pH, time, and temperature. For these reasons, a method which prescribes direct comparison of the blue color of the sample solution with that of a known solution simultaneously prepared under identical conditions [19] tends to be more reliable. In spite of the large number of published molybdenum blue methods, new modifications [20, 21] have continued to appear in the literature.

Other Methods. There are several gravimetric methods of determining phosphorus. The classical method involves precipitation of magnesium ammonium phosphate and ignition to produce $Mg_2P_2O_7$ which is weighed. A superior method [5, 22] consists of precipitating quinoline molybdophosphate which, upon heating at 200°C, conforms to the formula $(C_9H_7N)_3 \cdot H_3PO_4 \cdot 12MoO_3$ with a molecular weight of 2212.8.

Instead of the gravimetric finish, the quinoline molybdophosphate precipitate can be dissolved in standardized sodium hydroxide solution and back-titrated with standardized sulfuric acid [23]. Selig [24] has described a complexometric method in which phosphorus is precipitated as cerous phosphate after closed-flask combustion with alkaline hypobromite as absorbent; the excess cerous ion is then back-titrated at pH 6-7 by means of EDTA in the presence of methylthymol blue or xylenol orange.

According to Bishara and Attia [25], when phosphorus compounds are combusted in the empty tube for carbon and hydrogen analysis (Sec. 2.1B), the phosphorus is quantitatively retained in the capsule as P_2O_5; sulfur and nitrogen, if present, do not interfere. Determination of phosphorus by direct [26] and indirect [26a] polarography and by atomic absorption spectrometry [27] has been advocated.

Hayton and Smith [28] have reported that Karl Fischer reagent [29] quantitatively oxidizes tervalent phosphorus in organic phosphines, tertiary phosphonium salts, and metal complexes. An excess of pyridine is sometimes required. There is no reaction with quaternary phosphonium salts or derivatives of quinquevalent phosphorus. This provides a method to differentiate tervalent phosphorus from quinquevalent phosphorus in certain organic samples.

Arsenic

Whereas arsenic can be determined gravimetrically in the form of $Mg_2As_2O_7$ very precisely after decomposition of the organic material, the method is tedious because the precipitate of magnesium ammonium arsenate must stand for at least 6 h in order for a crystalline product to be obtained. Therefore titrimetric methods are generally employed. The iodometric procedure, which involves oxidation of iodide ions by arsenate in acidic medium,

$$H_3AsO_4 + 2HI \longrightarrow I_2 + H_3AsO_3 + H_2O$$

followed by titration of the liberated iodine with 0.01 $\underline{N}$ sodium thiosulfate, was first published by Wintersteiner [30]. It has been recommended by Steyermark [31] and other workers. On the other hand, Ingram [32] has pointed out the necessity of controlling the pH and excluding air in the operation; therefore he has advocated a more reliable method in which arsenic(III) is determined by titration with 0.01 $\underline{N}$ iodine solution. The experimental detail is given in Sec. 7.6.

Recently Griepink and Krijgsman [33] have proposed a method to titrate the arsenate solution, after flask combustion, by means of lead(II) nitrate solution electrometrically; the changes in pH during the titration are followed in a pH range from 6 to 7. The determination is not affected by the presence of a 10-fold excess of chloride, bromide, iodide, or sulfate. Bigois [34] has described a coulometric method for arsenic in which the decomposition products after HNO_3/H_2SO_4 digestion is further treated with hydrogen peroxide, and the AsO_4 is titrated with iodine generated by constant-current electrolysis of potassium iodide. Kopp [35] has suggested a procedure that involves reduction of the AsO_4^{3-} to arsine which is dissolved in chloroform and determined spectrophotometrically using silver diethyldithiocarbamate as the color-forming reagent. McCall and coworkers [36] have determined arsenic in organic compounds by X-ray fluorescence. According to Schreiber [37], atomic absorption spectrometry for the determination of arsenic is best carried out by reducing the HNO_3/H_2SO_4 digestion mixture with sodium borohydride in 4 $\underline{N}$ HCl and feeding the arsine thus formed into the flame. Pahil and Sharma [37a] have determined As(IV) in organic compounds by refluxing the sample with anhydrous methanol and subsequently titrating the solution with 0.1 $\underline{N}$ iodine.

Antimony

As reported by Schreiber [37], antimony in organic materials is conveniently determined by iodometric titration of Sb(III) which is obtained by reduction of Sb(V) in the H_2SO_4/HNO_3 digestion mixture. In order to prevent the precipitation of Sb_2O_3, tartrate ions are added to the solution to form the Sb(III) complex. This method has been recommended by Ingram [32]. The experimental procedure is given in Sec. 7.7.

Another titrimetric method described by Bigois [34] is as follows. After H_2SO_4/HNO_3 digestion, the mixture is further oxidized by means of potassium permanganate to produce SbO_4^{3-}. Iodine generated by constant-current electrolysis of potassium iodide contained in the solution is the titrating reagent, with sodium thiosulfate as intermediate reagent, and amperometric endpoint detection. The amount of iodine consumed is measured coulometrically.

Gravimetric determination [38] of antimony can be carried out by weighing the antimony chromium ethylenediamine complex, $[Cren_3]SbS_4 \cdot 2H_2O$. This is obtained by neutralizing the acid digestion mixture with ammonia, adding sodium sulfide to form sodium thioantimonate, and precipitating the complex by means of chromium ethylenediamine chloride. The product is collected in a filter tube and dried at room temperature. It may be mentioned that certain organic compounds containing the C–O–Sb linkage can be treated directly with hydrogen sulfide to precipitate the antimony, or titrated directly with iodine [39].

A spectrophotometric method to determine antimony has been described by Horwitz and Galan [40]. After H_2SO_4/HNO_3 digestion, the mixture is evaporated to fumes of SO_3; the residue is diluted with water and filtered. An aliquot is treated with an aqueous solution of KI and NaH_2PO_2 and the absorbance at 425 nm is measured against another aliquot containing no KI/NaH_2PO_2 reagent.

Bismuth

Ingram [32] has recommended the gravimetric method for determining bismuth in organic materials. The method consists of precipitating and weighing the bismuth salt of pyrogallol, $Bi(C_6H_3O_3)_3$, as described by Streibinger and Flaschner [41]. The experimental procedure is given in Sec. 7.8.

A titrimetric method has been proposed by Masino [42]. It involves the precipitation of bismuth sulfide [43]. The sulfide is subsequently dissolved in an excess of standardized silver nitrate solution which is back-titrated by means of potassium thiocyanate.

Determination of bismuth in biological materials by atomic absorption spectroscopy has been described by Willis [44] and by Hall and Farber [45]. The latter workers extracted the bismuth, after acid digestion, into isobutyl

methyl ketone as its complex with ammonium pyrrolidine-1-carbodithiolate.
The extract was then nebulized into the hydrogen-air flame and Bi was measured at 223.7 nm. Plank [46] has determined bismuth in pharmaceutical
preparations by direct-current polarography.

C. Interferences and Complications

It should be noted that all colorimetric methods for phosphorus analysis are
dependent on the reaction of the orthophosphate ion. Attention is called to
the fact that in some procedures of decomposition, phosphorus may not
end up as orthophosphate even though it is at the +5 oxidation state. In such
cases, the resulting solution should be diluted with water and heated for a
period of time to convert the whole of the phosphorus into orthophosphate.
When the sample is decomposed in a bomb under oxygen pressure, polymeric phosphates are formed and further treatment with $HNO_3/HClO_4$ is
necessary [46a].
    While the molybdenum blue methods are most commonly employed,
they are also beset with complications and interferences. Different procedures for the production of the color give different absorptions for the
same amount of phosphorus taken. Furthermore, silicon, arsenic, and
germanium, which form heteropoly acids with molybdic acid, are reducible
under conditions similar to those used for phosphomolybdic acid and produce similar blue colors [47]. When these elements are simultaneously
present, phosphomolybdic acid can be selectively extracted from the
aqueous solution with 20% isobutyl alcohol in chloroform [48]. Terent'ev
and co-workers [7] have determined phosphorus and silicon in separate
aliquots, after sodium peroxide fusion, as their respective heteropoly
acids by spectrophotometric procedures.
    When arsenic is determined by titration of iodine liberated by AsO_4^{3-},
there is no interference from halogens, sulfur, or phosphorus [31]. However, the solution must be adjusted to 3.3 $\underline{N}$ with hydrochloric acid [32].
If the acidity is too great, high results will be obtained owing to the action
of the acid on the potassium iodide present. It is also necessary to exclude
air during the titration in order to avoid oxidation of iodide by oxygen.
Despite the use of boiled solutions, a blank determination is necessary to
account for iodine liberated by dissolved air in the solution. Sandhu et al.
[32a] have described a method to determine As(V) in organic compounds
in the presence of Cu(II), Be(II), and Sn(IV), which involves reduction in
acetic acid-methanol and titration of the resultant As(III) after filtration.
    In the decomposition of antimony compounds, H_2SO_4/HNO_3 digestion
may produce the very resistant Sb(IV) state [49]; this difficulty can be
overcome by the use of a little perchloric acid, which oxidizes the whole

of the antimony to the Sb(V) state [50]. Iodometric titration of Sb(III) is
interfered by the presence of arsenic; in such cases, antimony can be determined gravimetrically as the pyrogallol salt [51]. During mineralization of the samples in open vessels, the volatility of the chlorides must be
kept in mind.

7.3 SURVEY OF RECENT DEVELOPMENTS

A. Mineralization and Determination

A selected bibliography on the analysis of phosphorus in organic materials
has been prepared by Laws [112]; it covers the literature up to 1962. The
literature on arsenic, antimony, and bismuth up to the same year has been
reviewed by Gorsuch [113]. Recent publications on the determination of
phosphorus and arsenic are summarized in Tables 7.1 and 7.2, respectively. A survey of the literature which has appeared during the past
decade indicates that the acid digestion and closed-flask combustion techniques are the favorite methods of decomposition, though the latter is less
frequently employed for arsenic than for phosphorus determinations.

Titrimetric finishes predominate in the analysis of arsenic. Colorimetric and spectrophotometric methods are popular for phosphorus, but
titrimetry still receives considerable attention. An iodometric method
described by Gawargious and Farag [114] for the determination of phosphorus in presence of nitrogen, sulfur, and halogens is of interest: After
combustion of the sample, aqueous bromine is added and the solution is
evaporated. After addition of acetone (10 ml), 0.1 $\underline{N}$ KI (2 ml), and 0.1 $\underline{N}$
KIO_3 (2 ml), the solution is titrated with $Na_2S_2O_3$ to a colorless endpoint.
Then 0.1 $\underline{N}$ NH_4Cl (1 ml), 0.2 $\underline{N}$ $ZnSO_4$ (1 ml), and solid KI and KIO_3 are
introduced; after 5 min, more acetone (10 ml) is added and the iodine lib-
erated is titrated with 0.01 $\underline{N}$ $Na_2S_2O_3$ without using an indicator. The
first addition of KI and KIO_3 causes liberation of iodine by acids of nitrogen, sulfur, chlorine, and fluorine, and by one proton of the H_3PO_4; the
addition of NH_4^+ and Zn^{2+} liberates two protons from $H_2PO_4^-$, which then
liberate iodine from the KI and KIO_3 subsequently added.

The modern instrumental methods such as atomic absorption and X-ray
spectrometry, and neutron activation have been applied to the determination
of Group V elements. We are currently investigating the conversion of
phosphorus, arsenic, and antimony in organic materials to PH_3, AsH_3,
and SbH_3, respectively, followed by gas chromatographic determinations.

TABLE 7.1. Summary of Recent Literature on the Determination of Phosphorus.

Decomposition method	Product determined	Finish technique	Reference
Acid digestion	PO_4^{3-}	Colorimetric	3, 52–53
	PO_4^{3-}	Titrimetric	54–58
	PO_4^{3-}	Spectrophotometric	59–60
	PO_4^{3-}	Gravimetric	61, 62
Closed-flask combustion	PO_4^{3-}	Colorimetric	3, 63–66
	PO_4^{3-}	Spectrophotometric	9, 67–70
	PO_4^{3-}	Titrimetric	24, 71–79
	PO_4^{3-}	Polarography	27
Metal bomb	PO_4^{3-}	Titrimetric	80–83
	PO_4^{3-}	Gravimetric	84, 85
Mg fusion	PH_3	Titrimetric	86
Sealed-tube combustion	PO_4^{3-}	Titrimetric	87–90
Empty-tube combustion	P_2O_5	Gravimetric	25
Karl Fischer reagent	P(III)	Titrimetric	29
Hydrogenation	PH_3	Gas chromatography	91
Flame	P	Atomic absorption	28, 92
X-ray	P	Fluorescence	93
Neutron activation	^{30}P, ^{32}P	Radiometric	94, 95

B. Evaluation of Methods

Steyermark [115] has conducted a collaborative study of methods for determining phosphorus in pure organic compounds. Johnson [116] has reported the collaborative study of an automated method [117] for phosphorus fertilizers in comparison with the gravimetric quinoline molybdophosphate method. Statistical evaluation showed that there was no difference between

TABLE 7.2. Summary of Recent Literature on the Determination of Arsenic.

Decomposition method	Product determined	Finish technique	Reference
Acid digestion	AsO_4^{3-}	Titrimetric	96-99
	AsO_4^{3-}	Colorimetric	97, 100
	AsO_4^{3-}	Spectrophotometric	35, 101, 102
	AsO_4^{3-}	Radiometric	103
Closed-flask combustion	AsO_3^{3-}, AsO_4^{3-}	Titrimetric	33, 104-106
	AsO_4^{3-}	Gravimetric	107
	$AsO_3^{3-} \rightarrow (C_6H_5)_3As$	Gas chromatography	108
Sealed-tube combustion	AsO_4^{3-}	Titrimetric	109
Acetylene flame	As	Emission spectrometry	110
X-ray	As	Fluorescence	36
Neutron activation	^{76}As	Radiometric	111

the mean results of the two methods, but that the results by the automated method were less reliable than those of the gravimetric method. Previously, Hoffman [118] used pure ammonium dihydrogenphosphate to compare the quinoline molybdophosphate methods, both gravimetric and titrimetric, with the vanadiphosphomolybdate spectrophotometric method and the magnesium pyrophosphate method, and found the quinoline molybdophosphate gravimetric procedure to be most precise.

Wilson [119] and Gorsuch [120] have compared the methods for the determination of arsenic in large samples; three procedures of decomposition are satisfactory, namely, dry ashing with magnesium nitrate, digestion with H_2SO_4/HNO_3, and digestion with $H_2SO_4/HNO_3/HClO_4$. On the other hand, Leblanc and Jackson [121] have found that the dry ashing technique for the determination of various seafoods gives much higher values for arsenic than does the H_2SO_4/HNO_3 digestion.

Plank [46] has published a report on the collaborative study of the determination of bismuth. It is concerned with bismuth in pharmaceutical dosage forms assayed by conventional direct-current polarography. The method was found to be quantitative and reliable.

7.4 EXPERIMENTAL: DETERMINATION OF PHOSPHORUS BY ACID
     DIGESTION AND PHOSPHOVANADOMOLYBDATE COLORIMETRY

A. Principle

When an organic compound containing phosphorus is oxidized, the phosphorus is converted to phosphorus pentoxide which dissolves in water to
yield phosphate ions. For samples containing 0.2-2 mg of phosphorus,
spectrophotometric determination of the yellow colored phosphovanadomolybdate complex is recommended (see Sec. 7.2B).

B. Equipment

     Micro Kjeldahl flasks, 10-ml capacity, and digestion rack (see Sec. 3.3).
     Spectrophotometer.

C. Reagents

     Phosphate standard solution (containing 0.500 mg P per ml): Dissolve
          2.1315 g of pure diammonium-hydrogen phosphate or 2.196 g of
          pure potassium-dihydrogen phosphate in a 1-liter volumetric flask
          and dilute to the mark.
     Ammonium vanadate solution (containing 0.20% vanadate): To 500 ml
          of boiling distilled water, cautiously add, first, 2.35 g of ammon-
          ium metavanadate, NH_4VO_3, CP, and then 100 ml of dilute sulfuric
          acid (1:12). Allow to cool and make up to 1 liter.
     Ammonium molybdate solution (containing 10% MoO_3): Dissolve 122 g
          of ammonium molybdate, $(NH_4)_6Mo_7O_{24} \cdot 4H_2O$, reagent grade, in
          880 ml of distilled water.
     Concentrated sulfuric acid, CP.
     Concentric nitric acid, CP.

D. Procedure

     Preparation of the Calibration Graph

     Measure 10 aliquot portions of the phosphate standard solution, cover-
ing the range from 0.10 to 2.00 mg P, into 100-ml volumetric flasks from
a microburet (accurate to 0.005 ml). Dilute each aliquot to 65 ml and treat
with 1.4 ml of concentrated sulfuric acid. Then slowly add to each batch,
with continuous swirling, 10 ml of the vanadate solution, followed by 10 ml
of the molybdate solution. Finally, fill the flasks to the mark with distilled

water, allow them to stand for 30 min, and compare the yellow color with
that of the blank. (With the Beckman spectrophotometer Model DU, set
the wavelength at 410 nm; use a slit width of 1.0 mm and 1.00-cm matched
Corex cells.) A straight line is obtained by plotting absorbance (A = - log
transmittance) against the amount of phosphorus (mg) in the solution (see
Fig. 7.1). From the slope and intercept of this line an equation can be
derived of the form

$$W = \alpha A + \beta$$

which gives the weight W of phosphorus in terms of absorbance A (Sec.
7.4F). For example, curve A in Fig. 7.1 is represented by

$$W = 1.4368 - 0.0129$$

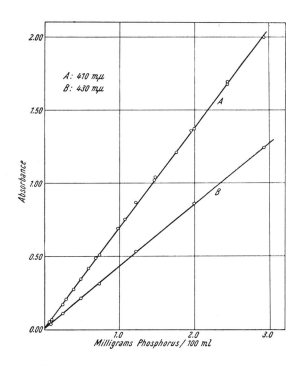

FIG. 7.1. Calibration graph for phosphorus determination. (From
Ref. 16, courtesy Mikrochim. Acta.)

Determination of Phosphorus

Accurately weigh a 3-10 mg sample (containing 0.2-2.0 mg P) into a dry 10-ml Kjeldahl flask. Add 1.5 ml of concentrated H_2SO_4 and four drops (0.2 ml) of concentrated HNO_3. Heat gently on the digestion rack until the reaction subsides. After cooling, add two more drops of concentrated HNO_3 and again heat with a medium flame until the solution is clear and sulfur trioxide fumes are evolved (20-30 min). Now cool the solution, dilute to about 10 ml, and transfer with several rinsings to a 100-ml volumetric flask. Develop the yellow color and measure the absorbance as described in the preparation of the calibration graph.

E. Calculation

Obtain the amount of phosphorus either from the graph or with the use of the equation mentioned above. Then,

$$\%P = \frac{mg\ P \times 100}{mg\ sample}$$

F. Note

It has been found that when new reagents are prepared, only the intercept β of the equation is changed. Therefore, it is not necessary to plot new calibration graphs from time to time. The new value of β may be obtained by analyzing a pure known phosphorus compound simultaneously with the unknown sample.

G. Comments

(1) In a recent collaborative study of the determination of phosphorus in organic compounds, Lalancette and co-workers [122] reported that there is no significant difference in either precision or accuracy between gravimetry and molybdenum blue or yellow colorimetry.

(2) Attention is called to the fact that the phosphate ions should be in the orthophosphate form for the formation of the colored complex. In case the sulfuric acid has been over-concentrated, the reaction mixture should be boiled for a few minutes after being diluted to about 10 ml.

7.5 EXPERIMENTAL: DETERMINATION OF PHOSPHORUS BY THE MOLYBDENUM BLUE METHOD

A. Principle

After the oxidation of phosphorus in organic material to form orthophosphoric acid, addition of ammonium molybdate followed by a reducing agent produces a blue color. Since the color is very intense, this method is recommended for the determination of phosphorus below the milligram level (see Sec. 7.2B).

B. Equipment

Micro Kjeldahl flask, 10-ml capacity, and digestion rack (see Sec. 3.3). Spectrophotometer, visual range (Sec. 7.5F, Note 1).

C. Reagents

Concentrated sulfuric acid, CP.
Hydrogen peroxide, 30%, phosphate-free.
Ammonium molybdate solution, 2.5% in 1 $\underline{M}$ H_2SO_4.
Reducing agent [19] (Note 2): Aqueous solution containing 0.25% 1,2,4-aminonaphtholsulfonic acid sodium salt, 15% sodium bisulfite, and 0.5% sodium sulfite.

D. Procedure

Accurately weigh a 2-10 mg sample containing about 0.2 mg P into the micro Kjeldahl flask. Add 2 ml concentrated sulfuric acid and several drops of 30% hydrogen peroxide. Heat gently until the reaction mixture chars. Cool, cautiously add more hydrogen peroxide, and reheat until the solution becomes clear. Upon cooling, add 5 ml of distilled water and transfer the solution to a 100-ml volumetric flask. Add 5 ml of ammonium molybdate solution and 2 ml of the reducing agent. Fill up to the mark and mix well. Place the volumetric flask in the boiling water bath for 20 min to develop the blue color. Cool to room temperature, and take an aliquot to measure the absorbance at 820 nm. Prepare a calibration graph under identical conditions for the range between 0.1 and 0.5 mg P.

E. Calculation

$$\%P = \frac{\text{mg P found x dilution factor x 100}}{\text{mg sample}}$$

F. Notes

Note 1: Owing to the intense blue color of the solution, a simple colorimeter or filter photometer may be used.

Note 2: Other reducing agents which have been advocated include ascorbic acid [2], amidol, metol [18], and hydrazine sulfate [47].

G. Comments

(1) The decomposition of the organic material may be carried out by other techniques. It is recommended to scan the absorption spectrum of the blue color developed under the specific experimental conditions in order to select the optimum wavelength for spectrophotometry.

(2) The interference due to silicon can be circumvented [70a] by controlling the conditions. Selenium used as a catalyst in Kjeldahl digestion of plant samples interferes with the automated analysis of molybdenum blue; this difficulty can be eliminated [123] by the addition of KCN.

7.6 EXPERIMENTAL: DETERMINATION OF ARSENIC

A. Principle

Organically bound arsenic is decomposed by oxidation with sulfuric acid and nitric acid. The arsenic acid obtained is then determined iodometrically (see Sec. 7.2B).

B. Equipment

Micro Kjeldahl flask, 10-ml capacity, and digestion rack (see Sec. 3.3). Iodine flask, 125-ml capacity.

C. Reagents

Concentrated sulfuric acid, CP.
Concentrated nitric acid, CP.

Hydrogen peroxide, 30%.

Concentrated hydrochloric acid, CP.

Potassium iodide solution, 10%: Dissolve 1 g of KI crystals in 9 ml of distilled water which has been boiled to remove dissolved oxygen.

Sodium thiosulfate standard solution, 0.01 $\underline{N}$: Dissolve 2.48 g of $Na_2S_2O_3 \cdot 5H_2O$ in 1 liter of distilled water. Transfer to an amber bottle. Add 1 ml of chloroform as preservative. Standardize against potassium biiodate, $KH(IO_3)_2$.

Starch indicator solution.

D. Procedure

Accurately weigh a sample containing 0.4-2 mg As into the Kjeldahl flask. Add 1 ml of concentrated sulfuric acid to cover the sample. Heat gently and add a few drops of concentrated HNO_3. If the contents of the flask do not clarify in 10 min, indicating incomplete destruction of the organic matter, remove the flask from the digestion rack. After cooling slightly, cautiously add several drops of 30% hydrogen peroxide. Again boil the reaction mixture gently. Repeat until the solution becomes colorless. Then evaporate the contents of the flask until sulfur trioxide fumes are evolved. Upon cooling, add 2 ml of distilled water, and transfer quantitatively into a 125-ml iodine flask, making the volume of solution approximately 20 ml. Add 12 ml of concentrated hydrochloric acid (Sec. 7.6F), followed by 1 ml of freshly prepared 10% KI solution. Stopper the iodine flask and mix thoroughly. Allow the flask to stand for 3 min, then titrate the liberated iodine with 0.01 $\underline{N}$ sodium thiosulfate. Add starch indicator when the yellow color of the solution becomes very faint; continue titration until the blue color disappears. Run a blank to correct for reagent errors.

E. Calculation

$$\%As = \frac{\text{ml } 0.01 \ \underline{N} \ Na_2S_2O_3 \times 0.3748 \times 100}{\text{mg sample}}$$

F. Note

The acidity of the solution should be approximately 3.3 $\underline{N}$ during the titration.

G. Comments

(1) According to Schreiber [37], the preferred method for determining arsenic in pharmaceuticals is to reduce the arsenic acid, after H_2SO_4/HNO_3

digestion, by means of sodium borohydride in 4 $\underline{N}$ HCl to liberate arsine which is determined by atomic absorption spectrometry.

(2) Recently Sandhu and co-workers [124] have described a method to reduce As(V) to As(III) by the action of zinc dust in boiling anhydrous acetic acid, followed by titration of the As(III) in the filtered solution with iodine, the endpoint being detected with starch as indicator or potentiometrically.

7.7 EXPERIMENTAL: DETERMINATION OF ANTIMONY

A. Principle

Organic material containing antimony is digested in sulfuric acid with the addition of hydrogen peroxide. The resulting Sb(V) is subsequently reduced to Sb(III) by means of sodium sulfite. The trivalent antimony, after being complexed with tartrate to prevent precipitation of Sb_2O_3, is determined by titration with 0.01 $\underline{N}$ iodine [37].

B. Equipment

Micro Kjeldahl flasks, 10-ml capacity, and digestion rack (see Sec. 3.3).

C. Reagents

Concentrated sulfuric acid.
Hydrogen peroxide, 30%.
Sodium sulfite.
Concentrated hydrochloric acid.
Tartaric acid.
Iodine standard solution, 0.01 $\underline{N}$: Dissolve 1.27 g of resublimed iodine crystals in a solution of 4 g of potassium iodide in 4 ml of distilled water. After 30 min, transfer this concentrated solution to a 1-liter volumetric flask and dilute to the mark with freshly boiled distilled water. Standardize against 0.01 $\underline{N}$ sodium thiosulfate solution immediately before use.
Sodium carbonate.
Sodium bicarbonate, saturated solution.
Starch indicator solution.

D. Procedure

Decomposition

Accurately weigh a sample containing 1-5 mg Sb into the Kjeldahl flask. Add 1 ml of concentrated sulfuric acid. Heat gently until the reaction mixture chars. Cool, cautiously add several drops of 30% hydrogen peroxide, and reheat until the solution is clear and sulfur trioxide fumes are evolved. Upon cooling, add 1 ml of distilled water and heat to a boil. Now add 100 mg of sodium sulfite to convert Sb(V) to Sb(III). Heat gently again (after adding 1 ml of distilled water) until the excess sulfur dioxide is completely removed (test with starch-iodide paper over the mouth of the flask).

Determination

Add 1 ml of concentrated hydrochloric acid and 1 ml of 5% tartaric acid solution into the Kjeldahl flask containing the Sb(III). Transfer quantitatively to a 100-ml Erlenmeyer flask, using distilled water for rinsing. Neutralize the solution with sodium carbonate and then add 1 drop of hydrochloric acid, followed by 1 ml of saturated sodium bicarbonate solution. Add a few drops of starch indicator and titrate with the standardized 0.01 $\underline{N}$ iodine solution to the blue endpoint.

E. Calculation

$$Sb(III) + I_2 = Sb(V) + 2I^-$$

$$\%Sb = \frac{ml\ 0.01\ \underline{N}\ I_2\ x\ normality^*\ x\ 0.6088\ x\ 100}{mg\ sample}$$

7.8 EXPERIMENTAL: DETERMINATION OF BISMUTH

A. Principle

The organic sample containing bismuth is decomposed by heating in nitric acid with the addition of sulfuric acid and hydrogen peroxide. The resulting Bi(III) is then precipitated by means of pyrogallol [41], which forms the phenate $Bi(C_6H_3O_3)$ that is crystalline and can be dried at 100°C and weighed as such.

---

*The normality factor is only required when the iodine solution is not exactly 0.01 $\underline{N}$.

B. Equipment

Micro Kjeldahl flask, 10-ml capacity, and digestion rack (see Sec. 3.3).
Filtration assembly and large test tube (see Sec. 5.9).

C. Reagents

Concentrated nitric acid.
Concentrated sulfuric acid.
Hydrogen peroxide, 30%.
Ammonium hydroxide, 0.5 $\underline{N}$.
Pyrogallol, CP.

D. Procedure

Accurately weigh a sample containing 1-3 mg Bi into the micro Kjeldahl
flask. Add 1 ml of concentrated nitric acid and two drops of concentrated
sulfuric acid. Heat gently for 30 min. Cool slightly, and cautiously add
several drops of 30% hydrogen peroxide. Reheat gently until the solution
is clear. Now add 1 ml of distilled water and evaporate the solution to
remove the volatile acids. Add 2 ml of distilled water and transfer quantitatively to the large test tube. Add 0.5 $\underline{N}$ ammonium hydroxide dropwise
until a faint opalescence appears. Then add 100 mg of pyrogallol and
swirl the test tube to mix the contents thoroughly. Place the test tube in a
boiling water bath for 15 min. After the yellow precipitate has settled,
move the test tube to the filtration assembly and transfer the precipitate
into the tared sintered glass filter tube. Use distilled water and benzene
alternately for rinsing. Dry the filter tube and yellow crystals at 100°C
to constant weight (see Figs. 5.14 and 5.15).

E. Calculation

$$\% Bi = \frac{mg\ Bi(C_6H_3O_3)\ x\ 0.6294\ x\ 100}{mg\ sample}$$

REFERENCES

1. Committee report on "Determination of busmuth in foods and biological
    materials", Analyst $\underline{58}$:607 (1933); $\underline{68}$:115, 217 (1945).
2. P. S. Chen, T. Y. Torribana, and H. Warner, Anal. Chem. $\underline{28}$:756
    (1956).

3. L. I. Diuguid and N. C. Johnson, Microchem. J. 13:616 (1968).
4. H. A. Williams, Analyst 66:228 (1941).
5. T. R. F. W. Fennell, M. W. Roberts, and J. R. Webb, Analyst
    82:639 (1957).
6. A. J. Christopher, T. R. F. W. Fennell, and J. R. Webb, Talanta
    11:1323 (1964).
7. A. P. Terent'ev, R. N. Potsepkina, and N. A. Grodskova, Zh.
    Anal. Khim. 28:391 (1973).
8. F. Bamford, Analyst 59:101 (1934).
8a. W. Pfirter, Mikrochim. Acta 1975(II):515.
9. F. G. Romer and B. Griepink, Mikrochim. Acta 1970:867.
10. W. Merz, Mikrochim. Acta 1959:456.
11. A. Dirscherl and F. Erne, Mikrochim. Acta 1960:775.
12. W. Merz, Mikrochim. Acta 1959:640.
13. M. Corner, Analyst 84:41 (1959).
13a. E. Celon, S. Degetto, G. Marangoni, and L. Sindellari, Mikrochim.
    Acta 1976(I):113.
14. R. Belcher, A. M. G. Macdonald, and T. S. West, Talanta 1:408
    (1958).
15. M. M. Tuckerman, J. H. Hodecker, B. C. Southworth, and K. D.
    Fleisher, Anal. Chim. Acta 21:463 (1959).
15a. S. S. Pahil, K. D. Sharma, S. S. Sandhu, and R. S. Sandhu, Z.
    Anal. Chem. 269:368 (1974); 273:32 (1975).
16. T. S. Ma and J. D. McKinley, Jr., Mikrochim. Acta 1953:4.
17. H. Wu, J. Biol. Chem. 43:189 (1920).
18. R. P. A. Sims, Analyst 82:584 (1961).
19. B. L. Horecker, T. S. Ma, and E. Haas, J. Biol. Chem. 136:775
    (1940).
20. L. H. Lazarus and S. C. Chou, Anal. Biochem. 45:557 (1972).
21. P. A. Drewes, Clin. Chim. Acta 39:81 (1972).
22. C. H. Perrin, J. Ass. Offic. Agr. Chem. 41:758 (1958); 42:567
    (1959).
23. H. N. Wilson, Analyst 79:535 (1954).
24. W. Selig, Z. Anal. Chem. 247:294 (1969).
25. S. W. Bishara and M. E. Attia, Microchem. J. 18:267 (1973).
26. F. Pottkamp and F. Umland, Z. Anal. Chem. 255:367 (1971).
26a. S. W. Bishara and M. E. Attia, Talanta 18:634 (1971).
27. G. F. Kirkbright and M. Marshall, Anal. Chem. 45:1610 (1973).
28. B. Hayton and B. C. Smith, J. Inorg. Nucl. Chem. 31:1369 (1968).
29. N. D. Cheronis and T. S. Ma, Organic Functional Group Analysts,
    Wiley, New York, 1964, p. 472.
30. O. Wintersteiner, Mikrochemie 4:155 (1926).
31. A. Steyermark, Quantitative Organic Microanalysis, 2nd Ed.,
    Academic, New York, 1961, p. 367.

32. G. Ingram, Methods of Organic Elemental Analysis, Reinhold, New York, 1962, p. 297.

32a. S. S. Sandhu, R. S. Sandhu, and K. D. Sharma, J. Indian Chem. Soc. 51:766 (1974).

33. B. Griepink and W. Krijgsman, Mikrochim. Acta 1968:574.

34. M. Bigois, Talanta 19:157 (1972).

35. J. F. Kopp, Anal. Chem. 45:1786 (1973).

36. J. M. McCall, Jr., D. E. Leyden, and C. W. Blount, Anal. Chem. 43:1324 (1971).

37. B. Schreiber, personal communication, August 1973.

37a. S. Pahil and K. D. Sharma, Indian J. Chem. 12:1316 (1974).

38. G. Spacu and A. Pop, Z. Anal. Chem. 111:254 (1938).

39. W. G. Christiansen, Organic Derivatives of Antimony, The Chemical Catalogue Co., New York, 1925.

40. G. Horwitz and M. Galan, Anal. Chim. Acta 61:413 (1972).

41. R. Streibinger and E. Flaschner, Mikrochemie 5:12 (1927).

42. G. Masino, Boll. Chim. Farm. 75:409 (1936).

43. C. L. Tseng and L. Wang, J. Chinese Chem. Soc. 5:3 (1937).

44. J. B. Willis, Anal. Chem. 34:614 (1962).

45. R. J. Hall and T. Farber, J. Ass. Offic. Anal. Chem. 55:639 (1972).

46. W. M. Plank, J. Ass. Offic. Anal. Chem. 55:155 (1972).

46a. H. Narasaki, K. Miyaji, and A. Unno, Japan Analyst 22:541 (1973).

47. D. F. Boltz and M. G. Mellon, Anal. Chem. 19:873 (1947).

48. C. Wadelin and M. G. Mellon, Anal. Chem. 25:1668 (1953).

49. T. H. Maren, Anal. Chem. 19:487 (1947).

50. P. F. Wyatt, Analyst 80:368 (1955).

51. F. Feigl, Z. Anal. Chem. 64:41 (1924).

52. T. S. Ma, in F. J. Welcher (ed.), Standard Methods of Chemical Analysis, 6th Ed., Vol. 2, Van Nostrand, Princeton, 1963, p. 399.

52a. A. M. Maksudov, Y. Tadzhibaev, and S. T. Akramov, Uzb. Khim. Zh. 1973:16.

53. T. D. Talbott, J. C. Cavognol, C. F. Smead, and R. T. Evans, J. Agr. Food Chem. 20:959 (1972).

54. K. Hozumi and K. Miauno, Japan Analyst 10:453 (1961).

55. J. Horacek, Collect. Czech. Commun. 27:1811 (1962).

56. H. Medzihradszky-Schweiger and S. Kutassy, Acta Chim. Acad. Sci. Hung. 41:265 (1964).

57. B. Griepink and J. Slanina, Mikrochim. Acta 1967:27.

58. J. Slanina, P. C. M. Frintrop, J. F. Mansveld, and B. Griepink, Mikrochim. Acta 1970:52.

59. W. J. Kirsten and M. E. Carlsson, Microchem. J. 4:3 (1960).

59a. R. R. Lowry and I. J. Tinsley, Lipids 9:491 (1974).

60. G. Lindner and I. Edmundson, Acta Chem. Scand. 21:136 (1967).

61. M. Kanand and H. Kashiwagi, Japan Analyst 10:789 (1961).
62. S. L. Erikson, Talanta 19:1457 (1972).
63. H. Kremsbrucker, Sci. Pharm. 27:294 (1959).
64. A. M. Ryadnina, Zavod. Lab. 27:405 (1961).
65. A. Nara, Y. Urushibata, and N. Oe, Japan Analyst 12:294 (1963).
66. H. Y. Yu and I. H. Sha, Chem. Bull. (Peking) 1965:557.
67. N. E. Gel'man and T. M. Shanina, Zh. Anal. Khim. 17:998 (1962).
68. E. Debal, Chim. Anal. 45:66 (1963).
69. L. H. Scroggins, Microchem. J. 13:385 (1968).
69a. W. I. Awad, S. S. M. Hassan, and S. A. I. Thoria, Mikrochim. Acta 1976(II):111.
70. T. M. Shanina, N. E. Gel'man, and T. V. Bychkova, Zh. Anal. Khim. 23:468 (1968).
70a. T. M. Shanina, N. E. Gel'man, V. S. Mikhailovskaya, and T. S. Serebvyakova, Zh. Anal. Khim. 27:1853 (1972).
71. R. Puschel and H. Wittmann, Mikrochim. Acta 1960:670.
72. E. Meier, Mikrochim. Acta 1961:70.
73. A. M. G. Macdonald and W. I. Stephen, J. Chem. Educ. 39:528 (1962).
74. K. Ziegler, H. Till, and H. Schindlbauer, Mikrochim. Acta 1963:1144.
75. T. L. Hunter, Anal. Chim. Acta 35:398 (1966).
76. M. F. Tysganova and K. F. Novikova, Zh. Anal. Khim. 24:272 (1969).
77. Y. A. Gawargious and A. B. Farag, Microchem. J. 16:333 (1971).
78. R. F. Sympson, Anal. Chim. Acta 61:148 (1972).
78a. L. Maric, M. Siroki, and Z. Stefanac, Microchem. J. 21:129 (1976).
79. V. Nuti, Farmaco, Ed. Sci. 27:179 (1972).
80. M. N. Chumachanko and V. P. Bulaka, Izv. Akad. Nauk SSSR 1962:560.
81. A. Kondo, Japan Analyst 9:416 (1960).
82. H. Buss, H. W. Kohlschulter, and M. Preiss, Z. Anal. Chem. 214:106 (1965).
83. E. A. Terent'eva and N. N. Smirnova, Zavod. Lab. 32:924 (1966).
84. T. R. F. W. Fennell and J. R. Webb, Talanta 2:105 (1959).
85. F. S. Malyukova and A. D. Zaitseva, Plast. Masey 1966:57.
86. V. M. Vladimirova and P. N. Fedoseev, Zavod. Lab. 39:8 (1973).
87. C. DiPietro, R. E. Kramer, and W. A. Sassaman, Anal. Chem. 34:586 (1962).
88. A. A. Abramyan, R. S. Sarkisyan, and M. A. Balyan, Izv. Akad. Nauk Arm. SSSR, Khim. Nauk 14:561 (1961).
89. Y. A. Mandelbaum, A. F. Grapov, and A. L. Itskova, Zh. Anal. Khim. 20:873 (1965).
90. W. Dindorf and R. Luckenbach, Mikrochim. Acta 1969:113.
91. A. D. Horton, W. D. Shults and A. S. Meyer, Anal. Lett. 4:613 (1971).

92. G. C. Toralballa, G. I. Spielholtz, and R. J. Steinberg, Mikrochim. Acta 1972:484.

93. B. Schreiber, reported at the International Symposium on Microchemical Techniques, University Park, Pa., August 1973.

93a. L. Beitz, Glas.-Instrum.-Tech. Fachz. Lab. 17:445 (1973).

94. I. P. Lisovski and L. A. Smakhtin, J. Radioanal. Chem. 8:75 (1971).

95. P. Fawcett, D. Green, and G. Shaw, Radiochem. Radioanal. Lett. 9:321 (1972).

96. T. S. Ma, in F. J. Welcher (ed.), Standard Methods of Chemical Analysis, 6th Ed., Vol. 2, Van Nostrand, Princeton, 1963, p. 401.

96a. M. R. Masson, Mikrochim. Acta 1976(I):399.

97. M. M. Tuckerman, J. H. Hodecker, B. C. Southworth, and K. D. Fleischer, Anal. Chim. Acta 21:463 (1959).

98. J. A. Perez-Bustmante and R. Burriel-Marti, Inform. Quim. Anal. Pura Apl. Ind. 22:25 (1968).

98a. Y. A. Gawargious, L. S. Boulos, and B. N. Faltaoos, Talanta 23:513 (1976).

99. B. F. A. Griepink, W. Krijgsman, A. J. H. E. Leenaers-Smeets, J. Slanina, and H. Cuijpers, Mikrochim. Acta 1969:1018.

100. R. M. Fournier, Mem. Poudres 40:385 (1958).

101. H. Kashiwagi, Y. Tukamoto, and M. Kan, Ann. Rep. Takeda Res. Lab. 22:69 (1962).

102. T. M. Shanina, N. E. Gel'man, and V. S. Mikhailovskoya, Zh. Anal. Khim. 25:358 (1970).

103. A. Arnold, S. Davis, and A. L. Jordan, Analyst 94:664 (1969).

104. Z. Stefanac, Mikrochim. Acta 1962:1115.

105. R. Puschel and Z. Stefanac, Mikrochim. Acta 1962:1108.

106. A. P. Terent'ev, M. A. Volodina, and E. G. Fursova, Zh. Anal. Khim. 22:640 (1967).

107. A. D. Wilson and D. T. Lewis, Analyst 88:510 (1963).

108. G. Schwedt and H. A. Russel, Chromatographia 5:242 (1972).

109. C. DiPietro and W. A. Sassaman, Anal. Chem. 36:2213 (1964).

110. J. A. Dean and R. E. Fues, Anal. Lett. 2:105 (1969).

111. G. Lunde, Acta Chem. Scand. 26:2642 (1972).

112. E. Q. Laws, in I. M. Kolthoff and P. J. Elving (eds.), Treatise on Analytical Chemistry, Part II, Vol. 11, Wiley, New York, 1965, p. 552.

113. T. T. Gorsuch, in I. M. Kolthoff and P. J. Elving (eds.), Treatise on Analytical Chemistry, Part II, Vol. 12, Wiley, New York, 1965, pp. 330, 344.

114. Y. A. Gawargious and A. B. Farag, Microchem. J. 16:342 (1971).

114a. J. Horacek, Collect. Czech. Commun. 38:1149 (1973).

115. A. Steyermark, private communication, 1974.
116. F. J. Johnson, J. Ass. Offic. Anal. Chem. 55:979 (1972).
117. C. W. Gehrke, J. Ass. Offic. Anal. Chem. 49:1213 (1966).
118. W. M. Hoffman, J. Ass. Offic. Anal. Chem. 44:779 (1961).
119. W. J. Wilson, Anal. Chim. Acta 22:96 (1960).
120. T. T. Gorsuch, Analyst 84:135 (1959).
121. P. J. Leblanc and A. L. Jackson, J. Ass. Offic. Anal. Chem. 56:383 (1973).
122. R. A. Lalancette, A. Steyermark, R. A. Lee, and D. M. Lukaszewski, J. Ass. Offic. Anal. Chem. 56:897 (1973).
123. P. J. Milham and C. C. Short, J. Ass. Offic. Anal. Chem. 56:882 (1973).
124. S. S. Sandhu, S. S. Pahil, and K. D. Sharma, Talanta 30:329 (1973).

Boron, Silicon, and Mercury

8.1 INTRODUCTORY REMARKS

In the previous several chapters we have discussed methods for the determination of nonmetallic elements in organic materials. We now turn to the
discussion of the techniques to determine metals and metalloids that are
integral components of organic substances. For reasons presented below,
the three elements boron, silicon, and mercury are treated in this chapter,
while the remaining metallic elements are covered in Chap. 9.

Organoboron, organosilicon, and organomercury compounds are the
most important and numerous groups of organometallic products. Organic
mercurials [1] have been employed as pharmaceuticals for a long time.
Silicon-containing polymers [2] are used extensively in many industries on
account of their water-repellent and fire-resistant properties. Borohydrides [3] are well known reducing agents and organoboranes [4] can be
utilized as reagents in a wide range of organic syntheses.

Boron, silicon, and mercury are toxic in certain forms. Severe poisoning may result from inhalation of small amounts of boronhydrides [5]
and their derivatives [6]. Silica, when inhaled as dust, produces the
disease known as silicosis [7], which is caused by irritation of the lung
tissue. Fortunately, because of the stability of organosilicon polymers
toward heat and oxidation, they can be used in industrial and commercial
products without great danger. In contrast, the instability of organomercury compounds poses very serious health hazards. Mercury is extremely
toxic to the nervous system. Since it was discovered that mercury in
polluted water eventually enters into organic matrices (e.g., methylmer-
cury) and is stored in aquatic plants and animals, much attention has been

given to the analysis of mercury in organic materials. For example, during the past few years, a steady stream of reports has appeared on the determination of mercury in fish [8-16], canned foods [17,18], eggs [19], cereals [20], beverages [21], drugs [22], cosmetics [23], petroleum oils [24], and biological fluids [25-31].

There are some aspects of analytical chemistry which differentiate boron, silicon, and mercury from the other elements discussed in Chap. 9. For instance, the titrimetric method for boron is unique (see Sec. 8.4). The spectrophotometric procedure for determining silicon in the form of its molybdate complex is not applicable to the metals. Unlike most metallic components in organic materials, mercury cannot be determined by dry ashing since it has substantial vapor pressure at the temperature of incineration. Similarly, boron will be lost as alkyl borates when the digestion mixture contains alcohols, and silicon will volatilize as SiF_4 in the presence of hydrofluoric acid.

8.2 CURRENT PRACTICE

A. Boron

Techniques of Decomposition

Various techniques can be applied to the decomposition of organoboron compounds. The choice is dependent primarily on the nature of the sample and the preference of the analyst. Generally speaking, the mineralization procedures may be divided into two categories: (1) digestion of the sample in a liquid medium, and (2) combustion of the sample at high temperatures.

Both acidic and basic digestion media have been employed. In the former case, the reagents can be (1) sulfuric acid with addition of catalysts [32] used in micro Kjeldahl digestion [33] (see Sec. 3.3); (2) a mixture containing nitric, sulfuric, and perchloric acids [34]; or (3) trifluoroperoxyacetic acid, CF_3CO_3H [35]. Basic digestion mixtures include alkaline potassium persulfate [36], and hydrogen peroxide in sodium hydroxide solution [37].

The closed-flask combustion technique has been utilized for the decomposition of boron compounds [38,39]. On the other hand, combustion in a metal bomb appears to be the most reliable procedure since it is applicable to decarboranes, carboranes, borazines, borazoles, boranes, borates, and borohydrides. The detailed experimental directions using KNO_3-sucrose and Na_2O_2 in a Parr microbomb are given in Sec. 8.4. Other fusion reagents such as alkali carbonate [40] and metallic potassium [41] (see Sec. 5.7) have been proposed. Some workers [42-44] have used

oxygen under 25-30 atmospheres of pressure. Decomposition by heating in
a combustion tube in the presence of a current of oxygen (see Sec. 2.7)
has been described [45,46]. For the complete destruction of polymeric
boron-containing materials [47,48], however, the use of oxyhydrogen combustion is recommended. The apparatus [49] is shown in Fig. 8.1. The
sample is heated with concentrated sulfuric acid in a flask through which a
stream of nitrogen leads to the oxyhydrogen flame. When the sample has
been decomposed, methyl alcohol is added to the flask, which is then
heated to 70°C. Methyl borate is formed and passes through the oxyhydrogen
flame where it is decomposed to boric acid, which is absorbed in water.

Methods of Determination

    After the conversion of organoboron compounds to boric acid, if the
amount of boron is between 0.3 and 1 mg, the prevailing techniques of finish

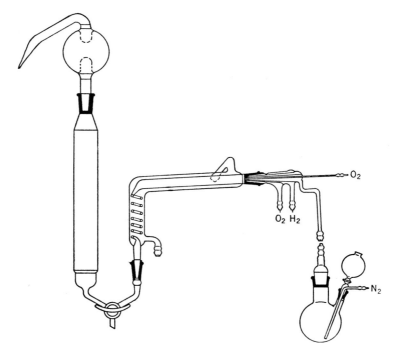

FIG. 8.1. Apparatus for oxyhydrogen combustion. (From Ref. 49,
courtesy J. Wiley & Sons, Inc.)

generally involve titrimetry by means of 0.01 $\underline{N}$ sodium hydroxide. Since boric acid is a very weak acid ($pK_1 = 9.2$), it cannot be titrated as such. However, boric acid can be transformed into stronger acidic species in the solution by complexation with cis-1,2-diols, as discovered by Biot [50] and depicted below:

$$
\begin{array}{c}
\text{-C-OH} \\
| \\
\text{-C-OH}
\end{array}
+
\begin{array}{c}
\text{HO} \\
\diagdown \\
\text{HO} \diagup
\end{array}\!\!\text{B-OH}
\xrightleftharpoons{-2H_2O}
\begin{array}{c}
\text{-C-O} \\
| \quad\diagdown \\
\text{-C-O} \diagup
\end{array}\!\!\text{B-OH}
\rightleftharpoons
\begin{array}{c}
\text{-C-O} \\
| \quad\diagdown \\
\text{-C-O} \diagup
\end{array}\!\!\text{B-O}^- + \text{H}^+
$$

$$
\begin{array}{c}
\text{-C-O} \\
| \quad\diagdown \\
\text{-C-O} \diagup
\end{array}\!\!\text{B-OH}
+
\begin{array}{c}
\text{HO-C-} \\
| \\
\text{HO-C-}
\end{array}
\xrightleftharpoons{-H_2O}
\left[
\begin{array}{c}
\text{-C-O} \quad\quad \text{O-C-} \\
| \quad\diagdown \quad\diagup \quad | \\
\text{-C-O} \quad\text{B}\quad \text{O-C-}
\end{array}
\right]^-
+ \text{H}^+
$$

Since these reactions are equilibrium reactions, the acidity of the solution is dependent on the concentration and nature of the reactants, and also on temperature. The formation constants of polyols with borate have been measured [51,51a]. Mannitol was first proposed as the analytical reagent [52] and is usually used, albeit other polyhydric compounds have been reported to be superior [53]. The dissociation constant of mannitol boric acid [54] is 6.3×10^{-6} to 8.4×10^{-6}.

One technique to determine boric acid by means of mannitol involves (1) neutralization of the strong acids or bases in the matrix after neutralization to the pH of 5.5 or methyl orange endpoint, (2) addition of a suitable amount of mannitol to produce the acidic complex, and (3) titration with standardized 0.01 $\underline{N}$ NaOH to the phenolphthalein endpoint. The equivalence point usually occurs around pH 8 and the solution is generally adjusted to about 0.35 $\underline{M}$ in mannitol. As mentioned above, the acidity of the complex varies with the mannitol content. When sufficient amount of mannitol is present (>1.5 $\underline{M}$), neutralization is complete [55] even at pH 6.0. Understandably, the solution containing the boric acid to be determined should be free from other weak acids and weak bases, as well as their salts. This condition usually cannot be met in analyzing organic materials; hence a separation step may be necessary.

An alternative technique, known as the "fixed pH" or "identical pH" method [55], is currently widely accepted since it eliminates the separation step. This technique involves (1) removal of carbon dioxide by boiling the solution, (2) exact adjustment to a selected pH value between 6 and 8, (3) addition of mannitol, and (4) titration by means of standardized 0.01 $\underline{N}$ NaOH back to the pH identical with the first adjustment, using an appropriate indicator or a pH meter. In view of the arbitrary feature of this method, each step should be carried out uniformly. The 0.01 $\underline{N}$ NaOH standardization must be performed against boric acid in exactly the same manner as it is used in the titration of the sample solution. The amount of organic material taken for analysis should contain nearly the same quantity of boric acid, and approximately the same amount of mannitol should be added for a given volume of solution. The fixed pH values of 7.6 [55], 7.10 [48], 6.9 [56], and 6.3 [35,57], respectively, have been recommended.

The mannitol titration technique is nearly specific for boron analysis in organic materials; only phosphate and germanium ions react with mannitol in a similar manner. It is apparent, however, that this method is not suitable for determining less than 0.1 mg of boron in the sample. Carlson and Paul [58] have described a method for the determination of boron as tetrafluoroborate with a liquid ion-exchange membrane electrode. Using columns of the boron-specific resin Amberite XE-243, boron can be separated from interfering ions (e.g., nitrate) and the tetrafluoroborate ion can be formed in less than 15 min with a small volume of 10% hydrofluoric acid. The tetrafluoroborate is eluted with sodium hydroxide solution and measured potentiometrically.

Boron in organic compounds has been determined by coulometric [39] and conductimetric [59] methods. For concentrations of boron ranging from 0.1 mg down to microgram quantities, colorimetric methods are most suitable, using reagents such as 1,1'-dianthrimide [60] and tumeric [61].

B. Silicon

Techniques of Decomposition

Decomposition of organosilicon compounds can be achieved by heating with oxidizing acids in a platinum crucible or Vycor digestion flask. Perchloric acid is frequently used [62], although a mixture of sulfuric and nitric acids is safer to handle [63]. Dry ashing or decomposition in the combustion tube [64] in an atmosphere of oxygen can be employed, provided that the organic material is heated slowly in order to prevent the formation of the extremely stable silicon carbide.

Fusion in metal bombs is recommended by some workers, particularly for more volatile organosilicon compounds [65]. Sodium peroxide is frequently used as the oxidizing agent [66,67]. Greive and Sporek [68] have reported that sodium hydroxide fusion is effective for the decomposition of polymeric siloxanes. Subsequently, Wetters and Smith [69] have found that potassium hydroxide is superior, and the usefulness of this fusion method is extended by employing an alcoholic alkali pretreatment step, in which siloxane bonds are converted to silanolates.

Methods of Determination

When organosilicon compounds are decomposed in a combustion tube, the resulting silicon dioxide remains in the microboat and can be determined as such. This method, however, is seldom employed when silicon alone is determined, because the fusion techniques are preferred for decomposition. After mineralization, the silicon in the fusion mixture can be determined gravimetrically, titrimetrically, or colorimetrically.

A gravimetric method for determining silicon involves the complexation of silica with molybdic acid; the silicomolybdic acid is then treated with a hydrochloric acid solution of 8-hydroxyquinoline, commonly known as oxine. An orange-colored oxine complex [70] is precipitated; it is composed of 12 parts of molybdic oxide and four parts of oxine to one part of silicon dioxide. Some workers [41] heat the precipitate at 110°C and weigh it; others [65, 71] ignite the dry precipitate at 500°C for 1 h and weigh the product as $SiO_2 \cdot 12MoO_3$. A blank determination should be carried out with each batch of reagent solutions, and the weight of the blank obtained is deducted from the weight of ignited precipitate. Christopher and Fennell [67] have recommended the use of quinoline in place of oxine; the yellow quinoline silicomolybdate is dried at 150°C and weighed.

Based on the precipitation of the oxine-silicomolybdate complex, McHard and co-workers [65] have described a titrimetric method as follows. A known excess of standardized oxine solution is added to the yellow silicomolybdic acid solution which is produced by the addition of ammonium molybdate to the acidic solution containing the silicate. The reaction mixture is heated at 65°C to coagulate the oxine complex and then made to a known volume. After filtration on a dry filter paper to separate the oxine complex, an aliquot is taken from the filtrate and the excess oxine is titrated with standardized bromate-bromide solution. It has been found that the theoretical factor calculated from the equations

$$KBrO_3 + 5KBr + 6HCl = 3Br_2 + 6KCl + 3H_2O$$

$$C_9H_7ON + 2Br_2 = C_9H_5ONBr_2 + 2HBr$$

gives slightly low silicon figures. Therefore an empirical factor should be determined using pure organosilicon compounds.

Another titrimetric method has been reported by Bartusek [72]. After mineralization of the sample in a flask with H_2SO_4/HNO_3, the silicon is precipitated with KHF_2 to produce K_2SiF_6. The latter is collected on paper pulp, dissolved and titrated at an elevated temperature with standardized sodium hydroxide solution using phenolphthalein as indicator.

The colorimetric methods for the determination of silicon are similar to those for phosphorus (see Sec. 7.2B). For 2-6 mg samples containing about 0.7 mg silicon, the yellow color of silicomolybdate is suitable; Debal [66] has recommended spectrophotometric measurement at 400 nm. For the analysis of organic materials containing less than 5% silicon, the yellow silicomolybdate should be reduced to the heteropoly blue with 1-amino-2-naphthol-4-sulfonic acid, and the absorbance of the solution is then measured at 800 nm at a pH between 1.2 and 1.7.

Morrow and Dean [73] have described a method to analyze organosilicon compounds by flame emission spectrometry without prior mineralization. The compound is aspirated as a solution in ethanol and the emission intensity at 251.6 nm is measured. Quantitation is made by using a working graph and by the method of standard addition.

C. Mercury

Techniques of Decomposition

A decomposition technique which is specific for organomercury compounds involves heating the organic sample in a combustion tube packed with calcium oxide. The mercury vapor liberated is conducted by a stream of air into an absorption tube packed with shredded gold foil, with which an amalgam is formed. The experimental procedure is given in Sec. 8.6. If the organic material contains sulfur, lead chromate is recommended in place of calcium oxide. Silver and copper wires may be incorporated in the combustion tube packing to remove halogens and nitrogen oxides, respectively, in an atmosphere of carbon dioxide [74].

Decomposition of organomercury compounds can be effected by wet oxidation with nitric acid [75], sulfuric acid and hydrogen peroxide [76], or potassium persulfate [77]. The most drastic oxidizing mixture consists of a combination of potassium permanganate, nitric acid, and sulfuric acid [78]; it should be used with caution because of the hazard of explosion (see Sec. 8.7). In another procedure [79], the sample is decomposed by heating with powdered potassium permanganate in a sealed glass tube at 400-500°C for 1-2 h; after cooling, the reaction mixture is leached with dilute nitric acid.

Compounds of the R-Hg$^+$ type can be decomposed by reduction with zinc in hydrochloric, acetic [80], or formic acid [81]. The amalgamated zinc is then separated and dissolved in nitric acid. Some organomercury compounds are directly titratable with ammonium thiocyanate in aqueous or nonaqueous media [82].

The decomposition of organic materials for the determination of low percentages or trace amounts of mercury deserves special attention. Since elemental mercury and some mercury salts have relatively high vapor pressures, dry combustion of the sample is seldom feasible. Furthermore, because of the extreme toxic effects of mercury, which is accumulative, government regulatory agencies specify very low tolerance limits for mercury in biological materials. Consequently, the destruction methods used must be very precise and accurate. The method generally accepted [83] was proposed by Klein [84]. It involves digestion of the sample with nitric acid and sulfuric acid in a special apparatus which allows part of the solution to be distilled into a reservoir; this enables the temperature of the reaction mixture in the flask to rise and the severity of the conditions to be increased. In the final stage, the digest is heated to the boiling temperature of sulfuric acid. Tölg and co-workers [85] have described the decomposition of biological materials by heating with HNO_3/HF under pressure in Teflon tubes. The recovery of mercury in the nanogram range is reported to be better than 98%.

Methods of Determination

When the decomposition is performed in a combustion tube, it is customary to determine the elemental mercury gravimetrically by collecting it as amalgam of gold. On the other hand, when wet digestion is employed for mineralization, the resulting mercury salts in the reaction mixture are usually determined by titrimetric procedures. Yeh [78] has recommended titration with 0.01 $\underline{M}$ potassium thiocyanate using an automatic second-derivative spectrophotometric titrator coupled to an automatic buret. Accurate endpoints are obtained even with very dilute solutions, and the acidity of the solution can be varied between 1 and 3 $\underline{M}$. Busev and Teternikov [86] have described a procedure in which the Hg(II) in the solution at pH 2-5 is precipitated by adding a 1.5-fold excess of 10^{-3} $\underline{M}$ nickel diethylphosphorodithioate. After the precipitate has been allowed to settle, the pH is adjusted to 6-7, and the unused precipitant is titrated with 10^{-3} $\underline{M}$ 4-dimethylaminophenylmercuric acetate, either visually or potentiometrically. This method cannot be used if bromine or iodine is present, although chlorine in the organic material does not interfere.

Understandably, the gravimetric and titrimetric methods are not suitable for mercury determinations at low levels. Thus, atomic absorption and other spectrophotometric procedures are generally employed for trace analysis of mercury. Holzbecher and Ryan [87] have described a simple fluorometric method for Hg(II) based on its oxidative reaction with thiamine to yield the highly fluorescent thiochrome. The fluorescence intensity is linear over a range of 10 to 200 ng/ml. The sample taken, however, should contain 10-500 μg of mercury to allow for the necessary dilution after acid digestion to lower the salt concentration. Sandi and co-workers [88] have determined mercury by an enzymatic method which is based on the inhibition of urease activity. Nishi and Horimoto [89] have determined microgram amounts of CH_3HgCl, C_2H_5HgCl, or C_6H_5HgCl in aqueous solutions by the following steps: (1) extraction with benzene; (2) reverse extraction with cysteine solution; (3) liberation of the organomercury compound by hydrochloric acid and, again, extraction into benzene; (4) finally, gas chromatographic determination using an electron capture detector.

D. Interferences and Complications

   Boron

   Analysis of boron in organic materials by sodium peroxide fusion and fixed pH titrimetry is relatively free from complications. When copper or aluminum is present, the former can be removed by filtration of CuO and the latter by separation of Al(OH)$_3$. The interference due to fluoride can be prevented by keeping the pH above 3 to eliminate HF. Germanium reacts like boron towards mannitol; hence both elements will be determined.

   Since acid digestion tends to cause sluggish titration, it is used only to decompose samples with low boron contents. When the closed-flask technique is employed, some investigators [90] have resorted to boron-free flasks which are shock sensitive. According to Strahm [91], borosilicate flasks can be used satisfactorily for routine analysis, especially if no alkaline material is introduced. The flammability and explosion hazards of some organoboron compounds must be considered.

   Hansen [92] has analyzed plant materials (e.g., leaves, fruits) for boron by the 1,1'-dianthrimide colorimetric method and reported that variation of drying temperature of the original sample (60-105°C), addition of lime before ashing, or treatment of the ash with ammonium persulfate and hydrazine sulfate does not affect the results. This method is also not vitiated by the presence of NO_3^-, Fe(II), Mn(II), Zn, and Mg ions.

   Silicon

   Attention is called to the ease of formation of silicon carbide when the organic material is decomposed by dry ashing. Once this occurs, it is almost impossible to bring the silicon into solution. Therefore a large excess of oxygen should be present during the combustion in order to produce SiO_2 quantitatively.

   Organo-tin-silicon compounds on combustion yield a residue consisting of SiO_2 plus SnO_2. Obtemperanskaya and co-workers [93] have described a procedure to determine the silicon as follows. The residue in the microboat is mixed with NH_4I and heated in a stream of nitrogen until evolution of vapor ceases. After cooling, the microboat containing only SiO_2 is weighed.

   Fluorine interferes with the determination of silicon by the oxine gravimetric method or the silicomolybdate spectrophotometric method. This interference can be removed by the addition of an excess of boric acid [41].

Since silicon and phosphorus behave very similarly toward molybdate, the presence of phosphorus causes severe interference. Horner [94] has described a laborious method in which the organic material is decomposed by sodium peroxide fusion; the silica is repeatedly precipitated, then treated with hydrofluoric acid, and determined by the loss of weight. Austin and co-workers [95] have reported that the molybdenum blue method [96] can be used for determining silicon in the presence of phosphorus by incorporating tartaric acid and sulfuric acid in the reaction mixture.

Mercury

Accurate determination of mercury in organic materials is often diffi-cult, especially at the trace level. The ready reduction of organomercury compounds to the metal, and the high volatility of both the metal and its compounds greatly increase the possibility of loss occurring during the de-composition stage. Heating in an open vessel at 200°C may cause losses of mercury from a few percent up to 100%.

Another complication in the analysis of mercury lies in the fact that this element amalgamates with many metals and forms stable complexes with organic species which are produced during the degradation of the or-ganic sample. Thus, when an extraction step is involved, the mercury may not go into the solvent layer as expected. In the determination of submicrogram amounts of mercury by the oxygen bomb combustion method, Bretthauer and co-workers [97] have employed a radiotracer of mercury to correct for unpredictable and unreproducible losses during analysis.

8.3 RECENT DEVELOPMENTS

A. Survey of Literature

The publications related to the determination of boron, silicon, and mercury are summarized in Tables 8.1, 8.2, and 8.3, respectively. This survey covers the period starting from about 1960. For earlier literature, the reader is referred to the excellent review on boron analysis written by Strahm [91] and that on silicon by Horner [94]. In 1965 Manger [152] pub-lished a review on the methods for the determination of mercury in organo-mercury compounds, citing 59 references. Table 8.3 attests to the recent activity of research on mercury analysis, due to the public concern about this toxic element in the environment.

B. Evaluation of Methods

Pierson [153] has evaluated six procedures for the determination of boron. Sharp endpoints in the final titration can be achieved with 0.05 $\underline{N}$ NaOH in

TABLE 8.1. Summary of Recent Literature on the Determination of Boron.

Decomposition method	Product determined	Finish technique	References
Closed flask	BO_3^{3-}	Titrimetric	39, 98–99
	BO_3^{3-}	Spectrophotometric	100
Acid digestion	BO_3^{3-}	Spectrophotometric	101
Sealed tube	BO_3^{3-}	Titrimetric	36, 102, 103
	BO_3^{3-}	Flame photometry	104, 105
Metal bomb	BO_3^{3-}	Titrimetric	106, 125a
	BO_3^{3-}	Spectrophotometric	107, 108
$CaCl_2$ fusion	$(CH_3)_3B$	Titrimetric	109
$Ca(OH)_2$	BO_3^{3-}	Fluorometric	110
Hydrolysis	BO_3^{3-}	Titrimetric	111, 112
Direct	B–O	Infrared spectrometry	113
Flame	B	Emission spectrometry	114
Neutron activation	^{116}In	Radiometric	115

TABLE 8.2. Summary of Recent Literature on the Determination of Silicon.

Decomposition method	Product determined	Finish technique	References
Acid digestion	SiO_2	Titrimetric	116
	H_2SiF_6	Titrimetric	117–119
	Si	Spectroscopy	119a
Alkali fusion	SiO_2	Colorimetric	66, 69, 120–126
	SiO_2	Titrimetric	127
	SiO_2	Gravimetric	67
Combustion tube	SiO_2	Gravimetric	128
Nickel flask	SiO_2	Spectrophotometric	129
Nickel-plated hood	SiO_2	Spectrophotometric	129a
$N_2O-C_2H_2$ flame	Si	Atomic absorption	130, 130a
Spark	Si	Spectroscopy	131, 131a

TABLE 8.3. Summary of Recent Literature on the Determination of Mercury.

Decomposition method	Product determined	Finish technique	References
Combustion tube	Hg	Gravimetric	132–137
	Hg	Spectrophotometric	23
	Hg	Atomic absorption	138
	Hg	u.v. photometric	139
Closed flask	Hg^{2+}	Gravimetric	140
	Hg	Hg-vapor absorption	24
Sealed tube	Hg^{2+}	Titrimetric	79
	Hg	Atomic absorption	141
Wet oxidation	Hg^{2+}	Titrimetric	78, 86, 142–143a
	Hg^{2+}	Spectrophotometric	8, 18, 144
	Hg^{2+}	Electrometric	145, 145a
	Hg^{2+}	Atomic absorption	9, 10, 20, 31, 146, 147
	^{197}Hg	Radiometric	16, 25–27, 148
Na_2O_2, Na_2CO_3	Hg	Gravimetric	149
Ozone	Hg	Atomic absorption	150
Oxyhydrogen flame	Hg^{2+}	Titrimetric	133
Neutron activation	$^{197}Hg, ^{203}Hg$	Radiometric	22, 151

the presence of excess mannitol provided that the initial oxidation process is controlled both as to the type and extent of reaction, depending on the structure of the organoboron compound. Kato and co-workers [106] have examined four oxidation methods, namely, Na_2O_2, alkaline H_2O_2 solution, trifluoroperoxyacetic acid, and $K_2S_2O_6$ solution, all involving reflux. Using the first method and NaOH as titrant, the errors are about 1%. On the other hand, employing $Ba(OH)_2$ as titrant and either Na_2O_2 or H_2O_2 for decomposition, the errors are about 0.1%. Debal and Levy [99] have compared the closed-flask technique with combustion-tube oxidation in the presence of moisted oxygen, and they recommended the former method in combination with titration at a fixed pH value of 7.5 using a pH meter. The sample size should be less than 1.5 mg.

A collaborative study on the determination of mercury in fish by flameless atomic absorption spectroscopy has been published [154]. The sample (5 g) is decomposed by digestion with $HNO_3/H_2SO_4/HClO_4$. The resulting Hg(II) is reduced by means of $SnCl_2$ to elemental Hg, the vapor of which is circulated in a closed system through a tubular cell, and the Hg is measured at 253.7 nm. The overall recovery of 0.3-0.8 ppm of Hg is 84%.

8.4 EXPERIMENTAL: DETERMINATION OF BORON

A. Principle

The organic sample containing boron is combusted in a Parr microbomb by means of sodium peroxide, sucrose, and potassium nitrate. The boron is converted to Na_3BO_3 which upon acidification gives H_3BO_3. The H_3BO_3 is titrated with 0.01 $\underline{N}$ NaOH in the presence of excess mannitol using the so-called fixed pH or identical pH method (see Sec. 8.2A). In the procedure described below, the pH of the boric acid solution is adjusted to 6.3, excess mannitol is added until the pH no longer decreases, and then the volume of 0.01 $\underline{N}$ NaOH standard solution which is required to return the pH to 6.3 is recorded.

B. Equipment

Parr microbomb, 2.5-ml capacity, with lead gaskets (see Sec. 5.7)
  (Parr Instrument Co., Moline, Illinois).
Nickel beaker (made of copper, nickel plated) with spout, 250-ml
  capacity (Macalaster Bicknell Co., New Haven, Connecticut).
pH Meter, Beckman Model H-2, equipped with the standard glass and
  calomel electrodes.
Microburet, 10-ml capacity, graduated in 0.02-ml intervals.

C. Reagents

Fusion mixture: Mix two parts of potassium nitrate with one part of
  sucrose and store in a glass-stoppered bottle. Use a scoop which
  holds about 100 mg of this mixture. Use a separate scoop (delivering about 1.5 g) for sodium peroxide, kept in another container.

Concentrated hydrochloric acid.

Bromocresol green-methyl red mixed indicator, see Sec. 3.3.

Sodium hydroxide solution, 6 $\underline{N}$, for rough pH adjustment (need not be standardized).

Sodium hydroxide standard solution, 0.01 $\underline{N}$: Standardize this solution against reagent grade boric acid.

Mannitol, CP.

D. Procedure

Decomposition

Accurately weigh a sample containing 2-4 mg B into the microbomb. Add the fusion mixture and combust the sample as described in Sec. 6.5. After cooling, rinse the microbomb with distilled water, open the lid, and place the microbomb and contents in a nickel beaker. Add distilled water so that the microbomb is just submerged, and boil the solution on a hot plate until effervescence ceases. Cool the solution to room temperature, remove the microbomb and transfer the solution quantitatively to a 100-ml volumetric flask. Add four drops of mixed indicator and carefully acidify the solution with concentrated hydrochloric acid (indicator changes from blue to pink). Fill up to volume with distilled water.

Determination

Withdraw an appropriately sized aliquot (usually 25 ml) and deliver into a 150-ml beaker. Add distilled water to make a total of 50 ml of solution and heat to boiling on the hot plate to remove carbon dioxide. Cool to room temperature and place on the magnetic stirrer. Submerge the electrodes and adjust the pH to 4-5 with 6 $\underline{N}$ NaOH, and then carefully with 0.01 $\underline{N}$ NaOH to exactly 6.30. Add excess mannitol (7-9 g) until the pH remains constant. Now titrate with 0.01 $\underline{N}$ NaOH standard solution (Sec. 8.4F). Record the volume of 0.01 $\underline{N}$ NaOH standard solution needed to bring the pH back to 6.30.

E. Calculation

$$\%B = \frac{N \times V \times 10.82 \times 100}{W}$$

where

N = exact normality of 0.01 $\underline{N}$ NaOH standard solution (against boric acid)

V = volume of the above standard solution required for titration

W = weight of sample in aliquot taken

F. Note

It is recommended that the tip of the microburet be positioned below the
surface of the solution during the titration.

G. Comments

(1) The described procedure calls for the titration of an aliquot of the total
bomb contents because it has been found that excessive salts tend to make
the subsequent titration sluggish.

(2) The advantage of the titration procedure described here (fixed pH
method) versus that recommended by other workers is the noninterference
of weak acids or bases which may be present.

(3) The procedure described here may also be used for the determination of germanium in organogermanium compounds. Both germanic acid
and boric acid liberate one H^+ upon the addition of mannitol. (It may be of
interest to note that germanic acid liberates $2H^+$ upon the addition of
catechol.)

(4) The only other interferences that have been found with the described
procedure are the presence of aluminum (which can be removed by precipitation) and fluorine in the sample. The interference due to F^- can be
circumvented by keeping the pH above 3.

(5) In determining low levels of boron (below 1% B) in organic materials,
a digestion procedure using nitric and sulfuric acids is recommended. The
amount of sulfuric acid used, however, should be kept to a minimum, since
the subsequent neutralization may lead to an excessive amount of salt,
which, as mentioned previously, gives a sluggish titration.

8.5 EXPERIMENTAL: DETERMINATION OF SILICON

A. Principle

After mineralization of the organosilicon compound to produce water-
soluble silicate, the reaction mixture is treated with ammonium molybdate
and 8-hydroxyquinoline. The complex formed is oxine-silicomolybdate
(see Sec. 8.2B) which can be collected on a filter, dried at 110°C, and
weighed. The procedure described below was developed by Schwarzkopf
and Heinlein [41]. It employs alkali fusion and hence can be used for
samples which contain fluorine. According to Dr. Otto Schwarzkopf [41a],
this method has been displaced by closed-flask combustion and weighing
of SiO_2 (see Sec. 8.5G) if applicable.

B. Equipment

   Nickel microbomb (manufactured by K. Krobe, Woodside, New York).
   Spark-plug gaskets, and bench jig for holding bomb.
   Nickel crucible.
   Polyethylene funnels.
   Polyethylene droppers.
   Filter tube, Pyrex fritted, porosity M.
   Muffle furnace.

C. Reagents

   Potassium metal (Sec. 8.5F, Note 1).
   Methanol.
   Sodium hydroxide solution, 1 $\underline{N}$.
   Dilute hydrochloric acid, 1:9 and 1:1.
   Ammonium molybdate solution, 10%.
   Boric acid.
   Oxine (8-hydroxyquinoline) solution: Dissolve 14 g of oxine in 20 ml
       of hydrochloric acid and dilute to 1 liter.

D. Procedure

   Decomposition

   Accurately weigh 2-20 mg of the sample into the microbomb. Use the
microweighing tube or platinum boat to weigh solids, and gelatin capsules
to weigh liquid samples. Add 150-200 mg of potassium metal. Place a
spark-plug gasket between the cup and lid of the bomb and close the bomb
tightly. (Apply graphite on the screw to facilitate reopening.) Heat the
bomb for 2 h at 650°C in the muffle furnace. Allow the bomb to cool, insert
it into the bench jig, and loosen the screw. Insert a bar into the hole in the
lid and loosen the lid by turning it a half turn. Remove the bar and un-
screw the bomb nut. Then remove the lid and rinse with methanol and 1 ml
of distilled water into a nickel crucible. Replace the lid on the cup. Clean
the outside of the cup with water and methanol; discard this rinse. Now
dry the bomb on a filter paper. Remove the lid and again rinse with 0.5
ml of methanol into the nickel crucible. Insert the cup into the crucible
and add 0.5 ml of methanol. Allow to stand for 5 min and mix the contents
with a spatula. Add three drops of distilled water and another three drops
3 min later. When the reaction is complete, place the nickel crucible and
contents in a water bath and boil for 10 min. Filter the contents of the
bomb using Whatman No. 40 filter paper and a polyethylene funnel. Use a

polyethylene dropper for the transfer and collect the filtrate in a 50-ml volumetric flask. Wash the nickel crucible and bomb three or four times with boiling water containing a few drops of 1 $\underline{N}$ NaOH. Filter the washings into the volumetric flask. Fill up to volume with distilled water.

Determination

Pipet a 5- or 10-ml aliquot from the volumetric flask into a 125-ml Erlenmeyer flask. Neutralize the solution with 1:9 dilute hydrochloric acid. Add 1 g of boric acid (for the purpose of avoiding the interference of fluorine), and acidify the mixture by adding 3 ml of 1:1 hydrochloric acid. Dilute the solution to 25 ml, add 5 ml of 10% ammonium molybdate solution and allow to stand for 15 min. Add 10 ml of 1:1 hydrochloric acid, followed by 15 ml of the oxine solution. Collect the oxine-silicomolybdate precipitate in the fritted Pyrex filter tube, rinse with oxine wash solution, dry at 110°C, and weigh.

E. Calculation

$$\%Si = \frac{\text{mg precipitate x 0.01167 (Note 2) x 100}}{\text{mg sample in aliquot}}$$

F. Notes

Note 1: See Sec. 5.7 for the preparation of the potassium capillary; use polyethylene tubings to prepare the capillary.

Note 2: This gravimetric factor should be checked against organosilicon compounds of known purity.

G. Comments

(1) If the sample does not contain fluorine, the easy way to convert organically bonded silicon to silicate is by means of sodium peroxide fusion in the Parr microbomb (see Sec. 8.4).

(2) Many organosilicon compounds can be decomposed by heating in a platinum crucible (see Sec. 9.4) with fuming HNO_3 and H_2SO_4. The residue is then weighed as SiO_2.

(3) The procedure for closed-flask combustion and weighing of SiO_2 is as follows [41a]. The sample is weighed in a gelatine capsule greased with Vaseline. The weighed capsule and a paper wick are fixed to the sample holder of the combustion flask. Concentrated sulfuric acid is added to the

flask and completely wets its walls. The flask is filled with oxygen and the sample is burned. The sulfuric acid dehydrates the silicic acid. After standing and diluting with water, SiO_2 is collected on a millipore filter, ashed, and weighed. For further details, see O. Schwarzkopf and F. Schwarzkopf, in M. Tsutsui (ed.), Characterization of Organometallic Compounds, Part I, p. 59, Wiley, New York, 1969.

8.6 EXPERIMENTAL: DETERMINATION OF MERCURY BY GRAVIMETRY

A. Principle

The organic mercury compound is decomposed by heating at high temperature to liberate mercury vapor. The latter is conducted through calcium oxide and then trapped in a tube packed with gold which forms the amalgam [132].

B. Apparatus

The combustion tube and gold tube are shown in Fig. 8.2. The combustion tube is packed with a column of granulated calcium oxide about 120 mm long, held between two pieces of glass fiber paper [155]. A 50-mm long glass rod with flattened ends provides an empty space in the combustion tube, between the combustion furnace which heats the section of calcium oxide, and the constricted tip of the combustion tube. The wide end of the combustion tube is connected to the air cylinder through a bubble counter (see Fig. 2.8, D). The absorption tube for mercury is 80 mm long and 8 mm in diameter. It is drawn out to a tip 3 mm in diameter and 20 mm long. A 40-mm layer of shredded gold leaf is packed near the constricted portion. The tip of the absorption tube is connected to the Mariotte bottle (see Fig. 2.8, K).

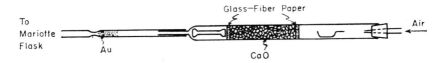

FIG. 8.2. Apparatus for the determination of mercury by gravimetry. (From Ref. 132, STANDARD METHODS OF CHEMICAL ANALYSIS, Vol. 2A, 6th Ed., edited by Frank J. Welcher © 1963 by Litton Educational Publishing, Inc., reprinted by permission of Van Nostrand Reinhold Company.)

C. Procedure

Accurately weigh a sample containing 1-5 mg Hg in a porcelain or silica
microboat. Push the boat into the combustion tube to within 50 mm of the
calcium oxide. With the section of calcium oxide maintained at 700°C,
pass a slow current of dry air through the combustion tube while the leveling tube of the Mariotte bottle is lowered in order to produce a suction effect on the absorption tube. (In this way the vapors coming out of the
combustion tube will be drawn through the gold-packed absorption tube,
which was previously weighed.)
    Now gradually move the burner under the microboat forward until the
vapors have passed into the calcium oxide section, where halogens and
sulfur are retained, while the mercury goes into the empty section of the
combustion tube. With a piece of clean, wet cloth placed over the absorption tube, carefully apply a small flame to the constricted tip of the combustion tube in order to drive all mercury into the absorption tube where
amalgam is formed. After aspirating 50 ml of air through the absorption
tube, disconnect the latter from the combustion train. Wipe the surface
of the absorption tube and then place it in the desiccator for 30 min. Reweigh the tube; the increase in weight is due to the mercury from the
sample.

D. Calculation

$$\%Hg = \frac{mg\ Hg\ x\ 100}{mg\ sample}$$

8.7 EXPERIMENTAL: DETERMINATION OF MERCURY BY AUTOMATED
      SPECTROPHOTOMETRIC TITRATION WITH POTASSIUM
      THIOCYANATE

A. Principle

The organic material containing mercury is oxidized in a micro Kjeldahl
flask by means of potassium permanganate in a mixture of concentrated
nitric and sulfuric acids. Any halogen, if present, is removed by volatilization of the halogen halide or the free element, while all mercury remains
in solution as Hg(II) ions. Utilizing the second derivative spectrophotometric technique [156] to detect the endpoint, Hg(II) ions can be titrated
with 0.01 $\underline{N}$ potassium thiocyanate with high precision, thus adapting the
time-honored Volhard [157] method to the micro scale. The procedure

described below is based on the report of Prof. Ching-Siang Yeh [78], Purdue University, West Lafayette, Indiana. The experimental details were verified by Prof. Yeh in October 1974.

B. Equipment

   Micro Kjeldahl flasks, 30-ml capacity, and digestion rack (see Sec. 3.3). Automated second-derivative spectrophotometric titrator and buret as-
      sembly (see Fig. 8.3): Use the Sargent-Malmstadt Automatic Ti-
      trator (E. H. Sargent Co., Chicago, Illinois).

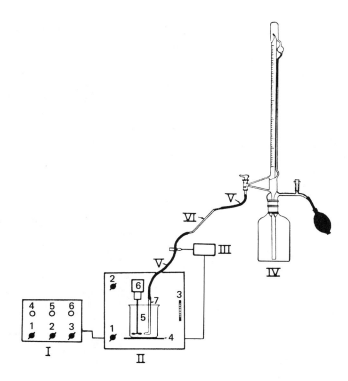

FIG. 8.3. Automated second-derivative spectrophotometric titrator and buret assembly. I, Controller; II, spectrophotometric titrator; III, solenoid valve; IV, automatic buret; V, amber tubing 1/8 x 1/32 in. latex; VI, glass capillary 5 x 0.5 mm. (Courtesy Microchem. J.)

C. Reagents

Concentrated sulfuric acid, specific gravity 1.84.
Concentrated nitric acid, specific gravity 1.42.
Hydrogen peroxide, 6%.
Potassium thiocyanate standard solution, 0.01 $\underline{M}$.
Ferric ammonium sulfate indicator solution: Dissolve 40 g of
    $Fe(NH_4)(SO_4)_2 \cdot 12H_2O$ in 100 ml of distilled water, add several drops
    of concentrated nitric acid, and filter the solution.
Triple-distilled mercury.

D. Procedure

Preparation of the Calibration Graph

Accurately weigh 100.3 mg of triple-distilled mercury into a Kjeldahl
flask. Add 5 ml of diluted nitric acid (1:1 by volume) and heat gently to
bring the mercury into solution. After the oxides of nitrogen have been
removed by boiling, cool the solution, transfer it to a 100-ml volumetric
flask and fill up to the mark with distilled water. Measure exactly 1-, 2-,
3-, and 10-ml aliquots, respectively, of this 0.005 $\underline{M}$ $Hg(NO_3)_2$ solution
into the 100-ml tall beaker of the titration assembly (see Fig. 8.3). Add
1 ml of the indicator and make up the volume of the solution to 50 ml with
distilled water. Titrate with the 0.01 $\underline{M}$ KCNS standard solution. Run a
blank and correct. Plot the calibration graph; it should be rectilinear
from 2 to 8 ml of the 0.01 $\underline{M}$ KCNS standard solution.

Determination

Accurately weigh 4-10 mg of the organic sample (estimated to consume
2-8 ml of 0.01 $\underline{M}$ KCNS) into the Kjeldahl flask. Add 1 ml of concentrated
nitric acid and 400 mg of potassium permanganate. Allow the reaction mixture to stand at room temperature behind a safety shield for 10 min
(caution: see Sec. 8.7F). Then add 3 ml of concentrated sulfuric acid.
Heat with a small flame for 5 min and with a stronger flame for 35 min or
until the solution is clear. Upon cooling, if permanganate is found in the
neck of the flask, add a few drops of 6% hydrogen peroxide to decompose
it and boil the solution for 3 min. Now cool the solution, cautiously add 5
ml of distilled water, and transfer to the 100-ml tall beaker (Fig. 8.3).
Carry out the automated titration in the same manner as described for the
preparation of the calibration graph.

E. Calculation

$$\%Hg = \frac{\text{mg Hg found x 100}}{\text{mg sample}}$$

F. Note

Potassium permanganate is extremely dangerous in concentrated acid medium. Add this reagent in small portions. If purple or red fumes (Mn_2O_7, highly explosive) evolve, let them subside before continuing the experiment.

REFERENCES

1. P. May and G. M. Dyson, Chemistry of Synthetic Drugs, 5th Ed., Longmans Green, London, 1959, p. 487.
2. W. Noll, B. Hazzard, and M. Landau, Chemistry and Technology of Silicones, Academic, New York, 1968.
3. S. H. Bauer, in R. E. Kirk and D. F. Othmar (eds.), Encyclopedia of Chemical Technology, Interscience, New York, 1957, Suppl. I, p. 103.
4. H. C. Brown, Boranes in Organic Chemistry, Cornell University Press, Ithaca, N.Y., 1972.
5. H. M. Rozendaal, Arch. Ind. Hyg. Occupat. Med. 4:257 (1951).
6. G. Roush, Jr., Occupat. Med. 1:46 (1959).
7. E. J. King, Biochem. J. 33:944 (1939).
8. A. Henrioul, F. Henrioul, and R. Henrioul, Anal. Falsif. Exp. Chim. 65:274 (1972).
9. J. G. Gonzalez and R. T. Ross, Anal. Lett. 5:683 (1972).
10. A. Fabbrini, G. Modi, L. Signorelli, and G. Simiani, Boll. Lab. Chim. Prov. 22:339 (1971).
11. J. F. Uthe, J. Solomon, and B. Grift, J. Ass. Offic. Anal. Chem. 55:583 (1972).
12. L. R. Kamps and B. McMahon, J. Ass. Offic. Anal. Chem. 55:590 (1972).
13. J. G. Saha and Y. W. Lee, Bull. Environ. Contam. Toxic. 7:301 (1972).
14. I. Skare, Analyst 97:148 (1972).
15. L. Magos, Analyst 96:847 (1971).
16. J. M. Rottschafer, J. D. Jones, and H. B. Mark, Jr., Environ. Sci. Technol. 5:336 (1971).
17. A. Cavallaro and G. Elli, Boll. Lab. Chim. Prov. 22:168 (1971).
18. E. Hauser, P. Holenstein, and M. Nussbaumer, Mitt. Geb. Lebensmittelunters. Hyg. 62:415 (1971).
19. K. Eichner, Lebensmitt. Gerichtl. Chem. 26:240 (1972).
20. M. Malaiyandi and J. P. Barretts, J. Ass. Offic. Anal. Chem. 55:951 (1972).

21. B. Legatowa, Roczn. Panst. Zakl. Hig. $\underline{23}$:429 (1972).
22. M. Margosis and J. T. Tanner, J. Pharm. Sci. $\underline{61}$:936 (1972).
23. C. Corvi, Mitt. Geb. Lebensmittelunters. Hyg. $\underline{63}$:135 (1972).
24. M. E. Hinkle, Prof. Pap., U.S. Geological Survey No. 750-B, B-171-H174 (1971).
25. H. Bader and E. Hedrich, Lebensmitt. Wiss. Technol. $\underline{5}$:178 (1972).
26. S. H. Omang, Anal. Chim. Acta $\underline{63}$:247 (1973).
27. H. Ruf and H. Rohde, Z. Anal. Chem. $\underline{263}$:116 (1973).
28. V. A. Thorpe, J. Ass. Offic. Anal. Chem. $\underline{54}$:206 (1971).
29. K. H. Schaller, P. Strasser, R. Woitowitz, and D. Szadkowski, Z. Anal. Chem. $\underline{256}$:123 (1971).
30. B. B. Mesman, B. S. Smith, and J. O. Pierce, Jr., Amer. Ind. Hyg. Ass. J. $\underline{31}$:701 (1970).
31. L. Magos and T. W. Clarkson, J. Ass. Offic. Anal. Chem. $\underline{55}$:966 (1972).
32. R. C. Rittner and R. Culmo, Anal. Chem. $\underline{85}$:1268 (1963).
33. T. S. Ma and G. Zuazuga, Ing. Eng. Chem., Anal. Ed. $\underline{14}$:280 (1942).
34. E. Abramson and E. Kahane, Bull. Soc. Chim. Fr. $\underline{1948}$:1146.
35. R. D. Strahm and M. F. Hawthorne, Anal. Chem. $\underline{32}$:530 (1960).
36. I. Dunstan and J. V. Griffiths, Anal. Chem. $\underline{33}$:1598 (1968).
37. H. R. Snyder, J. A. Kuck, J. R. Johnson, and V. Campen, J. Amer. Chem. Soc. $\underline{60}$:105, 121 (1938).
38. M. Corner, Analyst $\underline{84}$:41 (1959).
39. S. K. Yasuda and R. N. Rogers, Microchem. J. $\underline{4}$:155 (1960).
40. D. J. Pflaum and H. H. Wenzke, Ind. Eng. Chem., Anal. Ed. $\underline{4}$:392 (1932).
41. O. Schwarzkopf and R. Heinlein, Weight and Development Center Technical Report, 56-19, Part II (1957).
41a. O. Schwarzkopf, private communication, October 1974.
42. A. L. Conrad and M. S. Bigler, Anal. Chem. $\underline{21}$:585 (1949).
43. J. A. Kuck and E. C. Grim, Microchem. J. $\underline{3}$:35 (1959).
44. H. Allen, Jr. and S. Tannenbaum, Anal. Chem. $\underline{31}$:265 (1959).
45. P. Arthur and W. P. Donahoo, U.S. Atomic Energy Committee Report, CCC-1024-TR-221 (1957).
46. R. C. Rittner and R. Culmo, Anal. Chem. $\underline{34}$:673 (1962).
47. A. B. Burg and R. I. Wagner, J. Amer. Chem. Soc. $\underline{75}$:3872 (1953).
48. F. G. A. Stone and A. Burg, J. Amer. Chem. Soc. $\underline{76}$:386 (1954).
49. R. Wickbold and F. Nogel, Angew. Chem. $\underline{69}$:530 (1957); $\underline{71}$:405 (1959).
50. M. Biot, Compt. Rend. $\underline{14}$:49 (1842).
51. J. P. Lorand and J. O. Edwards, J. Org. Chem. $\underline{24}$:769 (1959).
51a. S. Friedman, B. Pace, and R. Pizer, J. Amer. Chem. Soc. $\underline{96}$:538 (1974).
52. F. A. Gooch and L. C. Jones, Amer. J. Sci. $\underline{7}$:23, 147 (1899).

53. F. L. Hahn, Anal. Chem. 33:316 (1961).
54. F. L. Hahn, Compt. Rend. 197:762 (1933).
55. F. J. Foote, Ind. Eng. Chem., Anal. Ed. 4:39 (1932).
56. J. R. Martin and J. R. Hayes, Anal. Chem. 24:182 (1952).
57. D. S. Taylor, J. Ass. Offic. Agr. Chem. 33:132 (1950).
58. R. M. Carlson and J. L. Paul, Anal. Chem. 40:1292 (1968).
59. W. J. Schule, J. F. Hazel, and W. N. McNabb, Anal. Chem. 28:505 (1956).
60. F. J. Langmyhr and O. B. Skaar, Anal. Chim. Acta 25:262 (1961).
61. F. W. Lima and C. Pagano, Analyst 85:909 (1960).
62. H. Gilman, R. N. Clark, R. E. Wiley, and H. Diehl, J. Amer. Chem. Soc. 68:2728 (1946).
63. J. C. B. Smith, Analyst 85:465 (1960).
64. E. G. Rochow, Chemistry of the Silicones, 3rd Ed., Wiley, New York, 1947.
65. J. A. McHard, P. C. Servais, and H. C. Clark, Anal. Chem. 20:325 (1948).
66. E. Debal, Talanta 19:15 (1972).
67. A. J. Christopher and T. R. F. W. Fennell, Talanta 12:1003 (1965).
68. W. H. Greive and K. F. Sporek, reported at the American Chemical Society Meeting, Phoenix, Ariz., 1966.
69. J. H. Wetters and R. C. Smith, Anal. Chem. 41:379 (1969).
70. M. I. Wolinetz, Ukr. Khim. Zhr. 11:18 (1936).
71. G. Ingram, Methods of Organic Elemental Analysis, Reinhold, New York, 1962, p. 363.
72. P. Bartusek, Textil. 28:51 (1973).
73. R. W. Morrow and J. A. Dean, Anal. Lett. 2:133 (1969).
74. G. Ingram, Methods of Organic Elemental Analysis, Reinhold, New York, 1962, p. 330.
75. L. T. Hallet, Ind. Eng. Chem., Anal. Ed. 14:956 (1942).
76. D. L. Tabern and E. F. Shelberg, Ind. Eng. Chem., Anal. Ed. 4:401 (1932).
77. H. A. Sloviter, W. M. McNabb, and E. C. Wagner, Ind. Eng. Chem., Anal. Ed. 13:890 (1941).
78. C. S. Yeh, Microchem. J. 14:279 (1969); private communication, October 1974.
79. A. A. Abramyan and M. A. Gevorkyan, Arm. Khim. Zh. 22:128 (1969).
80. V. H. Chambers, E. R. Cropper, and H. Crossley, J. Sci. Food Agr. 7:17 (1956).
81. British Pharmacopeia, the Pharmaceutical Press, London, 1958.
82. K. F. Sporek, Analyst 81:474 (1956).
83. W. E. Horwitz (ed.), Official Methods of Analysis of the AOAC, 12th Ed., Association of Official Analytical Chemists, Washington, D.C., 1975.

84. A. K. Klein, J. Ass. Offic. Agr. Chem. 35:537 (1952).
85. L. Kotz, G. Kaiser, and G. Tölg, Z. Anal. Chem. 260:207 (1972).
86. A. I. Busev and L. I. Teternikov, Zh. Anal. Khim. 23:1886 (1968).
87. J. Holzbecher and D. E. Ryan, Anal. Chim. Acta 64:333 (1973).
88. E. Sandi, K. Soos, R. Liebmann, and A. Hellvig, Chem. Tech.
     (Berlin) 22:557 (1970).
89. S. Nishi and Y. Horimoto, Japan Analyst 17:1247 (1968).
90. R. Belcher, A. M. G. Macdonald, and T. S. West, Talanta 1:408
     (1958).
91. R. D. Strahm, in I. M. Kolthoff and P. J. Elving (eds.), Treatise
     on Analytical Chemistry, Part II, Vol. 12, Wiley, New York, 1965,
     p. 169.
92. P. Hansen, Tidsskr. Plavl. 76:653 (1972).
93. S. I. Obtemperanskaya, T. K. Pham, D. H. Nguyen, and I. V.
     Karandi, Zh. Anal. Khim. 27:1421 (1972).
94. H. J. Horner, in I. M. Kolthoff and P. J. Elving (eds.), Treatise
     on Analytical Chemistry, Part II, Vol. 12, Wiley, New York, 1965,
     p. 241.
95. J. H. Austin, R. W. Rinehart, and E. Baill, Microchem. J. 17:670
     (1972).
96. E. J. King, B. D. Stacey, P. F. Holt, D. M. Yates, and D. Pickles,
     Analyst 80:441 (1955).
97. E. W. Bretthauer, A. A. Moghissi, S. S. Snyder, and N. W.
     Mathews, Anal. Chem. 46:445 (1974).
98. S. I. Obtemperanskaya and V. N. Likhosherstova, Vestn. Mosk.
     Univ. 1960:57.
98a. A. Mazzeo-Farina, Farmaco, Ed. Sci. 28:937 (1973).
99. E. Debal and R. Levy, Bull. Soc. Chim. Fr. 1969:79.
100. E. Celon and S. Bresadola, Gazz. Chim. Ital. 100:549 (1970).
101. A. Kaczmarczyk, J. R. Messer, and C. E. Peirce, Anal. Chem.
     43:271 (1971).
101a. M. Skalicky, H. Plotova, and M. Filip, Chem. Prum. 23:452
     (1973).
102. D. G. Shaheen and R. S. Braman, Anal. Chem. 33:893 (1961).
103. A. A. Abramyan, M. A. Gevorkyan, and R. S. Sarkisyan, Arm.
     Khim. Zh. 22:668 (1969).
104. T. Yoshizaki, Anal. Chem. 35:2177 (1963).
105. R. A. Shah, A. A. Qadri, and R. Rehana, Pakistan J. Sci. Ind.
     Res. 8:282 (1965).
106. S. Kato, K. Kimura, and Y. Tsuzuki, J. Chem. Soc. Japan, Pure
     Chem. Sect. 83:1039 (1962).
107. T. M. Shanina, N. E. Gel'man, and V. S. Mikhailovskaya, Zh.
     Anal. Khim. 22:782 (1967).

108. R. A. Moore, Proc. Soc. Anal. Chem. 9:35 (1972).
109. L. A. Fedotova and M. G. Voronkov, Zh. Anal. Khim. 22:1431 (1967).
110. D. Monnier, C. A. Menzinger, and M. Marcantonatos, Anal. Chim. Acta 60:233 (1972).
111. M. Wronski, Chem. Anal. (Warsaw) 8:299 (1963).
112. H. C. Kelley, Anal. Chem. 40:240 (1968).
113. F. Sefidvash, Anal. Chem. 40:1165 (1968).
114. J. C. M. Pau, E. E. Pickett, and S. R. Koirtyohann, Analyst 97:86 (1972).
115. A. Selecki and Z. Nowakowska, Radiochem. Radioanal. Lett. 1:247 (1969).
116. A. P. Terent'ev, S. V. Syatsillo, and B. M. Luskina, Zh. Anal. Khim. 16:83 (1961).
117. L. V. Myshlyaeva and V. V. Krasnoshchekov, Tr. Mosk. Khim.- Tekhnol. Inst. D. I. Mendeleeva 1963:178.
117a. P. Bartusek, Textil. 28:51 (1973).
118. A. P. Kreshkov, L. V. Myshlyaeva, O. B. Khachaturyana, and V. V. Krasnoschchekov, Zh. Anal. Khim. 18:1375 (1963).
119. A. P. Kreshkov, L. V. Myshlyaeva, and V. V. Krasnoshchkov, Plast. Massy 1962:51.
119a. G. V. Andreeva, Zavod. Lab. 39:1341 (1973).
120. J. Jenik and M. Jurecek, Collect. Czech. Chem. Commun. 26:967 (1961).
121. T. M. Shanina, N. E. Gel'man, and L. M. Kiparenko, Zh. Anal. Khim. 20:118 (1965).
122. A. P. Terent'ev, E. A. Bondarevskaya, N. A. Gradskova, and E. D. Kropotova, Zh. Anal. Khim. 22:454 (1967).
123. J. E. Burroughs, W. G. Kator, and A. I. Attia, Anal. Chem. 40:657 (1968).
124. A. P. Terent'ev, E. A. Bondarevskaya, and N. A. Gradskova, Zh. Anal. Khim. 24:753 (1969); 25:196 (1970).
125. H. Kadner and C. Biesold, Pharmazie 27:187 (1972).
125a. N. A. Gradskova and E. A. Bondarevskaya, Zh. Anal. Khim. 29:1237 (1974).
126. A. P. Terent'ev, N. A. Gradskova, E. A. Bondarevskoya, R. N. Potsepkina, and O. D. Kuleshova, Zh. Anal. Khim. 26:1850 (1971).
127. V. A. Klimova and M. D. Vitalina, Tr. Komis, Anal. Khim. Akad. Nauk SSSR 13:17 (1963).
128. V. Bilik, I. Jezo, and L. Stankovic, Chem. Zvesti. 18:688 (1964).
129. R. Reverchon and Y. Legrand, Chim. Anal. 47:194 (1965).
129a. A. Radecki and H. Lamparczyk, Chem. Anal. (Warsaw) 19:457 (1974).

130. V. Prey, H. Teichmann, and D. Bichler, Mikrochim. Acta
      1970:138.
130a. I. B. Peetre and B. E. F. Smith, Mikrochim. Acta 1974:301.
131. A. P. Kreshkov, L. V. Myshlyaeva, E. A. Kuchkarev, and T. G.
      Shatunova, Zh. Anal. Khim. 20:1325 (1965).
131a. N. A. Makulov, D. Y. Zhinkin, N. A. Gradskova, and O. P.
      Trokhachenkova, Zavod. Lab. 41:180 (1975).
132. T. S. Ma, in F. J. Welcher (ed.), Standard Methods of Chemical
      Analysis, 6th Ed., Vol. 2, Van Nostrand, Princeton, 1963, p. 403.
133. F. Martin and A. Floret, Bull. Soc. Chim. Fr. 1960:610.
134. A. Kondo, Japan Analyst 10:658 (1961).
135. V. Pechanec and J. Horacek, Collect. Czech. Chem. Commun.
      27:239 (1962).
136. T. Mitsui, K. Yoshikawa, and Y. Sakai, Microchem. J. 7:160
      (1963).
137. V. Pechanec and J. Horacek, Chim. Anal. 46:457 (1964).
138. D. H. Anderson, J. H. Evans, J. J. Murphy, and W. W. White,
      Anal. Chem. 43:1511 (1971).
139. R. J. Thomas, R. A. Hagstrom, and E. J. Kuchar, Anal. Chem.
      44:512 (1972).
140. R. Donner, Z. Chem. 5:466 (1965).
141. G. Cumont, Chim. Anal. 53:634 (1971).
142. A. A. Abd, E. Raheem, and M. M. Dokhana, Z. Anal. Chem.
      180:339 (1961).
143. S. Kinoshita, Microchem. J. 8:79 (1964).
143a. A. Warshawsky and S. Ehrlich-Rogozinski, Microchem. J. 22:362
      (1977).
144. T. Shibazaki and M. Koibuchi, J. Pharm. Soc. Japan 88:140 (1968).
145. S. Erlich-Rogozinsky and R. Sperling, Anal. Chem. 42:1089 (1970).
145a. T. J. Rohm, H. C. Nipper, and W. C. Purdy, Anal. Chem. 44:869
      (1972).
146. T. C. Rains and O. Menis, J. Ass. Offic. Anal. Chem. 55:1339
      (1972).
147. G. Lindstedt and I. Skare, Analyst 96:223 (1971).
148. K. K. S. Pillay, C. C. Thomas, Jr., J. A. Sondel, and C. M.
      Hyche, Anal. Chem. 43:1419 (1971).
149. H. Jerie, Mikrochim. Acta 1970:1089.
150. L. Lopez-Escobar and D. N. Hume, Anal. Lett. 6:343 (1973).
151. R. H. Filby, A. I. Davis, K. R. Shah, and W. A. Holler, Mikrochim.
      Acta 1970:1130.
152. L. Manger, Kenrija Ind. 14:317 (1965).
153. R. H. Pierson, Anal. Chem. 34:1642 (1962).
154. R. K. Manns and D. C. Holland, J. Ass. Offic. Anal. Chem. 54:202
      (1971).
155. T. S. Ma and A. A. Benedetti-Pichler, Anal. Chem. 25:999 (1953).
156. H. V. Malmstadt and C. B. Roberts, Anal. Chem. 28:1408 (1956).
157. J. Volhard, J. Prakt. Chem. 8:217 (1874).

The Metallic Elements

9.1 GENERAL CONSIDERATIONS

In this chapter we discuss the methods and techniques for the determination of metallic elements, excepting mercury, that are present in organic compounds and materials. The analysis of mercury is treated separately (see Chap. 8) because it is a liquid at ordinary temperatures, volatile, and hence not amenable to the general procedures described for the determination of metals.

Broadly speaking, organic elemental analysis that involves the determination of metallic elements can be classified into two categories. In one category, the metal is linked to the carbon atom through an oxygen, nitrogen, or sulfur atom. These organic compounds are salts (e.g., copper acetate, sodium phenolate) or addition products (e.g., amine chloroplatinate, aldehyde, sodium bisulfite) in which the metallic element retains its ionic property. In another category, the organic molecule possesses the direct metal-carbon bond. These compounds are known as organometallics. While organometallic reagents, notably the Grignard reagents, have been employed by organic chemists since the turn on the century, their analytical methods became important only during the past decade. The discovery of ferrocene in 1951 spurred the investigation of organometallics. The scientific interest and industrial uses have greatly expanded this field. Now there are periodicals [1, 2] specializing in organometallic chemistry. Treatises on organometallic compounds continue to appear [3, 4]. The annual reviews [5, 6] cite hundreds of research papers yearly. Monographs [7, 8] have been prepared on the analytical aspects of this class of organic compounds. It should be noted that compounds containing boron, silicon, and arsenic are also considered as organometallics. The methods for the determination of boron and silicon are discussed in Chap. 8 and those for arsenic are described in Chap. 7.

When the metallic element present in the organic compound is in the ionic form, it can be determined using the conventional inorganic analytical methods for the particular metal, provided that the organic portion of the molecule does not interfere with the procedure. Interestingly, however, this is not the common practice in the organic analysis laboratory. As a rule, the metal (e.g., sodium, platinum) is determined by mineralization of the organic sample and weighing the residue (see Secs. 9.4 and 9.5). On the other hand, analysis of organometallics usually involves destructive decomposition (see Sec. 9.2C) as a preliminary step, followed by the determination of the metallic element in question.

9.2 CURRENT PRACTICE

A. The Ashing Technique

The ashing technique for the determination of metals consists of heating the organic sample at high temperatures with or without a reagent. All volatile products (i.e., oxides of carbon, hydrogen, nitrogen, sulfur; halogens) are expelled, whereby the metallic element is recovered in the residue in a suitable weighable form. If the metal is stable in the free state, it is obtained as such. For instance, platinum in an amine chloroplatinate can be determined by heating in air, or with a little nitric acid:

$$\text{Amine chloroplatinate} \xrightarrow{\Delta} \text{Pt}$$

If the metal sulfate is stable, the sample is heated with sulfuric acid. This method is applicable to the alkali and alkaline earth metals. Thus, sodium is weighed as sodium sulfate:

$$\text{Sodium salt} \xrightarrow[\Delta]{H_2SO_4} Na_2SO_4$$

For metals which form stable oxides but not sulfates, the organic sample is heated with nitric acid. For example, copper is determined as copper oxide:

$$\text{Copper salt} \xrightarrow[\Delta]{HNO_3} CuO$$

The ashing method is recommended for the determination of the percent composition of a known metal in the pure organic compound. It has the advantage of being simple and not requiring special reagents and additional experimental operations to determine the metal ion. This method is also employed to measure the amount of inorganic matter in an organic sample [9].

The ashing technique can be conveniently performed in a microcrucible (see Sec. 9.4) or in a combustion tube using a microboat (see Sec. 9.5). The latter procedure permits passage of different gases in the heating process. For instance, nickel does not form an oxide of definite composition and hence cannot be weighed as the oxide residue. The determination can be accomplished, however, by first combusting the organic sample with nitric acid in the presence of oxygen and then, after cooling and displacing the atmosphere with hydrogen, by reducing the mixed nickel oxides to metallic nickel and weighing.

A closed-tube oxygen combustion apparatus is commercially available [10-10b]. As shown in Fig. 9.1, it is designed for continuous operation at 900°C. The combustion chamber is 32 mm in diameter and 200 mm long; its mouth is ground to take tapered Teflon plugs of the oxygen charging head and sample carriers. Recently the plasma ashing technique has been advocated, and the apparatus for this is now commercially available [10c].

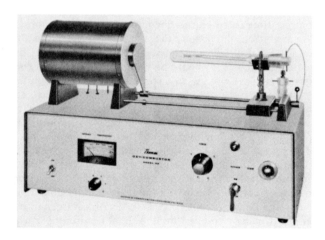

FIG. 9.1. Closed-tube combustion apparatus and sample carrier. (From Ref. 10, courtesy A. H. Thomas Co.)

B. Analysis of Specific Elements

It is apparent that the ashing techniques discussed in the previous section
do not differentiate the metals. Their value in the quantitative analysis of
metals is dependent on the organic compound being pure and containing
only one metallic element. Since organometallics are seldom obtained in
high purity, their metal contents are not determined by weighing the residue after ashing. In fact, ashing is not the recommended procedure for
the mineralization of organometallics (see Sec. 9.2C).

       In order to determine a specific metal in the organic sample, a suit-
able method should be selected which is applicable to the problem. Colorimetry, complexometric titration, and atomic absorption spectroscopy
are the common procedures, as indicated in Table 9.1 which was prepared
by Schwarzkopf and Schwarzkopf [11]. The reader is also referred to the
extensive list of chelatometric methods compiled by Reilley and Barnard
[12]. It is beyond the scope of this chapter to discuss the individual metals.
Needless to say, a separation operation should be incorporated in the
analytical procedure when more than one metallic element is present in
the organic sample, unless the respective metals can be analyzed without
mutual interference.

C. Decomposition Techniques for Organometallics

There is copious literature on the decomposition of organometallic compounds. For analytical purposes, the methods may be classified into three
groups, namely, acid digestion, pyrolysis, and hydrolysis; they are separately discussed below. It should be noted that the preferred decomposition technique varies from case to case and is dependent on the nature of
the sample, characteristics of the organic moiety, and the metallic element
to be determined. Organic samples containing osmium merit special attention in view of the fact that osmium tetraoxide is volatile. Therefore osmium compounds are decomposed by heating in a combustion tube in the
presence of oxygen, and OsO_4 is collected in ice water [11]. The resulting
solution is treated with potassium iodide, whereupon osmium(VIII) is reduced to osmium(IV), liberating four atoms of iodine. The iodine is extracted into carbon tetrachloride and the absorbance measured. According
to Schwarzkopf and Schwarzkopf [11], OsI_4 exhibits a green color in aqueous solution but it cannot be used for the direct colorimetric determination
of osmium.

TABLE 9.1. Methods for Determining Specific Metals.

Elements	Atomic absorption	Complexometric titration	Colorimetry	Other methods
Alkali metals (Li, Na, K, Rb, Cs)	x			
Be	x		x	
Mg, Ca, Sr, Ba	x	x		
Al	x	x	x	
Ga, In, Te	x	x		
Sn, Pb, Bi	x	x		
Ti	x	x	x	
Zr, Hf	x	x	x	
V	x	x	x	Hydroquinone, kinetic
Nb, Ta			x	
Cr	x	x	x	Hydroquinone
Mo	x	x	x	
W			x	
Mn	x	x		
Tc, Re			x	
Te	x	x	x	
Ru, Os			x	
Co	x	x	x	
Rh			x	
Ir			x	Hydroquinone
Ni	x	x	x	Gravimetry
Pd	x		x	Gravimetry
Pt			x	Gravimetry

TABLE 9.1. (Continued)

Elements	Atomic absorption	Complexometric titration	Colorimetry	Other methods
Cu	x	x	x	
Ag	x			Gravimetry
Au	x		x	Hydroquinone
Zn, Cd	x	x		
Hg	x	x	x	Carbamate titration
Rare earths	x	x	x	Hydroquinone (Ce)

(From Ref. 11, courtesy J. Wiley & Sons, Inc.)

Acid Digestion

Digestion of the organic substance in mineral acids is the technique most frequently employed as a preliminary step in the determination of metals. Nitric acid and sulfuric acid, either singly or together, are used as oxidizing agents. Sometimes hydrogen peroxide or perchloric acid is added to enhance the oxidizing power of the medium, while hydrochloric acid is incorporated to provide the required anion for the subsequent step of determination. The nonmetallic elements of the sample are oxidized to gaseous products, e.g., carbon dioxide, halogens, nitrogen oxides, and sulfur oxides.

When the boiling temperature of the acid medium is sufficient for the decomposition of the organometallic compound, the micro Kjeldahl flask (see Sec. 2.3) is a convenient reaction vessel. If the volume of solution is large, a conical flask may be employed. The latter can serve as the titration vessel subsequently, thus eliminating the need of transfer after digestion.

If the decomposition reaction requires high temperatures, or if the organometallic sample is a volatile liquid, the sealed-tube technique (see Sec. 5.1B) is employed. In Table 9.2 are listed the condition and acids recommended by Schwarzkopf and Schwarzkopf [11] for the decomposition of organic substances containing the various metallic elements. It should be noted that some metals may produce precipitates after the decomposition of the organic matter. For example, SnO_2 and TiO_2 are insoluble in nitric acid, while the platinum group of metals precipitate as the free metal in this medium. Then these decomposition products again should be brought into solution for the determination of the metal ions.

TABLE 9.2. Decomposition of Organometallics by Acid Digestion.

Elements	Acids used	
	Open tube	Sealed tube
Li, Na, K, Rb, Cs	H_2SO_4, HNO_3, $HClO_4$	Not required
Be	H_2SO_4, (HNO_3, $HClO_4$) and fuming	HNO_3 + H_2SO_4, followed by fuming
Mg, Ca, Sr, Ba	H_2SO_4 or HNO_3/HCl	
Al, Ga, Ir, Ti	H_2SO_4, HNO_3, $HClO_4$	
Sn	HNO_3 + HCl, HNO_3 + $HClO_4$	HNO_3 + HCl, HCl + $HClO_4$
Pb	HNO_3 + H_2SO_4	
Bi	HNO_3 + H_2SO_4	
Ti	HNO_3 + H_2SO_4 + $HClO_4$	HNO_3 + HCl
V	HNO_3 + H_2SO_4	HNO_3 + HCl
Cr, Mn, Fe, Co, Ni, Cu, Zn	HCl + $HClO_4$ + HNO_3	
Zr, Nb, Hf, Ta	HNO_3 + H_2SO_4 + HCl	HNO_3 + HCl
Ru, Ir	H_2SO_4	
Rh, Pd	H_2SO_4 - HNO_3	HNO_3/HCl, HCl/$HClO_4$, $KHSO_4$ + KCl
Ag, Cd	HNO_3	HNO_3
W, Re, Mo, Te	HNO_3 + H_2SO_4	HNO_3
Pt, Au	HNO_3 + HCl	HNO_3 + HCl($HClO_4$)
Rare earth	HNO_3	HNO_3

(From Ref. 11, courtesy J. Wiley & Sons, Inc.)

Sometimes an HNO_3/HCl mixture facilitates the solution of the metallic elements like platinum, tin, and titanium, but oxidizes incompletely the organic portion of the molecule. After the metal is dissolved by digestion in a sealed tube, the oxidation has to be pursued further with a stronger oxidizing agent, for example [11], in an open tube with H_2SO_4/HNO_3/$HClO_4$.

For the decomposition of organoberyllium compounds, after digestion with nitric acid, the residue should be treated with sulfuric acid and heated to fuming in order to remove the last traces of nitric acid and nitrous acid. This operation is necessary because the nitrogen acids interfere with the determination of beryllium by a method which is based on the measurement of the ultraviolet absorption of its salicylic acid complex [13].

Organoruthenium compounds must be digested with sulfuric acid only. The resulting RuO_4 is distilled into a solution of alkaline hypochlorite. Perruthenate containing ruthenium(VII) is formed and the color of this solution is measured [14]. If nitric or perchloric acid were used, nitrous and chloro complexes, respectively, would be produced and would interfere with the determination.

Combustion in an Oxygen Atmosphere

Combustion of organometallics in an atmosphere of oxygen serves to oxidize the organic moiety to gases and leave the metallic elements as solid residues which are then brought into solution with a suitable reagent. In view of the popularity of the closed-flask combustion technique (see Sec. 5.1B), it is natural that many analysts have employed this method for the decomposition of organometallic materials. This method, if applicable, has the advantage of being simple and rapid. Macdonald and co-workers [15-17], Schwarzkopf and Schwarzkopf [11], and Bishara and co-workers [18-20] have reported on their studies of the closed-flask combustion technique on organometallics. The consensus is that the closed-flask decomposition is effective for the analysis of many, but not all, metallic elements. For example, it has been found [16] that lead, bismuth, and iron form alloys with platinum in the ignition basket or adhere very strongly to other supports. Nickel and aluminum produce oxides which are exceedingly difficult to dissolve. Understandably, glass vessels cannot be used for the determination of alkali metals, whereupon Schöniger [21] has proposed combustion in cooled polyethylene flasks.

A serious limitation of the closed-flask combustion for the analysis of metals in organic materials lies in the fact that it is nearly impossible to control the temperature and duration of the oxidation reactions. Unlike halogen or sulfur compounds, most organometallics do not burn to produce a flame, hence it is difficult to assure complete combustion which is essential for quantitative analysis. For this reason, decomposition of the sample in the combustion tube (see Fig. 9.1 and Sec. 9.5) is recommended in certain cases. This technique can be performed in a static state of oxygen atmosphere or by passing a current of oxygen through the system during the combustion. The residue in the microboat can be weighed and returned for further heating if necessary.

Decomposition by Hydrolysis

For organometallic compounds that react readily with water, such as magnesium alkyl iodide and sodium triphenylmethyl, hydrolysis provides a simple and convenient method of decomposition. The metallic elements are thus separated from the organic portion of the molecule. Then the metal ions in solution can be analyzed directly, if the presence of the organic moiety does not interfere with the method of determination. The hydrolysis can be performed in dilute acidic or alkaline medium.

D. Difficulties and Complications

A characteristic phenomenon in the analysis of metal salts and organometallics is that the material does not disappear completely after combustion. When the decomposition is effected without a liquid reagent, the sample is usually charred and leaves a residue. If the metal produces a refractory substance like carbide or nitride, the product which forms on the surface of the sample will keep the oxygen from coming into contact with the center core. The decomposition is then incomplete.

Many organometallic compounds are sensitive to moisture, oxygen, and carbon dioxide. Difficulty is usually encountered while the sample is being quantitatively measured in the initial stage of the analytical procedure. In extreme cases, therefore, weighing and transfer of the sample should be carried out in a dry box [11, 22] filled with purified nitrogen.

Some organometallics exhibit high vapor pressures. In fact, a number of metal-carbon bonded compounds are specifically synthesized for this feature, and the volatility of the organometallic compound is utilized for practical purposes. When these compounds are subjected to quantitative analysis, close attention should be given to the decomposition procedure. This is particularly important in the closed-flask and combustion-tube methods. When the acid digestion technique is employed, it is prudent to attach a reflex condenser to the digestion flask. In this way the gaseous oxidation products like CO_2 can escape while the unchanged organometallic compound will be returned to the acid solution.

As to the methods of finish, the reader is referred to the treatises on inorganic analysis. For the determination of a particular metallic element, attention is called to the possible interference due to the presence of certain anions and extraneous cations.

9.3 SURVEY OF RECENT LITERATURE

The reader who wishes to cover the old literature on the determination of metallic elements in organic materials is referred to the reviews written

by Belcher and co-workers [23], Sykes [24], Gorsuch [24a], Schwarzkopf
and Schwarzkopf [11], and Ingram [25]. The last-mentioned reference also
describes the procedures for the determination of over 30 metals. Publications which have appeared during the past decade are summarized in
Table 9.3. It is worthy of note that a large portion of the current literature
is found in Russian journals. This seems to reflect the research activities
on organometallics in Russia. On the other hand, the paucity of American
papers does not indicate a lack of interest in metal analysis in the United
States. According to our personal knowledge, many studies have been
undertaken in this country. However, these investigations were not published because of the policies of the industrial management or government
agencies. It is apparent that many organometallic compounds have commercial potential or are related to national defense. In consequence, the
preparation of these compounds and the techniques for analyzing them are
frequently considered as classified information. Hence many research
reports are still not available for public dissemination.

A. Methods of Decomposition

As can be seen in Table 9.3, acid digestion is the preferred method recently reported for the decomposition of organic materials containing metallic elements. The reaction is usually performed in an open vessel.
Closed-flask combustion has been described for certain metals. In contrast to the decomposition techniques for nonmetals and metalloids, fusion
in a metal bomb is rarely used in the analysis of metals.

    Jenik [66] has reviewed the methods for mineralization of organome-
tallic sandwich-type compounds. These compounds may contain iron or
cobalt. For iron compounds, another metal such as aluminum, bismuth,
mercury, or titanium may be present.

    Goldstein [67] has investigated the methods for the decomposition of
metal naphthenates. It has been established that when the metallic element
is calcium, cobalt, lead, manganese, zinc, or zirconium, the naphthenate
can be completely hydrolyzed in the cold with 3 $\underline{N}$ nitric acid. The naph-
thenic acid is removed by filtration and the metal ions are then determined
in the aqueous solution.

    Quantitative decomposition of the organic sample containing trace
amounts of metallic elements is often confronted with difficulties. For
instance, Kaiser and co-workers [33] have reported that previously published methods for the decomposition of beryllium compounds always cause
significant losses at the nanogram level of beryllium. These samples,
however, can be decomposed by heating in a sealed Teflon tube containing
a mixture of about two parts of 70% nitric acid and one part of 40% hydrofluoric acid for 3 h. Alternatively, organic materials can be mineralized

TABLE 9.3. Summary of Recent Literature on the Determination of
Metallic Elements in Organic Materials.

Element	Decomposition method	Finish technique	Reference
Al	Acid digestion	Titrimetric	26–30
	Metal bomb	Titrimetric	31
	Metal bomb	Spectroscopy	32
Al, Cd, Co, Cu, Fe, Mg, Mn, U, Zn	Closed flask	Polarography	19, 20
Al, Bi, Cu, Ni	Closed flask	Polarography	19a
Al, Fe, Ni, Pb	HCl	Polarography	20
Ba, Ca, Co, Mg, Mn, Zn	Closed flask	Titrimetric	17
Ba, Ca, Zn	Closed flask	Polarography	20a
Be	$HNO_3 + HF$	Gas chromatography	33
Bi, Cu, Fe, Ni	Acid digestion	Titrimetric	17
Ca, Cu, Fe, Mn	Spark	Emission spectroscopy	34
Cd, Mg, U, Zn	Closed flask	Gravimetric, titrimetric	18
Co, Ni, Sb	Closed flask	Polarography	18a
Cr	Acid digestion	Spectrophotometric	35
Cr, Cu, Mn, Sb	Acid digestion	Coulometric	36
Cs	Ashing	Radiometric	37
Cu	Acid digestion	Spectrophotometric	38
	Acid digestion	Titrimetric	39
Fe	Acid digestion	Titrimetric	40, 41
	Acid digestion	Polarographic	41a
	Metal bomb	Spectrophotometric	42
Ge	Metal bomb	Titrimetric	43
	Metal bomb	Spectrophotometric	43a
	Closed flask	Spectrophotometric	44–44b

TABLE 9.3. (Continued)

Element	Decomposition method	Finish technique	Reference
Ge	Closed flask	Polarography	45
	Closed flask	X-ray fluorescence	46
K	Acid digestion	Titrimetric	47, 48
Li	Acid digestion	Titrimetric	49
Mn	Acid digestion	Polarography	50
	Acid digestion	Titrimetric	51
	Closed flask	Titrimetric	17
Na	Hydrolysis	Titrimetric	52
	Fusion with Mg	Titrimetric	53
	Fusion with Mg	Gravimetric	53
Pb	Acid digestion	Gravimetric	54, 55
	Metal bomb	Atomic absorption	55a
Pd	Metal bomb	Gravimetric	56
Pt	Acid digestion	Atomic absorption	56a
Re	Closed flask	Polarography	57
Sb	Acid digestion	Titrimetric	57
Sn	Acid digestion	Titrimetric	30, 58-60
	Acid digestion	Polarography	61
	Closed flask	Titrimetric	62
	Flame	Atomic absorption	62a
		X-ray fluorescence	62b
Ti	Acid digestion	Titrimetric	27, 30, 41, 63
	Spark	Spectroscopy	63
	Ashing	Gravimetric	63
Tl	Acid digestion	Polarography	64
Zr	Acid digestion	Titrimetric	65

by low-temperature ashing [68,69] which involves the application of oxygen plasma generated by microwave irradiation. The last-mentioned technique is expected to gain popularity as the instruments become generally available. According to Hozumi [69a], the chief drawback of the oxygen plasma decomposition technique is incomplete recovery of the metals.

B. Methods of Separation and Determination

Since most of the current demands on the determination of metallic elements in organic materials are concerned with impure substances, the analytical procedures by necessity involve methods which are discriminative with respect to the metals in question. When more than one metal is present in the matrix after decomposition of the sample, a separation step may be incorporated in order to remove the interfering cations. The ideal approach, of course, is to find a method which is specific or selective for the particular metal to be determined.

Perusal of the recent literature reveals that chelatometric titrimetry for the metal ions has received most attention. The reagent EDTA (ethylenediamine tetracetic acid, also called ethylene dicyanotetraacetic acid) is still the favorite. It forms useful complexes with many metals. By applying the technique of masking, certain metals may be determined in the presence of other metallic elements, thus avoiding the tedium of separation [70,71]. The technique of EDTA titration has been extended to the alkali metals which cannot be determined by direct complexation. For instance, de Sousa [72] has described a method to determine potassium by indirect chelometry as follows. The potassium ions are precipitated with perchlorate, and the precipitate is converted to chloride by heating with ammonium chloride. The potassium chloride is then dissolved, silver nitrate is added, and the silver chloride precipitate is caused to react with $K_2Ni(CN)_4$ in solution, liberating Ni^{2+} ions which are titrated with standardized EDTA solution using murexide as the indicator.

The electroanalytical methods proposed for metal determinations include polarography and coulometric titration. Germanium [45], manganese [50], and thallium [53] have been determined by measurement of the current at half-wave potentials of -1.34 V, -0.5 V, and -0.8 V, respectively. An indirect polarographic method proposed by Bishara and co-workers [19,20] is applicable to the determination of aluminum, cadmium, cobalt, copper, iron, lead, magnesium, manganese, nickel, uranium, and zinc. It is based on the quantitative precipitation of the metal oxinates by means of 0.025 $\underline{M}$ 8-hydroxyquinoline (oxine), followed by the polarographic determination of the unconsumed reagent. Coulometry has been suggested by Bigois [36] for the determination of antimony, chromium, copper, and manganese. After decomposition of the organic matter, copper is obtained as Cu^{2+} while antimony is further oxidized to $SbO_4{}^{3-}$ by means of potassium permanganate. Both ions are titrated with I^- which is electrolytically generated. For the determination of chromium and manganese, these elements are converted to $Cr_2O_7{}^{2-}$ and $MnO_4{}^-$ with perchloric acid and ammonium persulfate, respectively. The dichromate or permanganate is then titrated with Fe^{2+} which is produced at constant current from an Fe^{3+} solution.

Emission spectroscopy is a general method for the analysis of metallic elements. For instance, Ng and Lai [34] have described the determination of calcium, copper, iron, and manganese in natural rubber by this technique. It has the advantage of simultaneous measurement of the spectra of several elements with one sample (see Chap. 10). This is also true for atomic absorption [73] and atomic fluorescence [74]. For quantitative analysis of metals in organic materials, however, it is sometimes difficult to work out a reliable procedure based on spectroscopy.

C. Comparison of Methods

Considerable interest has been directed to the investigation of the decomposition techniques for organic samples containing metallic elements, and the findings are sometimes controversial. Using pure compounds for testing, Macdonald and Sirichanya [17] reported that barium, cobalt, manganese, and zinc presented no problems in the closed-flask combustion method. However, when compounds containing aluminum, copper, or nickel were burned, the oxides formed could not be readily dissolved. Addition of sodium carbonate, bicarbonate, hydroxide, or peroxide, or of potassium nitrate or bisulfate provided only slight improvements; only treatment with concentrated acid dissolved the oxides. Therefore, these workers have advocated acid digestion for aluminum, copper, nickel, and also for bismuth, iron, or lead compounds because the latter group of metals formed alloys [16] even when silica sample holders were used. Contrastingly, Terent'ev and co-workers [75] have recommended closed-flask combustion for the analysis of organocopper compounds. Bishara and co-workers [20] have confirmed this. Besides, the latter investigators also claimed that aluminum, iron, lead, or uranium compounds gave correct results by closed-flask combustion when a Pyrex glass spiral was employed.

Whereas the evaluation of analytical methods for determining specific organometallics has been conducted undoubtedly in the industrial laboratories, very few reports have appeared in the journals. As an example, Kreshkov and co-workers [63] have determined titanium in complex siloxanes by complexometry, spectroscopy, and gravimetry, and found that the three methods were of equal accuracy. Kovac et al. [64a] have compared the determination of calcium by atomic absorption spectroscopy, after closed-flask combustion, with the gravimetric and spectrophotometric methods after combustion in a stream of oxygen.

9.4 EXPERIMENTAL: DETERMINATION OF METALLIC ELEMENTS BY ASHING IN A CRUCIBLE

A. Principle

The organic material containing metallic elements is decomposed by oxidation in a microcrucible after the addition of sulfuric or nitric acid, depending on the nature of the metal present (see Sec. 9.2A). Strong heating of the reaction mixture removes the volatile matter, and the ash left behind is weighed as the free metal, metal oxide, or metal sulfate [76].

B. Equipment

Microcrucible, platinum or porcelain, 1- to 2-ml capacity.
Crucible lids, large and small (see Fig. 9.2).

C. Reagents

Dilute nitric acid (1:1).
Dilute sulfuric acid, 20%.

D. Procedure

Clean and accurately weigh the microcrucible together with the small crucible lid. Add the sample (containing 0.5-2.0 mg of metallic element) to the bottom of the crucible and reweigh. Place the crucible on the large crucible lid (Fig. 9.2) which rests on a silica triangle. Cautiously deliver about 20 μl (half a drop) of the oxidizing acid along the inner wall of the

FIG. 9.2. Determination of metal using a microcrucible. (From Ref. 76, STANDARD METHODS OF CHEMICAL ANALYSIS, Vol. 2A, 6th Ed., edited by Frank J. Welcher, 1963 by Litton Educational Publishing, Inc., reprinted by permission of Van Nostrand Reinhold Company.)

crucible and cover the crucible with its lid. Apply a very small flame to the side of the crucible as illustrated in Fig. 9.2. When no more acid fumes escape through the lid, shift the position of the burner so that it heats the bottom of the microcrucible from under the large lid. Raise the flame and heat strongly for 2 min. Upon cooling, reweigh the microcrucible and small lid, together. Repeat the process until the weight of the residue is constant.

E. Calculation

$$\% \text{ Metal} = \frac{\text{mg residue x factor* x } 100}{\text{mg sample}}$$

F. Comments

(1) If duplicate analysis is desired, another sample can be weighed into the microcrucible after the final weight of the residue has been determined, without cleaning the crucible. Contrastingly, this is not feasible when the decomposition is carried out in the microboat (see Sec. 9.5).

(2) The platinum microcrucible can be cooled much more rapidly than the porcelain crucible, when it is placed on the metal block inside the microdesiccator (Sec. 1.6).

(3) If the sample also contains boron, boric oxide will remain in the ash. In order to remove boron as the volatile BF_3, the experiment is modified as follows. After wetting the sample with sulfuric acid, add a drop of concentrated HF (48%). Heat to the appearance of SO_3 fumes and ignite until SO_3 is removed, but no further. Cool, add another drop of sulfuric acid and a drop of concentrated HF. Repeat the heating, then complete the ignition by placing the platinum crucible in the muffle furnace at 850°C for 1 h.

9.5 EXPERIMENTAL: DETERMINATION OF METALLIC ELEMENTS BY HEATING IN THE COMBUSTION TUBE

A. Principle

The organic sample is mixed with the oxidizing acid in a microboat and heated inside the combustion tube while a current of oxygen passes through

---

*Factor depends on the nature of the residue (oxide, sulfate, or free metal).

[76]. The residue in the microboat is then weighed as the free metal, oxide, or sulfate. For the determination of certain metals (e.g., nickel, cobalt), after the oxidative combustion, the current of oxygen is replaced by hydrogen, and subsequent heating converts the oxide to the free metal which is weighed (see Sec. 9.2A).

B. Equipment

Combustion tube, 10-mm i.d., and connecting tubes (see Fig. 9.3). Microboat, made of platinum, porcelain, or silica.

C. Reagents

Dilute nitric acid (1:1).
Dilute sulfuric acid, 20%.
Oxygen, small cylinder.
Hydrogen, small cylinder.

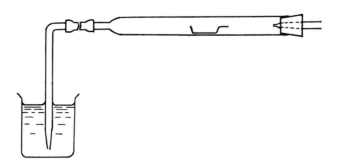

FIG. 9.3. Determination of metal using a microboat. (From Ref. 76, STANDARD METHODS OF CHEMICAL ANALYSIS, Vol. 2A, 6th Ed., edited by Frank J. Welcher, ©1963 by Litton Educational Publishing, Inc., reprinted by permission of Van Nostrand Reinhold Company.)

D. Procedure

Oxidation

Accurately weigh the clean microboat. Add a sample containing 0.5-2.0 mg of metallic element and reweigh. Carefully add half a drop of the oxidizing acid by means of a fine capillary, so that the sample is just wetted. Place the microboat and contents inside the combustion tube as shown in Fig. 9.3. Connect the wide end of the combustion tube to the oxygen cylinder through a control gauge, and the constricted end to a bent tube with the tip submerged in water. Pass a slow current of oxygen through the combustion tube. Gradually heat the microboat to prevent spattering of the sample. After the acid fumes have been driven off, heat the microboat at high temperature. Upon cooling, withdraw the microboat from the assembly and reweigh.

Reduction

If the metallic element in the organic material produces mixed oxides which can be reduced to the free metal, cool the combustion tube after oxidation. Exchange the oxygen cylinder with the hydrogen cylinder. Pass a current of hydrogen through the cold combustion tube for 5 min to displace all the air in the system. Then gradually heat the microboat and observe the color change of the residue. After the reduction is complete, cool the combustion tube, withdraw the microboat, and weigh it after allowing it to stand near the microbalance for 10 min.

E. Calculation

$$\% \text{ Metal} = \frac{\text{mg residue x factor* x 100}}{\text{mg sample}}$$

F. Comments

This experiment can also be carried out in the closed-tube combustion apparatus shown in Fig. 9.1, except that it is difficult to observe the changes which occur during heating.

---

*Factor depends on the nature of the residue (free metal, oxide, or sulfate).

9.6 EXPERIMENTAL: DETERMINATION OF METALLIC ELEMENTS BY ACID DIGESTION AND INDIRECT POLAROGRAPHY

A. Principle

Many organometallic compounds can be decomposed by digestion in hydrochloric or nitric acid in an open vessel. Some of the resulting metal cations can be quantitatively precipitated as oxinates. Bishara and co-workers [19, 20] have described a method to determine such metals by performing the precipitation with a known amount of oxine, followed by the measurement of the unreacted oxine polarographically. The procedures for the determination of aluminum, iron, lead, and nickel, respectively, are given below. The experimental details were verified by Dr. S. W. Bishara in November 1974.

B. Equipment

> Polarograph, Orion KTS Model 510, or equivalent.
> Electrolytic vessel: Use the universal U-shaped Kalousek cell, saturated calomel anode, capillary drop-time of 3-4 s under an open head of 68 cm of mercury.
> pH meter.

C. Reagents

> Hydrochloric acid, $1\underline{M}$ and 6 $\underline{M}$.
> Nitric acid, concentrated.
> Urea.
> Sodium bisulfite.
> Potassium sulfate.
> Oxine (8-hydroxyquinoline) standard solution, 0.025 $\underline{M}$, in 50% ethanol.
> Ammonia-ammonium chloride buffer of pH 10: Mix 142 ml of concentrated ammonium hydroxide with 17.5 g of ammonium chloride and dilute to 250 ml with distilled water.
> Methyl red indicator, 0.1% in H_2O.
> Pure nitrogen.

D. Procedure

Decomposition of Aluminum Compounds

Accurately weigh the sample (3-7 mg) into a 50-ml Erlenmeyer flask. Add 10 mg of sodium bisulfite followed by 5 ml of 6 $\underline{M}$ hydrochloric acid.

Heat gently to effect complete dissolution and allow to digest for 5 min.
Cool the solution, add two drops (0.1 ml) of methyl red indicator and neutralize with 6 $\underline{M}$ ammonium hydroxide to the yellow color.

Decomposition of Iron Compounds

Accurately weigh the organic material (2-5 mg) into a 50-ml Erlenmeyer
flask. Add 10 mg of sodium bisulfite, 5 ml of 6 $\underline{M}$ hydrochloric acid, and
two drops (0.1 ml) of concentrated nitric acid. After the sample dissolves,
allow it to digest for 5 min. Evaporate the solution to dryness; then add 2
ml of distilled water and 500 mg of urea to destroy the residual nitric acid.
Repeat this process twice. Finally dissolve the reaction mixture in 10 ml
of distilled water and cool to room temperature.

Decomposition of Lead Compounds

Accurately weigh the sample (2-5 mg) into a 50-ml Erlenmeyer flask.
Add 5 ml of 1 $\underline{M}$ hydrochloric acid. Heat gently to dissolve the material and
allow to digest for 5 min. Cool the solution to room temperature.

Decomposition of Nickel Compounds

Accurately weigh the organonickel compound (2-5 mg) into a 50-ml
Erlenmeyer flask. Add 20 mg of potassium sulfate and 5 ml of 1 $\underline{M}$ hydrochloric acid. Heat until complete dissolution is achieved and continue heating for 5 min. Cool to room temperature and add ammonia-ammonium
chloride buffer of pH 10 until the solution attains a pH of 9.25.

Precipitation and Polarography

For Ni(II): Precipitate the Ni(II) oxinate by adding, over a period of 3
min, 2 ml of 0.025 $\underline{M}$ oxine standard solution while the solution in the 50-ml
Erlenmeyer flask is being stirred. Transfer the reaction mixture to a
100-ml volumetric flask and fill to the mark with distilled water. For
Al(III), Fe(III), and Pb(II): Add 2 ml of 0.025 $\underline{M}$ oxine standard solution to
the 50-ml Erlenmeyer flask containing the respective metal cation (in order
to prevent the precipitation of hydroxides), and then adjust the pH of the
mixture to 9.25 by means of the ammonia-ammonium chloride buffer. Stir
for 5 min to insure quantitative precipitation of the metal-oxine complexes.
Transfer the reaction mixture to a 100-ml volumetric flask and add distilled water to the mark.

Mix the contents of the volumetric flask thoroughly. (If necessary,
allow the reaction mixture to stand and let the precipitate settle to the
bottom.) Measure an aliquot of the clear solution into the cathode compartment of the polarographic cell. Deaerate with pure nitrogen for 3 min.
Record the cathodic reduction wave of the oxine in the solution, starting at
a potential of -1.0 V vs. a saturated calomel electrode. Use a suitable
damping, a compensation of condenser current 2, and a sensitivity of 3 x 10^{-8}
$\mu A/mm$.

Carry out blank determinations using identical conditions but without the organic sample. The wave height obtained corresponds to the total amount of oxine in the 0.025 $\underline{M}$ standard solution.

E. Calculation

$$\% \text{ Metal} = \frac{(a - b) \times C \times M \times D \times 100}{b \times V \times W}$$

where a = wave height (mm) corresponding to total amount of oxine in 2 ml
      of solution
  b = wave height corresponding to the surplus oxine
  C = volume of oxine solution added (2 ml)
  M = molarity of the oxine solution
  D = atomic weight of the metal
  V = valence state of the metal
  W = weight of sample in mg

9.7 EXPERIMENTAL: DETERMINATION OF NICKEL, COBALT, OR PALLADIUM IN ORGANOMETALLIC COMPLEXES BY INDIRECT EDTA TITRATION

A. Principle

Organometallic complexes containing nickel, cobalt, or palladium are decomposed by boiling with a mixture of sulfuric acid and nitric acid. The resultant metallic ion is then determined by adding an excess of standardized EDTA solution and back-titrating with standardized zinc chloride solution in the presence of Eriochrome Black T indicator at a pH between 9 and 10 (see Sec. 9.7G).

B. Equipment

    Micro Kjeldahl flasks, 10-ml capacity (see Sec. 3.3).
    Digestion stand.
    pH meter equipped with glass-calomel electrodes.
    Magnetic stirrer.
    Microburets, 10-ml capacity, graduated in 0.02-ml intervals.

C. Reagents

    Concentrated sulfuric acid.
    Concentrated nitric acid.

EDTA standard solution, 0.01 $\underline{M}$: Weigh 3.723 g of $Na_2EDTA \cdot 2H_2O$
   into a 1-liter volumetric flask and make up to volume with distilled
   water.

Zinc chloride standard solution, 0.01 $\underline{M}$: Weigh 0.6538 g of zinc dust
   (AR) into a 1-liter volumetric flask. Dissolve in a minimum amount
   of concentrated hydrochloric acid and dilute to volume with distilled
   water.

Buffer solution containing ammonia: Add 67.6 g of NH_4Cl to 572 ml
   of concentrated NH_4OH and dilute to 1 liter with distilled water.

Buffer solution containing no ammonia: Dissolve 70 g of Na_2CO_3 in
   500 ml of distilled water, adjust the pH to 10 with concentrated HCl,
   and then dilute to 1 liter with distilled water.

Eriochrome Black T indicator: Mix thoroughly 1 g of Eriochrome
   Black T with 90 g of NaCl.

D. Procedure

Decomposition

Accurately weigh the sample (containing 0.05-0.09 milliequivalents of
metal to be determined) into the micro Kjeldahl flask. Add 1 ml of concentrated H_2SO_4, 10 drops of concentrated HNO_3 and allow to digest until
the solution is clear (about 30 min, see Sec. 9.7F, Note 1). Then transfer
the contents of the flask to a 250-ml beaker, and dilute with distilled water
to 150 ml.

Titration

From a 10-ml buret, add exactly 10.0 ml of 0.01 $\underline{M}$ EDTA standard
solution to the sample solution in the beaker. Adjust the pH of the solution,
by means of the buffer, to 9-10. For nickel or cobalt, use the ammonia
buffer. For palladium, use the carbonate buffer (Note 2). Add a few
milligrams of Eriochrome Black T indicator (the solution now turns blue
or green), and titrate the excess EDTA with standardized 0.01 $\underline{M}$ $ZnCl_2$
solution until the color changes to purple.

E. Calculations

$$\% \text{ Metal} = \frac{(V_{EDTA} - V_{Zn}) \times E \times 100}{W}$$

where V_{EDTA} = volume of 0.01 $\underline{M}$ EDTA added

$\quad\quad V_{Zn}$ = volume of 0.01 $\underline{M}$ $ZnCl_2$ needed to titrate excess EDTA

$\quad\quad$ E $\quad$ = equivalent weight of metal divided by 100 (E_{Ni} = 0.5871,

$\quad\quad\quad\quad$ E_{Co} = 0.5893, E_{Pd} = 1.064)

$\quad\quad$ W $\quad$ = weight of sample in mg

F. Notes

Note 1: The color of solution containing nickel is green, that of a solution containing cobalt is pink, and a solution containing palladium is orange or yellow.

Note 2: Solutions containing palladium cannot be buffered with an NH_3 solution because palladium tends to form very stable ammine complexes.

G. Comments

(1) Metals other than the alkali metals, if present, will interfere with this titrimetric method.

$\quad$ (2) The standard deviation of this method is ±0.3%.

9.8 EXPERIMENTAL: GRAVIMETRIC DETERMINATION OF TIN AND OTHER METALS BY MEANS OF CUPFERRON

A. Principle

Cupferron, the ammonium salt of nitrosophenyl hydroxylamine, reacts with a number of metallic elements (see Sec. 9.8F) to form insoluble cupferrates. The reaction for tin may be depicted as follows:

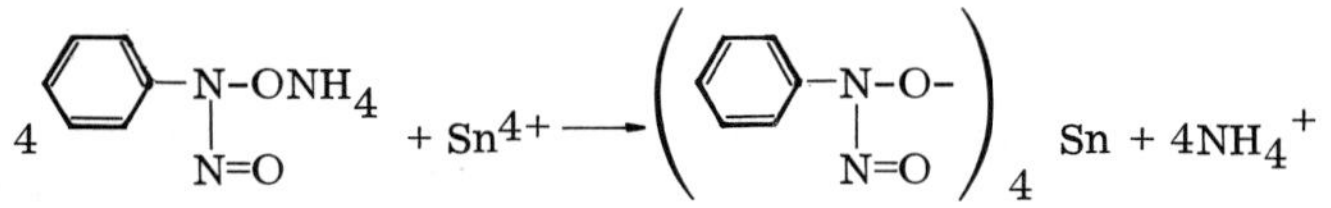

For quantitative analysis, the organic sample is combusted by the metal bomb technique. The melt is dissolved in water and the solution is made acidic. The cupferron reagent is added; the resulting precipitate is collected on a filter and then ignited to the metal oxide which is weighed.

B. Apparatus

   Parr microbomb (see Sec. 5.6).
   Filtering crucibles, Selas, 20-ml capacity.
   Muffle furnace.

C. Reagents

   Fusion mixture (see Sec. 5.6).
   Concentrated hydrochloric acid.
   Concentrated sulfuric acid.
   Cupferron reagent solution, 10%, aqueous: Weigh the cupferron in a
      beaker. Add a portion of the water to make a paste; then add the
      remaining amount of water. (The cupferron should be entirely
      dissolved, but if it is not, the solution should be filtered.) This
      solution can be kept up to a week in the refrigerator. If the solution
      turns dark, it should be discarded. The bottle of cupferron flakes
      should contain a desiccant and the material should give a colorless
      or light yellow solution.
   Wash solution: Use a 10% HCl or H_2SO_4 solution containing 1% cup-
      ferron.

D. Procedure

   Decomposition

   Accurately weigh a sample (estimated to give 20-30 mg of the metal
oxide) into the microbomb. Add 0.1 g of 2:1 KNO_3-sucrose mixture and
1.5 g of Na_2O_2. Mix well and carry out the combustion as described in
Sec. 5.6.

   Precipitation

   Dissolve the fusion product in about 40 ml of distilled water in a covered nickel beaker on a hot plate, and rinse the bomb thoroughly. Cool the
solution and neutralize with hydrochloric or sulfuric acid; then make the
solution 5-10% acid. Keep the sample solution and the cupferron reagent
solution, respectively, at 5 to 15°C in an ice bath. Add 5 ml of the cupferron reagent, dropwise, into the sample solution with stirring. Let the
reaction mixture stand 20-30 min in the ice bath before filtering through a
tared Selas crucible, washing the precipitate with cold wash solution.

Ignition

Heat the crucible containing the precipitate carefully (preferably with a heat gun), since the precipitate decomposes at the temperature required to dry it. When the gaseous compounds cease evolving, ignite the crucible with a burner until the carbon residue is burned off. Then ignite in a muffle at 800-900°C for 1 h to obtain the metal oxide.

E. Calculations

$$\% \text{ Metal} = \frac{\text{mg metal oxide x factor x 100}}{\text{mg sample}}$$

$$\text{factor} = \frac{\text{atomic weight of metal}}{\text{molecular weight of metal oxide weighed}}$$

For tin,

$$\text{factor} = \frac{Sn}{SnO_2} = \frac{118.69}{150.69}$$

F. Comments

(1) The following metals can also be precipitated completely or very nearly so by cupferron: antimony, bismuth, gallium, hafnium, iron, molybdenum, niobium, palladium, polonium, tantalum, titanium, tungsten, vanadium, and zirconium.

(2) The following metals may be partially precipitated: actinium, copper, lanthanum and lanthanum series, protactinium, thallium, thorium.

(3) Uranium is precipitated only if present in the quadrivalent state.

(4) The interfering elements are nitrogen (nitric acid or other oxidizing agents destroy cupferron), phosphorus (forms insoluble phosphates with titanium or zirconium), and silicon (taken down to some extent by the precipitates).

(5) Boron does not interfere. Hence this method is suitable for the determination of certain metals in organoboron compounds.

REFERENCES

1. Journal of Organometallic Chemistry, Elsevier, Amsterdam.
2. Synthesis and Reactivity in Inorganic and Metalorganic Chemistry, Marcel Dekker, New York.

3. M. Dub and R. W. Weiss (eds.), Organometallic Compounds, Springer-Verlag, New York, Vols. 1-3, 1968-1972.

4. A. K. Sawyer (ed.), Oragnotin Compounds, Marcel Dekker, New York, Vols. 1-3, 1971-1972.

5. R. B. King and D. Seyferth, Annual Survey of Organometallic Chemistry, Elsevier, Amsterdam.

5a. New York Academy of Sciences, Conference on Horizons in Organometallic Chemistry, August 1973.

6. J. P. Candlin, K. A. Taylor, and A. W. Parkins, "Organometallic Compounds of the Transition Elements," in Annual Reports 68, The Chemical Society, London, 1972, Sect. B, p. 273.

7. M. Tsutsui (ed.), Characterization of Organometallic Compounds, Wiley, New York, Vol. 1, 1969; Vol. 2, 1971.

8. T. R. Crompton, Analysis of Organoaluminum and Organotin Compounds, Pergamon, London, 1968.

9. J. Polesuk, A. Amadeo, and T. S. Ma, Mikrochim. Acta 1973:507.

10. A. H. Thomas Co., Philadelphia.

10a. W. J. Kirsten and J. E. Fildes, Microchem. J. 7:34 (1963); 9:411 (1965).

10b. M. E. Fernandopulle and A. M. G. Macdonald, Microchem. J. 11:41 (1966).

10c. International Plasma Corp., Hayward, California; Yanaco, Kyoto, Japan.

11. O. Schwarzkopf and F. Schwarzkopf, in M. Tsutui (ed.), Characterization of Organometallic Compounds, Vol. 1, Wiley, New York, 1969, p. 35.

12. C. N. Reilley and A. J. Barnard, Jr., in L. Meites (ed.), Encyclopedia of Analytical Chemistry, McGraw-Hill, New York, 1963, Sect. 3.77.

13. H. V. Meek and S. V. Banks, Anal. Chem. 22:1512 (1950).

14. R. P. Larsen and L. E. Ross, Anal. Chem. 31:176 (1959).

15. R. Belcher, A. M. G. Macdonald, and T. S. West, Talanta 1:408 (1958).

16. A. M. G. Macdonald, in C. N. Reilley (ed.), Advances in Analytical Chemistry and Instrumentation, Vol. 4, Wiley, New York, 1965, p. 75.

17. A. M. G. Macdonald and P. Sirichanya, Microchem. J. 14:199 (1969).

17a. J. Penic, I. Bregovec, Z. Stefanac, and Z. Sliepcevic, Microchem. J. 18:596 (1973).

18. A. B. Sakla, S. W. Bishara, and S. A. Abo-Taleb, Microchem. J. 17:436 (1972).

18a. S. W. Bishara, Y. A. Gawargious, and B. N. Faltaoos, Anal. Chem. 46:1103 (1974).

19. S. W. Bishara, Mikrochim. Acta 1973:25.

19a. A. B. Sakla, S. W. Bishara, and R. A. Hassan, Anal. Chim. Acta 73:209 (1974).

20. S. W. Bishara, A. B. Sakla, M. E. Attia, and H. N. A. Hassan, Mikrochim. Acta 1974:257.

20a. Y. A. Gawargious, S. W. Bishara, and B. N. Faltaoos, Indian J. Chem. 12:1113 (1974).

21. W. Schöniger, Z. Anal. Chem. 181:28 (1961).

22. N. D. Cheronis and T. S. Ma, Organic Functional Group Analysis, Wiley, New York, 1964, p. 400.

23. R. Belcher, D. Gibbons, and A. Sykes, Mikrochemie 40:76 (1952).

24. S. Sykes, Mikrochim. Acta 1956:1155.

24a. T. T. Gorsuch, in I. M. Kolthoff and P. J. Elving (eds.), Treatise on Analytical Chemistry, Part II, Vol. 12, Wiley, New York, 1965, p. 295.

25. G. Ingram, Methods of Organic Elemental Analysis, Reinhold, New York, 1962, p. 290.

26. L. V. Myshlyaeva and T. G. Shatunova, Tr. Mosk. Khim. Tekhnol. Inst. 1965:48.

27. L. Vojda, Magy. Kem. Lapja 19:497 (1964).

28. D. F. Hagen, D. G. Biechler, W. D. Leslie, and D. E. Jordan, Anal. Chim. Acta 41:557 (1968).

29. D. E. Jordan and W. D. Leslie, Anal. Chim. Acta 50:161 (1970).

30. L. V. Myshlyaeva and T. G. Maksimova, Zh. Anal. Khim. 23:1584 (1968).

31. M. P. Strukova and T. V. Kirillova, Zh. Anal. Khim. 21:1236 (1966).

32. A. P. Terent'ev, E. A. Bondarevskaya, N. A. Gradskova, and E. D. Kropotova, Zh. Anal. Khim. 22:454 (1967).

33. G. Kaiser, E. Grallath, P. Tschöpel, and G. Tölg, Z. Anal. Chem. 259:257 (1972).

34. S. K. Ng and P. T. Lai, Appl. Spectrosc. 26:369 (1972).

35. E. A. Kalinovskaya and L. S. Sil'vestrova, Zavod. Lab. 34:30 (1968).

36. M. Bigois, Talanta 19:147, 157 (1972).

37. J. Benes and M. Tomasek, Atompraxis 14:259 (1968).

38. Society for Analytical Chemistry, Analytical Methods Committee, Analyst 88:253 (1963).

39. T. V. Reznitskaya and E. I. Burtseva, Zh. Anal. Khim. 21:1132 (1966).

40. H. D. Graham, J. Food Sci. 28:440 (1963).

41. J. Jenik and F. Renger, Collect. Czech. Chem. Common, 29:2237 (1964).

41a. E. A. Terent'ev and T. M. Malolina, Zh. Anal. Khim. 19:353 (1964).

42. M. P. Strukova and V. N. Kotova, Zh. Anal. Khim. 22:1239 (1967).

43. V. A. Klimova and M. D. Vitalina, Zh. Anal. Khim. 19:1254 (1964).

43a. A. P. Terent'ev, E. A. Bondarevskaya, R. N. Potsepkina, and O. D. Kuleshova, Zh. Anal. Khim. 27:812 (1972).

44. S. I. Obtemperanskaya, I. V. Dudova, and G. F. Dikaya, Zh. Anal. Khim. 23:784 (1968).

44a. T. M. Shanina, N. E. Gel'man, and T. V. Bychkova, Zh. Anal. Khim. 28:2424 (1973).

44b. M. R. Masson, Mikrochim. Acta 1976(I):385.

45. K. Remtova and V. Chvalovsky, Collect. Czech. Chem. Commun. 33:3899 (1968).

46. M. Schlunz and A. Koster-Pflugmacher, Z. Anal. Chem. 232:93 (1967).

47. E. O. Schmalz and G. Geiseler, Z. Anal. Chem. 190:233 (1962).

48. E. Bladh and P. Gedda, J. Clin. Lab. Invest. 12:274 (1960).

49. B. Sarry and H. Grossman, Z. Anorg. Allg. Chem. 359:2341 (1968).

50. E. A. Terent'eva and T. M. Malolina, Zh. Anal. Khim. 23:1070 (1968).

51. R. Riemschneider and K. Petzoldt, Z. Anal. Chem. 176:401 (1960).

52. D. Ceausescu, Z. Anal. Chem. 176:1 (1960).

53. P. N. Fedoseev and M. E. Grigorenko, Zh. Anal. Khim. 16:100 (1961).

54. British Standards Institution, B. S. 2878 (1968).

55. G. B. Kyriakopoulos, J. Inst. Petrol. 54:376 (1968).

55a. R. E. Mansell and T. A. Hiller, Anal. Chem. 45:975 (1973).

56. M. P. Strukova and T. V. Kirillova, Zh. Anal. Khim. 22:1110 (1967).

56a. J. P. Macquet and T. Theophanides, Anal. Chim. Acta 72:261 (1974).

57. E. A. Terent'eva and I. M. Pruslina, Zh. Anal. Khim. 28:2352 (1973).

57a. C. H. Stapfer and R. D. Dworkin, Anal. Chem. 40:1891 (1968).

58. G. Tagliavina, Anal. Chim. Acta 34:24 (1966).

59. R. Geyer and H. J. Seidlitz, Z. Chem. 4:468 (1964).

60. V. Chromy and J. Vrestal, Chem. Listy 60:1537 (1966).

61. R. Geyer and H. J. Seidlitz, Z. Chem. 7:114 (1967).

62. R. Reverchon, Chim. Anal. 47:70 (1965).

62a. I. B. Peetre and B. E. F. Smith, Mikrochim. Acta 1974:301.

62b. V. M. Glazov, Tr. Khim. Khim. Tekhnol. 1973:151.

63. A. P. Kreshkov, L. V. Myshlyaeva, E. A. Kuchkarev, and T. G. Shatunova, Zh. Anal. Khim. 20:1325 (1965).

64. E. A. Terent'eva, E. N. Vinogradova, and N. P. Akimov, Zavod. Lab. 34:414 (1968).

64a. V. Kovac, M. Tonkovic, and Z. Stepanac, Microchem. J. 19:37 (1974).

65. E. A. Terent'eva and M. V. Bernatskaya, Zh. Anal. Khim. 21:870 (1966).

66. J. Jenik, Chem. Listy 60:783 (1966).

67. M. Goldstein, Chim. Peint. 29:177 (1966).

68. G. Kaiser, P. Tschöpel, and G. Tölg, Z. Anal. Chem. 253:177 (1971).

69. K. Omemoto, M. Hutoh, and K. Hozumi, Mikrochim. Acta 1973:301.

69a. K. Hozumi, private communication, September 1976.

70. G. Schwarzenbach and H. Flaschka, Die Kompleximetrische Titration, 5th Ed., Enke, Stuttgart, 1965.

71. Y. Tsuchitani, Y. Tomita, and K. Ueno, Talanta 9:1023 (1962).

72. A. de Sousa, Mikrochim. Acta 1961:644.

73. J. W. Robinson, Atomic Absorption Spectroscopy, 2nd Ed., Marcel Dekker, New York, 1975.

74. R. M. Dagnall, M. D. Silvester, K. C. Thompson, and T. S. West, Talanta 14:557 (1967); 18:1103 (1971).

75. A. P. Terent'ev, S. I. Obtemperanskaya, and V. N. Likosherstova, Zh. Anal. Khim. 15:748 (1960).

76. T. S. Ma, in F. J. Welcher (ed.), Standard Methods of Chemical Analysis, 6th Ed., Vol. 2, Van Nostrand, Princeton, 1963, p. 402.

Simultaneous Determination
of Several Elements

10.1 INTRODUCTORY REMARKS

This chapter deals with the principles and techniques whereby two or more elements present in the organic material can be determined simultaneously by taking only one sample of the substance to be analyzed. For convenience, we also include in this chapter discussion of some methods in which several elements can be determined using the same instrument and operation but different samples (see Sec. 10.2D).

As mentioned in Chap. 2, carbon and hydrogen determinations have been carried out simultaneously from the very beginning of quantitative organic analysis. The need for simultaneous analysis of other elements, however, was not apparent until recently. The early textbooks [1-3] on organic microanalysis did not discuss the possibility of simultaneous determinations, while the manuals [4-6] published in the 1960s only made scanty mention of this subject. One explanation of the long neglect is that heretofore elemental analysis was concerned primarily with the verification of the composition of pure organic compounds. Since it was a general practice for the research organic chemist to publish only values of C, H, or N, or some other element to establish the identity of a new compound, it was not necessary to make a complete analysis even though the new compound was known to contain several elements. Nevertheless, the microanalyst usually reweighed the microboat after combustion and such operation could provide additional analytical data. The following experiences of ours may serve as typical examples: (1) In the determination of chlorine in an amine chloroplatinate by the catalytic combustion technique, the residue in the microboat gave the percentage of platinum. (2) After the C,H analysis of organocobalt or organonickel compounds, the microboat with its contents was transferred to another combustion tube and heated in an atmosphere of hydrogen (see Sec. 9.5), giving the weight of cobalt and nickel respectively. (3) The percentage composition of C, H, S, and Na were obtained simultaneously for the sodium bisulfite addition products of carbonyl compounds by weighing the residue (Na_2SO_4) in the microboat after the C,H analysis [7].

After the development of the empty-tube technique for C,H analysis (see Sec. 2.1B), Belcher and Spooner [8] proposed a method to determine sulfur simultaneously by collecting the sulfur oxides on hot silver gauze in the combustion tube and subsequently weighing the silver gauze. Later Ingram [9] reported a procedure to determine carbon, hydrogen, and chlorine in one sample by incorporating an absorption tube packed with manganese dioxide which retains chlorine quantitatively. During the 1950s, an intensive effort was made by Korshun [10] and her co-workers in the Institute of Elemento-Organic Compounds, Academy of Sciences, USSR, to investigate the feasibility of simultaneous determinations. This included procedures for the determination of carbon and hydrogen plus one or more heteroelements, as well as methods for determining two or more heteroelements by using only one weighed sample. The impetus of their studies was due to the fact that they had to analyze a large number of organic compounds containing heteroelements. The advantages of simultaneous determinations attracted the attention of other groups of microanalysts who adopted the Russian methods and also devised new ones. The reader is referred to the annual reviews [11, 12] for the developments. A number of the papers which appeared originally in the Russian language have been translated into English [13, 14]. Some established procedures are described in detail later in this chapter (see Secs. 10.2 and 10.5).

A new approach toward simultaneous determinations emerged in the 1960s when gas chromatography was utilized to separate the products of combustion in organic elemental analysis, and physical methods were employed to measure the individual components. This has led to the development of the simultaneous determination of carbon, hydrogen, and nitrogen, and the automated elemental analyzers [15-20]. Further extension of this development opens the way for the determination of other elements using the same automatic apparatus (see Sec. 10.2D).

Recently several groups of investigators have applied the techniques of neutron activation and X-ray fluorescence to multielement analysis in organic materials. Thus, Nadkarni and Morrison [20a] have determined up to 25 elements in human blood serum, while Gaudy et al. [20b] have determined 24 elements in bovine liver by means of neutron activation. Yang et al. [20c] have used Ge(Li) spectrometry to determine 13 elements in tobacco leaves. Hutson et al. [20d] have demonstrated the determination of C, N, and O in tissues using muonic X-rays. Anderson et al. [20e] have carried out the determination of S, P, Cl, and a metal in organometallic compounds by X-ray fluorescence spectrometry.

10.2 CURRENT PRACTICE

A. Simultaneous Determination of Carbon and Hydrogen Together With
    Other Elements

  Carbon, Hydrogen, and Halogens

    Simple procedures are available for the simultaneous determination of
carbon, hydrogen, and any one element of the halogen group. Chlorine,
bromine, and iodine can be quantitatively retained by silver and weighed
at the end of the experiment. The device used by Korshun and Sheveleva
[21] is shown in Fig. 10.1. The organic sample is placed in a quartz capsule and is combusted at 900°C with a stream of oxygen in the presence of
platinum as catalyst. Silver gauze retains the halogen while carbon dioxide
and water vapor pass through. Later these workers [22] found that silver
deposited on pumice is a better reagent because it can be used at 425°C
without the need of the platinum contacts. As shown in Fig. 10.2, the silvered pumice is placed in the quartz sleeve to which the capsule containing
the sample is inserted. Since quartz is attacked by halogens, Klimova and
Merkulova [23] have recommended the use of a platinum boat 70-80 mm
long containing 0.4-0.7 g of electrolytically deposited silver. The apparatus is illustrated in Fig. 10.3.
    Absorption of chlorine in an external tube has been described by Ingram
[9], using manganese dioxide as the reagent. Bromine is not retained quantitatively, and poisons the reagent for further use. Recently Hadzija [24]
has employed lead dioxide as an external absorbent for chlorine, bromine,
and iodine. It should be noted that neither reagent can serve the purpose
when the organic material also contains nitrogen or sulfur.

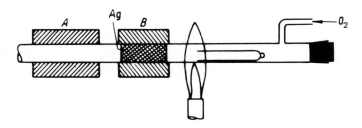

    FIG. 10.1. Combustion tube for simultaneous determination of car-
bon, hydrogen, and halogen, after Korshun and Sheveleva. A, length 110
mm for 900°C; B, length 60 mm for 600°C. (From Ref. 21, courtesy Zh.
Anal. Khim. and by permission of Gordon and Breach Science Publishers.)

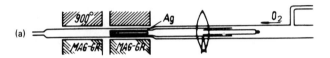

FIG. 10.2. (a) Apparatus for simultaneous determination of carbon, hydrogen, and halogen, after Korshun and co-workers. (b) Sleeve 8.5-mm and 11-mm o.d., 9-mm i.d. (From Ref. 22, courtesy Zh. Anal. Khim. and by permission of Gordon and Breach Science Publishers.)

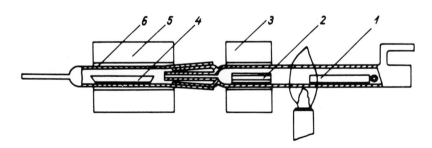

FIG. 10.3. Apparatus for simultaneous determination of carbon, hydrogen, and halogen, after Klimova and Merkulova. 1, Boat with sample; 2, platinum contact; 3, electric furnace with temperature 950°C; 4, boat with electrolytically deposited silver; 5, electric furnace with temperature 420-440°C; 6, ground-joint apparatus. (From Ref. 23, courtesy Izvt. Akad. Nauk SSSR and by permission of Gordon and Breach Science Publishers.)

As mentioned in Chap. 5, the methods for the determination of fluorine are different from those for chlorine, bromine, or iodine. Hence, for the simultaneous determination of carbon, hydrogen, and fluorine, the reagents described above cannot be used to retain fluorine. Gel'man and Korshun [25] have found that magnesium oxide mixed with the sample in a

quartz capsule can quantitatively retain fluorine as magnesium fluoride
while carbon and hydrogen are oxidized to carbon dioxide and water, respectively. The granulated magnesium oxide employed should be conditioned immediately before use by heating at 1100°C for 20-30 min [26]. Air-
dry magnesium oxide contains considerable amounts of water and carbon
dioxide at room temperature, and it gradually gives these up on heating.
It loses its chemical activity, however, after calcining at 1600°C and becomes inert toward fluorine. The quartz capsule (15 cm long) and its contents are positioned in the empty combustion tube as illustrated in Fig. 10.4.
In the early attempts, Gel'man and co-workers determined fluorine gravimetrically by weighing the quartz capsule before and after the combustion,
and calculated the %F based on the equation

$$2MgO + 2F_2 = 2MgF_2 + O_2$$

Recently, however, these investigators have recommended the liberation of
fluorine from the magnesium oxide-magnesium fluoride mixture by means
of superheated steam [27-29]. The hydrogen fluoride is distilled through a
platinum or silica condenser, collected in aqueous solution, and determined by titration with thorium nitrate (see Sec. 5.7).

Since the boiling points of organic fluoro compounds decrease with their
fluorine content [30], the analyst is faced with the problem of handling low-
boiling liquids and gaseous substances. It is apparent that the arrangement
shown in Fig. 10.4 cannot be used for such samples, because some of the
fluorine compound will escape before reacting with the magnesium oxide in
the quartz capsule. In order to circumvent this difficulty, Gel'man and co-
workers [31] have constructed an apparatus for analyzing liquid samples
which can be weighed in quartz capillaries. As depicted in Fig. 10.5, the
essential feature is an attachment which fits into the combustion tube and
permits oxygen to enter in two places. One end of the sealed quartz capillary is inserted through the hook of the stopcock. Thus the capillary is
cut open inside the combustion tube by turning the stopcock, and the sample
can be expelled from the capillary and led through the hot magnesium oxide

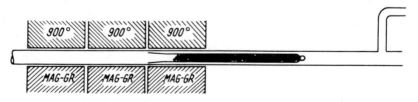

FIG. 10.4. Apparatus for simultaneous determination of carbon, hydrogen, and fluorine, after Gel'man et al. (From Ref. 26, courtesy Zhr.
Anal. Khim. and by permission of Gordon and Breach Science Publishers.)

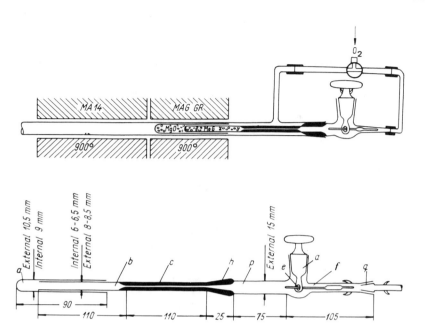

FIG. 10.5. Apparatus for simultaneous determination of carbon, hydrogen, and fluorine in liquid samples, after Gel'man et al. (From Ref. 31, courtesy Zhr. Anal. Khim. and by permission of Gordon and Breach Science Publishers.)

by a slow stream of oxygen. For materials which boil at temperatures below 20°C, these workers treat the samples as gases and employ a gasburet (Fig. 10.6) to conduct a measured quantity of the sample into the combustion tube. The gasburet has a capacity of 2 ml and is connected to the attachment (Fig. 10.5) for liquid samples. By means of the crescent stopcock B, it is possible to introduce either the gas to be analyzed or oxygen into the combustion tube. The stopcocks C and D allow the gasburet to be filled with the gaseous material without disconnecting it from the combustion train. The water jacket E is connected to a thermostat to maintain a constant temperature for the gas to be analyzed. When sampling, the gas passes into the gasburet either directly from a storage cylinder, or from a gaspipet. The leveling bulb F containing mercury is fixed in a position above stopcock B, the stopcock is turned to position I, and, by opening clamp G, the graduated part of the gasburet is gradually filled with mercury;

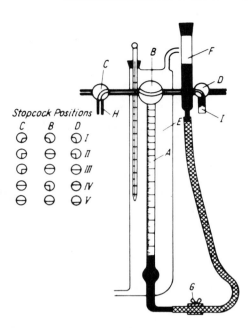

FIG. 10.6. Gasburet for measuring gas sample into combustion tube,
after Gel'man et al. (From Ref. 31, Zhr. Anal. Khim. and by permission
of Gordon and Breach Science Publishers.)

then stopcock B is turned to position II and clamp G is closed. The gas
inlet tube of the cylinder or gaspipet is connected to the outlet H and gas
is passed through stopcocks C, B, and D for 30 s. During this operation
the tiny drop of mercury which has entered into the opening in stopcock B
is blown out by the gas stream into outlet I and is removed. The leveling
bulb F is then lowered, clamp G is opened slightly, and, by careful rotation of stopcock B into position I, the gas is slowly delivered into the gasburet A, after which stopcock B is returned to position II, and oxygen is
driven through all three stopcocks. The volume of gas is read off at atmospheric pressure. Now the stopcocks are set in position IV and the gas
to be analyzed is conducted into the combustion tube at the rate of about
0.2 ml per min by controlling clamp G. When the requisite amount of gas
has been passed into the apparatus, the stopcocks are turned to position V
for 10 min and the gas supply line is swept with oxygen.

In view of the fact that fluorine reacts differently from the other halogens, it is possible to devise methods for the simultaneous determination of carbon, hydrogen, fluorine, and another halogen. Zimin and co-workers [32] have described a gravimetric method in which the combustion products are absorbed as follows: water by means of 98% sulfuric acid, chlorine by metallic silver, fluorine (as SiF_4) by potassium fluoride, and carbon dioxide by Ascarite, successively. Gel'man and co-workers [31] have used magnesium oxide to retain fluorine and silver to absorb the other halogens in an arrangement shown in Fig. 10.7. The combustion tube has the usual length, 550 mm, with 14 mm o.d. and 12 mm i.d. The quartz capsule holding the sample and magnesium oxide is 90 mm long, with 8-8.5 mm o.d. and 6-6.5 mm i.d.

Carbon, Hydrogen, and Sulfur

Simultaneous determination of carbon, hydrogen, and sulfur can be performed using a principle and apparatus similar to that for the analysis of carbon, hydrogen, and chlorine, bromine or iodine, since sulfur, like the halogens, is quantitatively absorbed by metallic silver. It should be noted, however, that sulfur is always retained in its highest oxidation state as silver sulfate when the combustion of the organic substance takes place in an atmosphere of oxygen [33], whereas the halogens are recovered at their lowest valence state as silver halides. In the method proposed by Korshun and Sheveleva [34], the absorption of sulfur oxides is effected in an external quartz tube placed between the combustion tube and the water absorption tube, as shown in Fig. 10.8. The sulfur oxide absorption tube is filled with silver wool or ribbons and is kept in an electrical furnace maintained at 650-800°C. During the combustion, the rate of oxygen intake is adjusted to 35-40 ml/min. Low sulfur values will be obtained if the furnace temperature is reduced to 420-600°C, indicating that the oxidation of

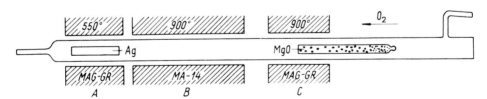

FIG. 10.7. Apparatus for simultaneous determination of carbon, hydrogen, fluorine, and another halogen, after Gel'man et al. (From Ref. 31, courtesy Zhr. Anal. Khim. and by permission of Gordon and Breach Science Publishers.)

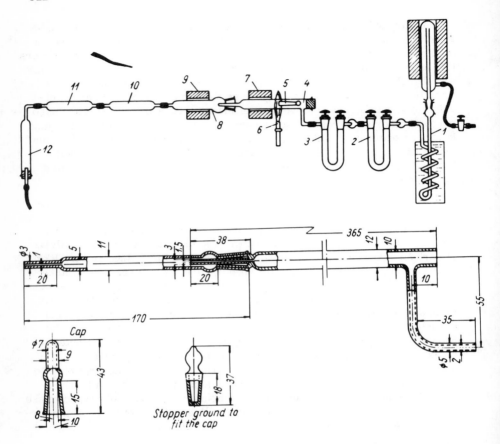

FIG. 10.8. Apparatus for simultaneous determination of carbon, hydrogen, and sulfur, after Korshun and Sheveleva. (top), 1, gas scrubber with electric furnace; 2, U-tube with Ascarite; 3, U-tube with Anhydrone; 4, quartz combustion tube; 5, quartz capsule for the weighed sample; 6, burner; 7 and 9, electric furnaces; 8, sulfur oxides absorption tube; 10, water absorption tube; 11, carbon dioxide absorption tube; 12, guard tube with Anhydrone. (From Ref. 34, courtesy Zhr. Anal. Khim. and by permission of Gordon and Breach Science Publishers.)

sulfur dioxide is incomplete under these conditions. Sokolova [35] has described a more effective absorbent which is prepared by mixing pumice with silver nitrate and heating at 800°C. According to Margolis and Egorova [36] the quartz tube filled with silver corrodes badly at high temperatures; these workers have proposed strontium silicate as absorbent which need be heated to only 600-650°C.

Klimova and co-workers [37] have described a procedure for the simultaneous determination of carbon, hydrogen, and sulfur by flask combustion in a closed chamber filled with oxygen (see Fig. 10.9). The device
is similar to the ignition combustion apparatus of Ingram [38] for carbon
and hydrogen analysis (Fig. 10.10). Ingram [39] has reported that sulfur

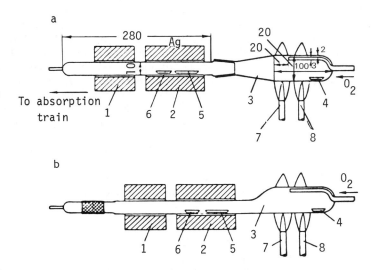

FIG. 10.9. Apparatus for simultaneous determination of carbon, hydrogen, and sulfur or halogen, after Klimova et al. (a) Ground-joint connection; (b) Vacuum-tubing connection. 1 and 2, electric furnaces; 3, combustion tube; 4, boat for sample; 5 and 6, boats with finely dispersed silver;
7 and 8, gas burners. (From Ref. 37, courtesy Izv. Akad. Nauk SSSR and
by permission of Gordon and Breach Science Publishers.)

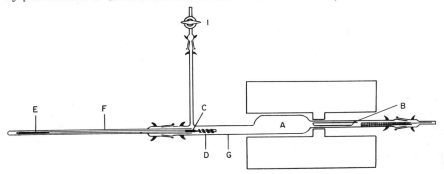

FIG. 10.10. Ignition combustion apparatus of Ingram. A, ignition
chamber; B, reagent section; C, sample holder; D, platinum wire coil for
sample container; E, iron core; F, sleeve for sample holder; G, inlet tube;
I, stopcock for cutting off oxygen supply. (From Ref. 38, courtesy J.
Wiley & Sons, Inc.)

and halogen oxidation products are absorbed quickly by their reagents as soon as the ignition has occurred. This is attributed to the diffusion of the gases to the hot reagents under the slightly reduced pressure created in the closed chamber. In the steup shown in Fig. 10.9, two quartz boats of 80 mm length each are filled with electrically precipitated or molecular silver and placed in the exit tunnel beyond the combustion chamber. All sulfur is retained in the first boat, the other boat being used as a control. Understandably, this apparatus also can be employed for the simultaneous determination of carbon, hydrogen, and chlorine, bromine, or iodine [37].

Other reagents have been proposed for the retention of sulfur. For instance, Fedorovskaya and Zakharova [40] have used empty-tube combustion and an external tube containing lead chromate at 380-400°C to absorb the sulfur oxides. The sulfur content is then determined gravimetrically. Abramyan and Kocharyan [41] have employed cadmium silicate as absorbent. These workers [42] have also published a method in which the organic substance is burned at 850-900°C in a stream of oxygen. The oxides of sulfur are absorbed at 400-450°C on the thermal decomposition product of potassium permanganate. Carbon and hydrogen are determined gravimetrically while sulfur is determined by extraction of the sulfate followed by titration with barium nitrate using thorin-methylene blue as indicator.

Carbon, Hydrogen, Sulfur, and Halogen

Taking advantage of the similarities and differences between the properties of sulfur and those of the halogens, it is feasible to perform four-element analysis using one weighed sample. Several schemes have been devised for this purpose. In the procedure described by Klimova and Mukhina [43] the organic material is pyrolyzed in a quartz capsule in a rapid stream of oxygen. The combustion train is illustrated in Fig. 10.11. A quartz boat containing cobaltic oxide is placed in the combustion tube, while an external quartz tube which houses a quartz boat containing electriolytically precipitated silver is connected between the combustion tube and the water absorption tube. Oxides of sulfur are absorbed by cobaltic oxide, and halogen is retained by the silver. Interestingly, the temperature to which the section of cobaltic oxide should be heated is dependent on the halogen present in the sample. If it contains chlorine or iodine, the furnace is set at 470-500°C, whereas C, H, S, Br combinations require a lower temperature range of 410-430°C. At the end of the combustion, which requires 25-30 min, the furnaces heating the quartz boats are removed. Carbon, hydrogen, and halogen are determined gravimetrically by weighing after 15 min the Anhydrone and Ascarite tubes, and the quartz boat containing silver halide, respectively. Sulfur is determined titrimetrically in the following manner.

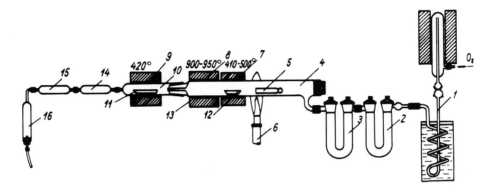

FIG. 10.11. Combustion train for simultaneous determination of carbon, hydrogen, sulfur, and halogen, after Klimova and Mukhina. 1, gas-purification apparatus with electric furnace; 2, U-tube with Ascarite; 3, U-tube with Anhydrone; 4, quartz combustion tube; 5, quartz test tube for sample; 6, burner; 7, 8, 9, electric furnaces; 10, quartz apparatus connected through ground joint to combustion tube; 11, quartz boat with electrolytically precipitated silver for absorption of halogen; 12, quartz boat with cobaltic oxide for absorption of oxides of sulfur; 13, platinum catalyst; 14, water absorption tube; 15, carbon dioxide absorption tube; 16, final tube with Anhydrone. (From Ref. 43, courtesy Izv. Akad. Nauk SSSR and by permission of Gordon and Breach Science Publishers.)

The contents of the quartz boat with cobaltic oxide are transferred to a 50-ml beaker and 5 ml of distilled water is added. After standing overnight, the solution is filtered through a sintered glass filter into the titration flask and the residue in the beaker is washed by decantation with hot water. The combined volume of the solution and washings should not exceed 20 ml. Five drops of 0.5 $\underline{N}$ NaOH and 1 ml of 25% acetic acid are added; then the mixture is heated for 1 min to remove carbon dioxide. After cooling, 20 ml of ethanol and three drops of 1% sodium alizarinsulfonate are added, and the mixture is titrated with 0.02 $\underline{M}$ Ba(NO$_3$)$_2$ until the yellow color changes to pink.

An all-gravimetric scheme has been proposed by Abramyan and Atashyan [44]. The organic material is burned in a stream of oxygen at the rate of 25 ml/min in a quartz combustion tube at 900–950°C. Sulfur oxides are absorbed by cobaltic oxide at 400–500°C; then water is absorbed by Anhydrone, halogen is retained by antimony metal at room temperature, and finally carbon dioxide is absorbed by Ascarite. For samples containing nitrogen, an absorbent for oxides of nitrogen is placed after the halogen absorber. A single analysis takes about 45 min.

Another gravimetric procedure has been published by Wang and Hsu [45]. These workers employed empty-tube combustion and absorbed sulfur and halogen together on silver wires. After weighing, the silver wires were washed with water to dissolve silver sulfate and then reweighed. The halogen content was calculated by difference. During analysis of nitrogen-containing compounds, an absorption tube filled with manganese dioxide was placed after the Anhydrone tube.

Carbon, Hydrogen, and Phosphorus or Metalloids

Organic materials containing phosphorus and metalloids like silicon and boron are conveniently discussed here since all such elements tend to remain behind in the sample container during carbon and hydrogen analysis. Complete oxidation of these compounds usually requires mixing the sample with some solid reagents. Klimova and co-workers [46] first carried out the simultaneous determination of carbon, hydrogen, and phosphorus by adding chromium sesquioxide (Cr_2O_3) on asbestos to the sample in the quartz capsule; they then employed empty-tube combustion at 900°C with a rapid stream of oxygen. Phosphorus pentoxide was retained in the capsule which was weighed before and after the combustion. Subsequently, these workers have found that pumice is a superior reagent [47]. Pumice has high porosity, large surface area, and is not hygroscopic like the fluffy fibrous asbestos. When finely ground, pumice pours easily. It quickly attains constant weight upon heating, and it does not corrode quartz. This method, however, is not suitable for the analysis of samples which contain sulfur because the oxides of iron and alkali and alkaline earth metals in the pumice will also retain sulfur oxides.

Korshun and Terent'eva [48] have found that crushed quartz after being treated with a strong alkali does not react with sulfur oxides but it retains phosphorus pentoxide. Thus a gravimetric method presents itself for the simultaneous determination of carbon, hydrogen, phosphorus, and sulfur. The apparatus shown in Fig. 10.8 is utilized and phosphorus is determined in the quartz capsule. A different method for the simultaneous determination of these four elements has been proposed by Papay and Mazor [49] using a modified Pregl technique (see Chap. 2). The oxidation of the pyrolysis products and the absorption of oxides of sulfur and phosphorus occur in an auxiliary tube coated with minium (lead orthoplumbate, Pb_3O_4). After combustion, the lead oxide layer is dissolved. Sulfate in the solution is determined by chelometric titration; then phosphate is precipitated as ammonium phosphomolybdate and titrated with alkali. It should be noted that lead orthoplumbate also retains fluorine [50] and arsenic [51] if these elements are present in the organic sample.

Chromium sesquioxide on asbestos has been recommended as the added reagent for the simultaneous determination of carbon, hydrogen, and silicon [53]. It serves to prevent the formation of silicon carbide which is resistant to oxidation and vitiates the results of carbon and silicon. The reagent is prepared as follows [53]: About 2 g of ammonium chromate or dichromate is heated by a small flame in a 50-ml crucible until all ammonium is expelled. The resulting Cr_2O_3 is mixed with fibrous asbestor previously calcined at 1200°C for 6 h. The mixture is screened on a sieve with 2-3 mm openings in order to separate it from the asbestos dust and excess chromic oxide. It is then calcined in a combustion tube at 900-950°C in a stream of oxygen for 4 h to remove the last traces of moisture. The apparatus for the analysis is illustrated in Fig. 10.12. Understandably, the added reagents may be omitted in cases where there is no danger of silicon carbide being produced [54].

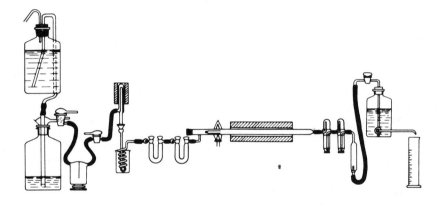

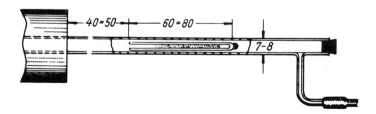

FIG. 10.12. (top) Apparatus for simultaneous determination of carbon, hydrogen, and silicon. (bottom) Location of capsule with sample and catalyst inside the combustion tube. (From Ref. 53, courtesy Zhr. Anal. Khim. and by permission of Gordon and Breach Science Publishers.)

Platonov [55] has proposed a method for the simultaneous determination of carbon, hydrogen, and silicon in organosilicon compounds which is based on the electro deposition of finely dispersed silicon dioxide. A high-frequency generator (12 000-15 000 V) is employed. Aranyi and Erdey [56] have described a procedure for the total analysis of organosilicon or organogermanium compounds with one weighed sample. The sample is covered with a 100-fold amount of powdered chromium trioxide at 900°C. The water and carbon dioxide produced are absorbed and weighed in the absorption tubes; the content of silicon or germanium is obtained from the change in weight of the ignition capsule. The packing of the added reagent is critical; loose packing allows escape of the silicon dioxide formed, whereas too-dense packing can lead to explosion.

Taking advantage of the finding that chromium sesquioxide does not retain sulfur oxides, Klimova and Bereznitskaya [57] have published a procedure for the simultaneous determination of carbon, hydrogen, silicon, and sulfur. The organic material is covered with chromium sesquioxide on asbestos, as mentioned above. Sulfur is retained in an external tube by means of silver heated at 700°C (see Fig. 10.8); silicon is determined by the change in weight of the quartz capsule. In contrast, the same authors [58] have performed simultaneous determinations of carbon, hydrogen, silicon, and halogen without the use of added reagents in the quartz capsule but with a platinum contact (made from platinum foil) placed in the combustion tube. The external tube packed with metallic silver is heated at 500-550°C. It was pointed out, however, that chromium sesquioxide on asbestos was necessary when the organosilicon compound contained less than 3% hydrogen. Furthermore, a drop of water (3-5 mg) should be added on top of the asbestos and the capsule reweighed. This amount of water is then subtracted from the weight of the Anhydrone tube for the calculation of the hydrogen content.

Simultaneous determination of carbon, hydrogen, and boron has been described by Rittner and Culmo [59]. The organic material is decomposed in the Pregl combustion train (see Sec. 2.6). Boric oxide left in the platinum boat is dissolved in water and determined by the identical pH method. The pH of the solution is adjusted to 7.10 using approximately 0.01 N NaOH and 10-12 g of mannitol is added until the pH remains relatively stable, being between 3.5 and 4.0. Then the solution is titrated with standardized 0.01 N NaOH until the pH is brought back to 7.10.

Carbon, Hydrogen, and Metals

The reader is referred to Chap. 9 on the general discussion of analysis of metallic elements in organic materials. In the present section, we shall confine ourselves to the methods which can be employed for the simultaneous determination of carbon, hydrogen, and one or more metallic elements

in one weighed sample. These methods may be divided into two categories: (1) collection of a volatile product, and (2) retention of a residue.

For Metals Which Form Volatile Products. During the decomposition
process for carbon and hydrogen analysis, some metallic elements, if
present, .form products which have sufficient vapor pressures at the temperature of the combustion furnace to move along the train together with
water and carbon dioxide. If the volatile products can be separately recovered quantitatively, then they can be simultaneously determined. Among
these metals may be mentioned osmium and rhenium whose tetroxides are
volatile, while mercury is usually liberated as the free metal.
    Simultaneous determination of carbon, hydrogen, and mercury can be
carried out in the apparatus shown in Fig. 10.2 with slight modifications.
Since condensation of the mercury vapor should take place at the lowest
possible temperature, the tip of the quartz sleeve should be cooled instead
of being heated by the electric furnace. One technique is to cover this
part of the quartz combustion tube with a wet flannel dipped into ice water
(see Sec. 8.6). A more elaborate technique is to use the cooling device
described by Trutnovsky [60]. The silver illustrated in Fig. 10.2 is preferably replaced by shredded gold leaves.
    Lebedeva and Kramer [61] have recommended the decomposition of the
organic material in a stream of oxygen at the rate of 18-20 ml per min at
900-950°C and continued combustion of the products over Co_3O_4 at 600-
650°C. Mercury is collected on silver-impregnated pumice while water
and carbon dioxide are absorbed in Anhydrone and Ascarite, respectively.
Pechanec [62] has described a procedure in which the sample is combusted
in a tube (heated at 550-600°C) packed with a layer of the decomposition
product of $AgMnO_4$, a layer of copper granules, and another layer of
$AgMnO_4$. Mercury is trapped in an external tube containing silver sponge.
    Using two quartz sleeves inside the combustion tube, Korshun and
co-workers [63] have demonstrated the feasibility of simultaneously determining carbon, hydrogen, mercury, and halogen or sulfur. The products
of pyrolytic combustion, after leaving the capsule, pass first over heated
silver in the first quartz sleeve to absorb the halogen or sulfur oxides, and
then over cooled gold leaves in the second quartz sleeve which retains the
mercury. An improved version of the apparatus has been published by
Gel'man and is described in detail in Sec. 10.5.

For Metals Which Leave Residues. Since many metallic elements
leave residues in the sample container after thermal decomposition of the
organic substance, it is a simple matter to determine the metal simultaneously with carbon and hydrogen, as mentioned in Sec. 10.1. Understandably, the residue should not interfere with the quantitative production of

water and carbon dioxide. For instance, alkali carboxylates cannot be
analyzed in this manner because a portion of the carbon in the compound
will be converted to the thermally stable alkali carbonate. According to
Zabrodina and Miroshina [65], if the sample is mixed with ground quartz,
then lithium, potassium, or sodium can be determined simultaneously with
carbon and hydrogen since carbon dioxide is thus prevented from combining
with the alkali metal.

Simultaneous gravimetric determination of carbon, hydrogen, and a
noble metal presents no difficulty since the latter is obtained in the free
state. For metallic elements which exhibit more than one valence state
but are known to form certain oxides, such as U_3O_8 for uranium [66] and
V_2O_5 for vanadium [67], the composition of the oxide should be previously
established. Detailed directions are given in Sec. 10.5.

Added reagents and oxidants are required for the determination of
some metals. For example, Lebedeva and co-workers [68] have described
a method for the simultaneous gravimetric determination of carbon, hydrogen, and thallium as follows. The sample is placed in a silica tube 9
cm long and is covered with a layer of specially treated powdered silica.
The tube is then heated in a stream of oxygen (12 to 15 ml per min), and
the pyrolysis products are passed over Co_3O_4 at 680-700°C. Thallium is
retained in the residue, probably as the silicate.

Gel'man and Bryushkova [69] have determined carbon, hydrogen,
aluminum, and halogen in compounds which are inflammable in air by performing the thermal decomposition in argon or nitrogen. Aluminum is
determined gravimetrically as Al_2O_3 in an atmosphere of oxygen to prevent
absorption of water.

B. Simultaneous Determination of Nonmetals Other Than Carbon and
    Hydrogen

Combinations of Chlorine, Bromine, and Iodine

Binary Mixtures. Simultaneous determination of binary mixtures of
halogens present in organic materials can be effected in several ways.
Olson [70] has recommended the following titrimetric techniques. (1) For
bromine-iodine combinations: transfer two suitable aliquots of the sample
solution obtained from the closed-flask combustion. Treat one aliquot with
bromine as in the iodometric determination of iodine (see Sec. 5.5) and
calculate the milliequivalents (meq) of iodine in the sample. Treat the
other aliquot with hypochlorite and determine the total milliequivalents of
bromine and iodine. Calculate the bromine content by difference. (2) For
chlorine-bromine and chlorine-iodine mixtures: argentimetrically titrate
one aliquot of the sample solution for total halogen (see Sec. 5.4). Determine bromine (or iodine) by the iodometric method.

A sequential coulometric titrimetric method for the simultaneous determination of chlorine and iodine has been published by Cassani [71].
After closed-flask combustion and absorption in solution containing potassium hydroxide and hydrogen peroxide, the flask is set aside for 1 h and
the solution is transferred to the titration cell (Fig. 10.13) with 60 ml of
$1 \underline{N} H_2SO_4$ and a few crystals of $NiSO_4$ which are added to destroy the residual peroxide prior to the sequential titration.   Korshun and Chumachenko
[72] have carried out the decomposition in a metal bomb by means of potassium fusion and recommended mercurometric titrimetry for the resulting chloride (or bromide) and iodometry for iodide.   Hassan and Elsayes
[73] have employed closed-flask combustion with acidic solution of sodium
nitrite as absorbent for halogens.   Then iodine is extracted by carbon tetrachloride and determined iodometrically, while the aqueous chloride (or
bromide) solution is analyzed mercurimetrically after the excess nitrite
has been removed.   A method for simultaneous determination of free bromine and chlorine using methyl orange has been described by Laitinen and
Boyer [74], which is based on the fact that the ring bromination reaction
produces an increase in absorbance at 317 nm, whereas the azo link cleavage by chlorine produces a decrease in absorbance at 317 nm.   By utilizing

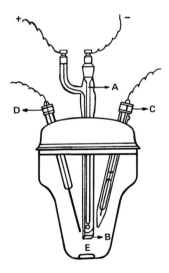

FIG. 10.13.   Coulometric titration cell, after Cassani.   A, platinum
electrode in 1 $\underline{N}$ H_2SO_4; B, Ag^+-generating electrode; C, Hg/Hg_2SO_4 reference electrode in saturated K_2SO_4; D, silver measuring electrode; E,
magnetic stirring bar.   (From Ref. 71, courtesy Chim. Ind.)

the changes in the ultraviolet-visible spectrum of aqueous methyl orange solutions upon reaction with gaseous mixtures of bromine and chlorine in nitrogen, the total halogen content and individual halogen mole fraction are determined. It should be noted that the formation of BrCl limits the total halogen concentration to less than 5 microequivalents/liter in nitrogen.

The Ternary Mixture. In contrast to the binary halogen combinations, simultaneous analysis of the ternary mixture containing chlorine, bromine, and iodine in organic materials is a difficult problem. None of the published simultaneous titration methods is entirely satisfactory. For instance, Shiner and Smith [75] have described a procedure which involves direct potentiometric titration of the halide mixture in a solution of acetate buffer containing a few drops of liquid nonionic detergent. We [76] have found, however, that application of this technique to the analysis of organic materials gave erratic results for chlorine and bromine while the iodine values were generally acceptable. It appeared that the potentiometric chloride break was overlapping that of the bromide. Interestingly, there was no overlap of these two breaks when iodine was absent in the sample. Based on these observations, a method for simultaneous determination of chlorine, bromine, and iodine has been developed as follows. The organic substance is decomposed in the closed flask and the halides are absorbed in hydrazine sulfate solution. One aliquot is used for iodine determination by potentiometric titrimetry. Another equal aliquot is taken and, after removal of the iodide by oxidation to iodine and boiling, titrated potentiometrically for chloride and bromide, respectively. The experimental procedure is given in Sec. 10.4.

Combinations of Other Nonmetals

A method for simultaneous determination of sulfur and iodine has been proposed by Nuti and Ferrarini [77]. The decomposition is performed in a closed flask which is fitted with a stopcock at its base in order to serve subsequently as a separatory funnel (see Fig. 5.1). The absorbent consists of carbon tetrachloride (1.5 ml), water (4 ml), 3% H_2O_2 (0.1 ml), and 3% $(NH_4)_6Mo_7O_{24} \cdot 4H_2O$ (0.1 ml). The flask is shaken for 8 min after combustion. Then 2 ml of carbon tetrachloride is added and the lower phase is drained off. Iodine is determined first by titration with 0.01 $\underline{N}$ $Na_2S_2O_3$. Then acetone is added and sulfate is determined by titration with 0.01 $\underline{M}$ $Ba(ClO_4)_2$ using sulfonazo III as the indicator.

Simultaneous determination of oxygen and chlorine, bromine, or iodine has been described by Korshun and Bondarevskaya [78]. The organic substance is burned in a quartz capsule and the decomposition products are caused to react with 3 cm of platinized carbon at 900°C where hydrogen halide and carbon monoxide are formed. The vapors are then conducted by a stream of nitrogen through the absorption apparatus filled with Ascarite to retain hydrogen halide. The carbon monoxide is driven into a tube containing copper oxide at 100°C, and the resulting carbon dioxide is absorbed by Ascarite and weighed. For organic materials which have low hydrogen contents the authors recommended adding paraffin to the sample to supply hydrogen for the formation of hydrogen halide.

A combustion train is also employed by the Korshun school for the simultaneous determination of fluorine and nitrogen [79]. The sample is placed in a quartz tube 9 cm long and filled with granulated nickel oxide which contains 15-20% of magnesium oxide. This tube is inserted in a large combustion tube and combustion is performed at 900-950°C in a carbon dioxide atmosphere. At the end of the combustion, nitrogen is collected in the microazotometer (see Chap. 3) and measured. The contents of the quartz tube are poured into the pyrohydrolytic tube [27-29] to liberate hydrogen fluoride which is determined by thorium nitrate titration.

The metal fusion technique has been utilized to determine two or more elements with one weighed sample. For instance, Chumachenko and coworkers [80] have used potassium fusion for the simultaneous determination of silicon and the halogens. The sample is placed in the steel test tube of the microbomb, together with 25-100 mg of metallic potassium. The bomb is heated at 850-900°C for 5 min. After cooling, the fusion mixture is dissolved in water (see Sec. 5. 7 about precautions in handling metallic potassium). The solution is quantitatively transferred to a polythene beaker, heated on a boiling water bath, and filtered into a volumetric flask. Aliquots are taken for the determination of silicon colorimetrically or amperometrically. Chloride and bromide are determined mercurometrically, iodide iodometrically, and fluoride thoriumetrically. Fedoseev and Ivashova [81] have described a procedure for determining nitrogen and sulfur in one sample by means of magnesium fusion. The material is heated with magnesium powder at 550-650°C for 30-40 min, whereby magnesium sulfide and nitride, respectively, are formed. Acidification of the fusion product with dilute hydrochloric acid in an atmosphere of carbon dioxide liberates hydrogen sulfide which is collected and determined iodometrically. The ammonium salt produced is determined by the Kjeldahl method (see Sec. 3.3).

Wet oxidation by digestion in an acidic medium, if applicable, provides a simple and convenient method for the simultaneous determination of several elements. For example, the feasibility of determining arsenic, phosphorus, and tellurium by digestion with nitric acid and sulfuric acid has been reported [82]. Faithful [83] has published a scheme for the simultaneous determination of nitrogen, phosphorus, calcium, and potassium by digestion of the organic material in concentrated sulfuric acid, followed by colorimetric determinations of the respective elements using aliquots. A method for the simultaneous determination of boron, nitrogen, and phosphorus [84] after sulfuric acid digestion of the sample is given in detail in Sec. 10.6.

C. Simultaneous Determination of Metallic Elements

Since the preliminary step in the determination of metallic elements in organic materials involves either ashing or digestion in solution (see Chap.
9), the feasibility of using a single sample to determine two or more components is evident. When the direct ashing technique is employed, direct
gravimetric determination of two metals is possible if one metal yields a
volatile product while the other metal remains in the sample container.
The experimental procedure is described in Sec. 10.5. In case the ash
contains more than one metallic element, it is brought into solution and the
respective metal ions are determined by appropriate methods.

Emission and atomic absorption spectrometric procedures can be employed for the simultaneous determination of several metallic elements
and these methods are usually applicable to the organic substance directly.
For instance, Ng and Lai [85] have used emission spectroscopy for the
simultaneous determination of calcium, copper, iron, and magnesium in
natural rubber. The accuracy of this method is reported to be parallel to
that of atomic absorption spectrometry [86]. Kopp and Kroner [87] have
described a direct-reading spectrochemical procedure to determine 16
metals plus arsenic, boron, and phosphorus.

When the metallic components in the organic material have been converted into ionic forms (e.g., by wet decomposition, closed-flask combustion, etc.) or when metal ions are present in the original organic material,
there are numerous ways to determine the metallic elements simultaneously. Sequential determinations (e.g., polarography [87a, b]) using one
sample solution is feasible if there is no mutual interference due to the
metallic ions present. Alternatively, aliquots of the solution are taken for
the determination of the respective metals. The simultaneous determination of calcium and potassium in green plants after acid digestion [83],
and the simultaneous determination of potassium and sodium in biological
fluids using ion-selective electrodes [88], are typical examples. Selective
solvent extraction, such as the method for Ge, Si, As, and P, also can be
applied to organic analysis after mineralization [88a].

D. Determination of Oxygen or Sulfur Using the C, H, N Analyzer

After the acceptance of automated C, H, N analyzers in the microanalytical
laboratory, it is logical to utilize the electronic systems and gas-separation
units in these instruments to determine other nonmetals in organic materials. The most versatile device is probably the apparatus designed by
Dugan and Aluise [89] for the determination of carbon, hydrogen, nitrogen,
sulfur, and oxygen. The schematic diagram is illustrated in Fig. 10.14.

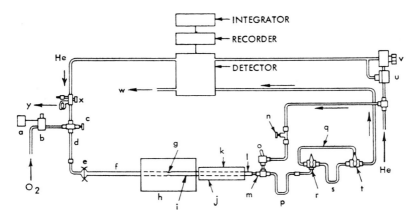

FIG. 10.14. Apparatus for determination of C, H, N, S, and O, after
Dugan and Aluise. a, electric timer; b, solenoid; c, cross pattern fine
metering valve; d, gas inlet tube; e, O-ring joint; f, quartz combustion
tube; g, indentations in combustion tube; h, high-temperature combustion
furnace; i, quartz wool; j, reduction furnace; k, copper; l, copper oxide;
m, Swagelok connections; n, needle valve; o, toggle valve; p, carbowax
column; q, by-pass tube; r and t, three-way valves; s, molecular sieve
column; u, constant differential-type flow controller; v, micro adjustable
restrictor; w, florator and soap film flowmeter; x, rotary gas sampling
valve. (From Ref. 89, courtesy Anal. Chem.)

Carbon, hydrogen, nitrogen, and sulfur are determined simultaneously by
combusting the sample (0.5-3.0 mg) at 1060-1080°C in an atmosphere of
40% oxygen in helium. The gaseous products are subsequently retained
and separated in two cold traps, one being packed with Carbowax 20M on
Teflon to retain CO_2, SO_2, and H_2O, and the other packed with molecular
sieve 5A to retain N_2. Sequential heating of the traps releases the trapped
gases which then are measured by means of a thermal conductivity detector. Oxygen is determined by dynamic flash-pyrolysis using a tube containing pelletized carbon black [90] at 1120°C in an atmosphere of helium.
The carbon monoxide formed is retained and separated in the same trapping system and is also measured by thermal conductivity. Later these
workers [91] constructed a dual-channel analyzer shown in Fig. 10.15A.
This instrument has two independent flow controls, two combustion and
reduction furnaces, and two trapping systems, all of which operate from
one integrator-recorder-detector system. Simultaneous determination of

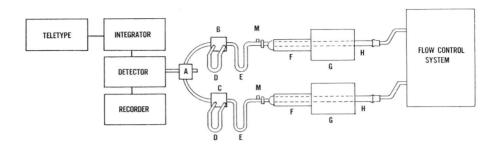

FIG. 10.15A.   The dual-channel analyzer for C, H, N, S, and O, after
Dugan and Aluise.   (top) A block diagram of the dual-channel C, H, N, S,
and O analyzer.   A, B, C, four-way valves; D, D', 1-foot molecular sieve
(5A) columns; E, E', 1-foot Carbowax 20M columns (20% on Teflon); F, F',
reduction furnaces; G, G', combustion furnaces; H, H', quartz tubes; M, M',
backflush valves.   (bottom) The dual-channel analyzer.   From right to left
are the balance, flow-control panel, quartz tubes, combustion furnaces,
reduction furnaces, trapping columns, detector, integrator, and recorder.
The operator is typing the input data for the computer program.   (From
Ref. 91, courtesy Hercules Chemists and G. Dugan.)

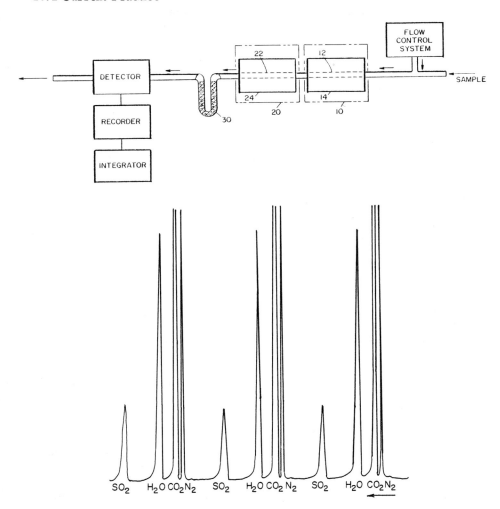

FIG. 10.15B. Simultaneous determination of C, H, N, and S, after Dugan. (top) Analyzer. (bottom) Typical chromatogram determination of sulfanilamide. (From Ref. 91a, courtesy Hercules Chemists and G. Dugan.)

carbon, hydrogen, nitrogen, and sulfur is performed in one channel with provisions to switch instantaneously to the other channel for the determination of oxygen. An analysis for C, H, N, S, and O can be made in 20 min. Simultaneous analysis of C, H, N, and S requires 13 min, while a single oxygen determination takes 9 min. The presence of bromine, chlorine, and fluorine does not interfere with the analysis or change the calibration; reaction with metallic copper in a combustion train effectively removes these materials.

Recently Dugan [91a] undertook a thorough investigation of the combustion of sulfur compounds and found that (1) solid oxidant or catalyst is not required when excess oxygen is used, and (2) sulfur is exclusively converted into SO_2 in the presence of metallic copper at 840°C. Thus an improved C, H, N, S analyzer has been developed which is based on the gas chromatographic determination of CO_2, H_2O, N_2, and SO_2 produced from the sample. Figure 10.15B (top part) shows the schematic diagram of the main features of this analytical system. In the apparatus, the combustion zone 10 consists of a combustion chamber 12 (maintained at 1080°C) and furnace 14. The combustion chamber is a quartz tube containing quartz wool. The reduction zone 20 consists of a reduction chamber 22 (maintained at 840°C) and furnace 24. The reduction chamber is a quartz tube packed with copper wire (30-60 mesh). The gas chromatographic column 30 consists of a 1/4-in. tube (2 m long) filled with Porapak QS (80-100 mesh) and maintained at 130°C. A thermal conductivity detector is used. The flow-control system provides a continuous flow of helium (30 ml per min) through the system. The chromatogram shown in the bottom part of Fig. 10.15B illustrates the analysis of sulfanilamide. The percentages of C, H, N, and S may be obtained by using calibration graphs or response factors prepared from microchemical standards.

Several commercial C, H, N analyzers can be modified to perform oxygen determinations. Culmo [92] has described the changes in the combustion tube packing when the Perkin-Elmer apparatus (see Sec. 2.3) is converted for such purposes. The organic substance is pyrolyzed in a helium atmosphere with subsequent formation of carbon monoxide by contact of the pyrolysis products with platinized carbon at 950°C. The carbon monoxide then passes through an oxidation tube containing cupric oxide at 670°C whereby carbon monoxide is oxidized to carbon dioxide which is measured by the detection system of the C, H, N analyzer. In the Carlo Erba [93] apparatus (see Sec. 2.5), a separate reactor is built into the system to perform pyrolysis and conversion of oxygen to carbon monoxide by activated carbon at 1050°C. The combustion products are separated by a gas chromatographic column tailored for carbon monoxide determination. The time-sharing design makes it possible to measure C, H, and N on one side or oxygen on the other side without interference, with a single hot wire thermal conductivity detector. The Heraeus [94] instrument uses a molecular sieve to separate carbon monoxide which is measured by thermal conductivity. While it is possible to convert the Heraeus C,H,N set-up to oxygen analysis, the use of a single instrument for two operating modes is not recommended by the manufacturer.

Conversion of the Perkin-Elmer C,H,N analyzer into a device for sulfur analysis has been reported by Culmo [95]. In the combustion area, tungstic oxide is used as the oxidation catalyst, calcium chloride is used to remove water, 8-hydroxyquinoline to remove residual halogen byproducts, and copper to remove excess oxygen and halogens. In the detection area, a sulfur dioxide absorber filled with silver oxide and heated at 220°C is inserted between the two sides of a differential thermal conductivity detector, thus producing a signal proportional to sulfur content. Known compounds are first analyzed to calibrate the detector voltage in terms of micrograms of sulfur. This calibration factor is then used to measure the unknown samples.

10.3 SURVEY OF RECENT DEVELOPMENTS

A. Summary of Publications

The literature on methods for the simultaneous determination of the elements in organic materials published during the past 15 years is summarized in Tables 10.1 and 10.2. Procedures which deal with the determination of one or more elements in addition to carbon and hydrogen are tabulated in Table 10.1. It will be noted that a majority of the studies utilized the gravimetric technique in the finish step. One reason is that most of these investigations were carried out before the advent of automated carbon and hydrogen analyzers. Secondly, if the heteroelement forms a nonvolatile product which is determined by gravimetry, it is convenient to determine carbon and hydrogen also by weighing the respective carbon dioxide and water absorption tubes. On the other hand, if all elements to be analyzed can be converted to suitable gaseous products, the recent approach is to separate the gases by appropriate means and measure the quantity of each gas by methods other than gravimetry. Gas chromatography is a favorable technique for separation, while selective absorption and molecular sieves also have been employed. Parallel to the development of gas chromatography, thermal conductivity measurement is the preferred technique of determination and seems to be displacing the methods based on volumetry and manometry. Thus Liebman and coworkers [196] have proposed a scheme for the on-line elemental analysis of gas-chromatographic effluents which can perform simultaneous determination of carbon, hydrogen, and nitrogen or oxygen in 10 μg to 1 mg of organic substance. Further discussion of this device is presented in Chap. 11.

TABLE 10.1. Summary of Publications on Simultaneous Determination of Several Elements Including Carbon and Hydrogen

Elements	Products	Finish technique	Reference
C, H, N	CO_2, H_2O, N_2	Gravimetric, gasometric	96–100
	CO_2, H_2O, N_2	Manometric	101, 102
	CO_2, H_2O, N_2	Gasometric	103, 104
	CO_2, H_2O, N_2	Gas chromatographic	105–118
C, H, halogen	CO_2, H_2O, X^-	Gravimetric, titrimetric	23, 28, 119–125
	CO_2, H_2O, X^-	Gravimetric	24, 126–127a
C, H, S	CO_2, H_2O, SO_3	Conductometric, gravimetric	40, 128–133
	CO_2, H_2O, SO_3	Gas chromatographic	134
	CO_2, H_2O, SO_2	Infrared spectroscopy	135
C, H, P	CO_2, H_2O, P_2O_5	Gravimetric	136
C, H, Si	CO_2, H_2O, SiO_2	Gravimetric	55, 56
C, H, Bi	CO_2, H_2O, BO_3^{3-}	Gravimetric	59
C, H, Se	CO_2, H_2O, SeO_2	Gravimetric	137, 138
C, H, metal	CO_2, H_2O, metal ion	Gravimetric	56, 66, 68, 138, 138a
C, H, N, S	CO_2, H_2O, N_2, SO_2	Gas absorption	89, 91
C, H, halogen, S	CO_2, H_2O, X^-, SO_3	Gravimetric	42–44, 139–142
C, H, halogen, metal	CO_2, H_2O, X^-, metal oxide	Gravimetric	143–145
C, H, halogen, Al	CO_2, H_2O, X^-, Al^{3+}	Gravimetric, titrimetric	69

TABLE 10.1.   (Continued)

Elements	Products	Finish technique	Reference
C, H, halogen, Hg	CO_2, H_2O, X^-, Hg	Gravimetric	61, 63, 146, 147
C, H, halogen, S, P	CO_2, H_2O, X^-, SO_3, P_2O_5	Gravimetric, titrimetric	49, 148
C, H, N, halogen, S, Hg	CO_2, H_2O, N_2O_3, X^-, SO_3, Hg	Gravimetric	64

Table 10.2 summarizes the publications in which carbon and hydrogen analysis is not a part of simultaneous determination, although the elements to be determined may include either carbon or hydrogen. A glance at this table will reveal the preponderance of titrimetric procedures. The halogen group of elements has been extensively investigated. It should be noted, however, that there is no simple method for the simultaneous determination of chlorine, bromine, and iodine (see Sec. 10.4). The great variety of binary combinations of elements is shown in the tabulation. The feasibility of simultaneously determining three or more elements using one sample of the organic material is also indicated.

B.   Evaluation and Improvement of Methods for Simultaneous
     Determinations

Although a wide range of possible combinations for the simultaneous determination of two or more elements has been proposed (see Tables 10.1 and 10.2), the reports on the evaluation of these methods are practically all concerned with automated procedures. Thus the C, H, N and C, H, N, O analyzers have received considerable attention and have been discussed intensively among microanalysts [197, 198].

As mentioned in the previous section, the development of gas chromatography opened the way for simultaneous determination of several elements like carbon, hydrogen, nitrogen, oxygen, and sulfur, which can be easily converted into measurable gaseous products. It should be noted, however, that quantitative gas chromatography restricts severely the

TABLE 10.2. Summary of Publications on Simultaneous Determinations
Not Including Carbon and Hydrogen Analysis.

Elements	Products	Finish technique	Reference
Halogens	Cl^-, Br^-	Titrimetric	73, 149-151
	Cl^-, Br^-	Coulometric	71
	Cl^-, I^-, IO_3^-, BrO_3^-	Polarographic	152
Halogen, C	X_2, CO_2	Titrimetric, gas chromatographic	153-157
Halogen, N	X^-, N_2	Titrimetric, gasometric	157a
	X^-, NH_3	Titrimetric	158, 158a
Halogen, S	X^-, S^{2-}	Titrimetric	77, 159-171
	X, SO_3	Gravimetric	171a
Halogen, Si	X^-, SiO_3	Titrimetric	172
Halogen, Cu or Fe	X^-, metal ion	Titrimetric	173
Cl, Hg	Cl^-, Hg	Gravimetric	174
F, O	HF, CO	Gravimetric	174a
F, N	F^-, N_2	Titrimetric, gasometric	79, 175
F, Si	F^-, SiO_2	Titrimetric, gravimetric	175a
F, P	F^-, PO_4^{3-}	Titrimetric	176
F, S	F^-, SO_4^{2-}	Titrimetric	176a
S, H	^{35}S, ^3H	Radiometric	177
S, P	SO_4^{2-}, PO_4^{3-}	Titrimetric	177a
S, B	SO_4^{2-}, BO_3^{3-}	Titrimetric	178
S, Fe	SO_4^{2-}, Fe^{3+}	Titrimetric, spectrophotometric	179
N, C	NH_4^+, CO_3	Titrimetric	180
N, H	H_2, N_2	Gasometric	181
N, O	N_2, CO	Gas chromatographic	182, 182a
N, S	N_2, SO_4^{2-}	Gasometric, titrimetric	182b

TABLE 10.2. (Continued)

Elements	Products	Finish technique	Reference
N, P	NH_4^+, PO_4^{3+}	Titrimetric, colorimetric	183, 183a
P, Si	PO_4^{3-}, SiO_2	Colorimetric, gravimetric	184-187
Si, Fe	SiO_3^{2-}, Fe^{3+}	Spectrophotometric	187a
P, Al	PO_4^{3-}, Al^{3+}	Titrimetric	188
Ca, Mg	Ca^{2+}, Mg^{2+}	Titrimetric	189
Halogen, S, C	X^-, SO_3, CO_2	Titrimetric	190
Cl, S, N	Cl^-, H_2S, NH_3	Titrimetric	191
N, P, B	NH_4^+, PO_4^{3-}, BO_3^{3-}	Titrimetric	84
Si, Al, C	SiO_3^{2-}, Al^{3+}, CO_2	Titrimetric, gravimetric	192-194
Halogen, N, C, metal	X_2, NH_4^+, CO_2, metal ion	Gravimetric, titrimetric	195
N, S, P, C	Hydrides	Flame ionization	195a

sample size of the organic material to be analyzed. In order to obtain accurate results, the devices for simultaneous determinations based on gas chromatographic separation usually demand that the organic sample should lie within a narrow range of 0.2 to 0.5 mg. While the possibility of analyzing organic substances below the milligram region is commendable (see Chap. 11), such procedures impose a condition of complete homogeneity of the preparation and delicate techniques to prevent contamination during transfer, especially for moisture- and air-sensitive samples. For this reason, a number of investigators [18, 89, 199] have proposed methods which are not dependent on gas chromatography. Thus, in the apparatus described by Wachberger and co-workers [199] for the simultaneous determination of carbon, hydrogen, and nitrogen, the sample weight is of the order of 3-5 mg. The organic substance is combusted in pure oxygen, the combustion products are separated in a stream of helium by means of selective adsorption and desorption, and the single components are determined successively by means of thermal conductivity measurements.

Several groups of workers have reported on their experience with the commercial C, H, N, O analyzers and suggested improvements. Cousin [200] has modified the Technicon apparatus (see Chap. 2) by using two electric furnaces to heat the combustion tube, one being 360 mm long and kept at 950°C and the other 80 mm long and kept at 1050°C. It is claimed that this arrangement gives correct C, H, N values for refractory compounds and long life to the combustion tube packing. Ono and co-workers [201] have made intensive studies on the F & M Model 185 C, H, N analyzer. They reported that (1) cupric oxide was not a suitable oxidant during such a short combustion period (25 s), (2) cobaltic oxide and manganese dioxide gave high blank values for both carbon and hydrogen even after heat treatment, (3) potassium chlorate showed the smallest blanks, and (4) silver permanganate crystals prepared by mixing silver nitrate solution with potassium permanganate solution showed zero blank for nitrogen and very small blanks for carbon and hydrogen. These workers have developed an automatic sampler to be coupled with the C, H, N analyzer and, by obtaining the integrated results of peak areas with the Hewlett-Packard 3370A integrator, have carried out 50 determinations in 5 h. They have also proposed a scheme to perform C, H, N analysis without weighing the samples, by keeping constant the gas chromatographic conditions.

Shimizu and Hozumi [202] have published a statistical study of the errors of Corder MT-2 C, H, N analyzer which involves a detector system of differential thermal conductometry. Considering the possible sources of error including (1) sample weight, (2) bar-gram reading, (3) atmospheric pressure, (4) temperature of the detector oven, (5) base signal, (6) bridge supply voltage, and (7) other functions, it was shown that the contributions of (6) and (7) were several times higher than the sum contribution of (1) to (5).

Stoffel [203] has evaluated the Carlo Erba Model 1102 C, H, N, O analyzer using 23 test samples. By coupling this apparatus with an electronic microbalance and a microcomputer, either 23 C, H, N, O determinations or 40 C, H, N determinations can be carried out automatically during a workday. A similar scheme has been described by Hozumi and co-workers [204] who use an automatic charger to hold 12 quartz capsules containing weighed organic samples, a double action pump, and a data printing system. The samples are sequentially charged into the combustion tube. The gaseous products with a carrier gas (helium) are withdrawn into one side of the double-action pump during a period of 5 min and the mixture is kept standing for 2.5 min for homogenizing. The gas mixture is then pushed out of the pump toward a series of three differential thermal conductometers successively removing or retaining water, carbon dioxide, and nitrogen. While the double-action pump is pushing out the combustion products toward the thermal conductometers, the next sample can be burned and the gas mixture is withdrawn into the opposite side of the pump. Charging of the samples thus takes place at an interval of 7.5 min. An analog-to-digital converter is provided to print out the successive signals from the thermal conductometers. The error functions of this system have been evaluated.

C. Limitations of Simultaneous Determinations

While simultaneous determinations have many advantages such as conservation of valuable samples and saving of labor, these methods are not yet the panacea of organic elemental analysis. In quantitative analysis, the precision and accuracy of the determinations are of primary concern. Hence, when the results of several elements are obtained from a single operation, the reliability of each result should be established. It is obvious that simultaneous determination of all elements is not suitable if some components are present in the organic material at low levels. For instance, the current device for C, H, N, S, and O is not recommended for total analysis of petroleum which has very little nitrogen, oxygen, and sulfur.

A method for simultaneous determination is desirable if it is simple and more convenient than the combination of two methods otherwise required. Conversely, it has little advantage except saving the sample if the operation becomes rather complicated. For example, in the procedure for the simultaneous determination of carbon, hydrogen, and fluorine [26, 119a] (see Sec. 10.2A) and that for nitrogen and fluorine [79] (see Sec. 10.2B), magnesium oxide is used to retain fluorine which is subsequently liberated. This entails special reagents and equipment besides difficult experimental operations.

In order to attain total destruction of the organic substance, the sample is sometimes mixed with solid oxidants or other additives, among which may be mentioned chromium oxides [46,52,56,58] (see Sec. 10.2A). This necessitates extremely careful preparation of the additives, especially for methods which include the elements carbon and hydrogen. It is very difficult to obtain and keep reagents entirely free from moisture and carbon-containing contaminants. Since the additives are not a permanent filling of the combustion tube, they cannot be purified in situ as in the classical carbon and hydrogen procedure (see Sec. 2.6).

In order to recover all combustion products, the empty-tube technique is frequently employed. Consequently the combustion time is greatly shortened. Thus it does not permit a gradual or slow decomposition of the organic material and increases the danger of explosion.

Notwithstanding the above-mentioned limitations, simultaneous determinations will be welcomed by organic analysts. Continuing research in this area will put forward better and generally useful methods.

10.4 EXPERIMENTAL: SIMULTANEOUS DETERMINATION OF
      CHLORINE, BROMINE, AND IODINE WITH ONE SAMPLE

A. Principle

The organic material containing chlorine, bromine, and iodine is decomposed by closed-flask combustion. Using as absorbent a solution of hydrazine sulfate, all three halogens are converted to the corresponding halide
ions. After being diluted to a definite volume, two equal aliquots are taken.
One aliquot is titrated for iodide only, and the other is treated to remove
the iodine present and then analyzed for chlorine and bromine in the same
potentiometric titration operation (see Sec. 10.2B).

B. Equipment

    Combustion flask, 300-ml capacity (see Sec. 5.3).
    Metrohm Potentiograph Model E 336-A, titration stand, and combined
        silver-calomel electrode with KNO_3 reference solution (see Sec. 5.4).

C. Reagents

    Hydrazine sulfate solution, 2% in H_2O.
    Potassium dichromate solution, 0.5 $\underline{N}$.
    Thyodene (starch indicator) (Fisher Scientific Co., Fair Lawn, New
        Jersey).
    Silver nitrate standard solution, 0.02 $\underline{N}$: Dissolve 3.40 g of $AgNO_3$ in
        1 liter of deionized water; standardize against a known amount of
        pure sodium chloride.
    Nitric acid, concentrated.
    Methyl red indicator, 1% in ethanol.
    Buffer solution: Prepare a solution of acetic acid by diluting 17.2 ml
        of glacial acetic acid to 500 ml with distilled water, and another
        solution by dissolving 24.6 g of anhydrous sodium acetate in water
        and diluting to 500 ml. Then mix the two solutions together.

D. Procedure

    Decomposition

    Accurately weigh the organic material containing about 0.1 meq of
chlorine, bromine, and iodine for closed-flask combustion as described
in Sec. 5.3. Place 4 ml of 2% hydrazine sulfate solution and 4 ml of distilled water in the flask and carry out the combustion. Then quantitatively
transfer the contents to a 100-ml volumetric flask and make up to volume
with distilled water.

Preparation of Solutions for Titration

Pipet two equal and appropriately sized aliquots (10-40 ml) into 150-ml beakers. Treat the contents of one beaker (solution A) in the following manner. Add 5 ml of 0.5 $\underline{N}$ potassium dichromate solution, one drop (0.05 ml) of concentrated nitric acid, and 0.5 g of Thyodene, and stir for 5 min. Then boil for 10-15 min (until the blue color of the starch-iodine complex disappears) to remove the iodine in the solution. Add 1 ml of 0.5 $\underline{N}$ potassium dichromate solution to confirm the absence of iodine.

Potentiometric Titrations

Refer to Sec. 5.4 for the potentiometric titration procedure. Set the Potentiograph dials as follows:

Range switch: 1000.
Compensating potential: 750 (for Cl^- and Br^-, solution A), 500 (for I^-, solution B).
Titration speed: 4.

Before proceeding with the titrations, add several drops of 1% methyl red indicator and a few drops of concentrated nitric acid to the resultant solution (solution A, now containing only Cl^- and Br^-). Do the same to the contents of the other beaker (solution B, containing Cl^-, Br^-, and I^-). Introduce 20 ml of the acetate buffer to each beaker and dilute the volume to 100 ml.

Now titrate solution B with the 0.02 $\underline{N}$ $AgNO_3$ standard solution by means of the Potentiograph. Just record the first break, which gives the value of I^-.

Then titrate solution A, which contains Cl^- and Br^-, using the same titrant. The first break indicates Br^- and the second break indicates Cl^-.

Determine the endpoint of each titration graphically by drawing the best straight line through the region of maximum $\Delta mV/vol$. The midpoint of the line between the points where the drawn line separates from the curve is the equivalence point.

E. Calculation

$$\%X = \frac{N \times v \times E_X \times 100}{w}$$

where N = normality of the 0.02 $\underline{N}$ $AgNO_3$ standard solution
  v = volume of the 0.02 $\underline{N}$ $AgNO_3$ standard solution required
  E_X = equivalent weight of the halogen (Cl = 35.45, Br = 79.91, I = 126.90)
  w = weight (mg) of sample in aliquot

F. Comments

This method can be applied to the simultaneous determination of any binary combination of Cl, Br, and I in organic materials. After closed-flask combustion, the solution obtained is handled like solution B, except that two potentiometric titration breaks are recorded in order to get the values of both halides in the sample.

10.5 EXPERIMENTAL: SIMULTANEOUS GRAVIMETRIC DETERMINA-
      TION OF CARBON, HYDROGEN, AND THE OTHER ELEMENTS
      WITH ONE SAMPLE

A. Principle

The method for simultaneous gravimetric determination of carbon, hydrogen, and the other elements (referred to below as heteroelements) in one
weighed sample of organic material is dependent on pyrolytic combustion
of the sample in a quartz container placed in an empty combustion tube.
Carbon dioxide and water vapor are collected in the respective external
absorption tubes (see Sec. 2.6). Heteroelements, in the form of volatile
compounds, are retained in previously weighed quartz tubes placed inside
the combustion tube. Heteroelements forming nonvolatile residues are
determined by the change in weight of the quartz sample-container (see
Sec. 10.2A).
     The procedure described below is based on the information supplied by
Dr. N. E. Gel'man of the Institute of Elemental Organic Compounds,
Academy of Sciences, Moscow, USSR. More than 40 elements in various
combinations can be determined simultaneously with carbon and hydrogen
[64]. The experimental details were verified by Dr. Gel'man in January
1975.

B. Apparatus and Reagents

    Combustion tube and electric furnaces, as depicted in Fig. 10.16.
    Quartz insert tubes.
    Quartz sample containers, 60-90 mm long.
    Absorption tube for water, filled with Anhydrone (see Fig. 2.11).
    Absorption tube for nitrogen oxides, filled with silica gel impregnated
        with saturated potassium dichromate solution in concentrated sul-
        furic acid.
    Absorption tube for carbon dioxide, filled with Ascarite and Anhydrone.
    Oxygen cylinder.
    Aspirator.

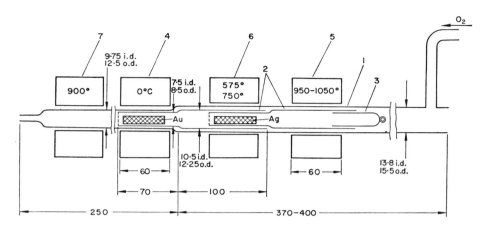

FIG. 10.16. Apparatus for simultaneous determination of carbon, hydrogen, and several heteroelements. Dimensions in millimeters. 1, combustion tube; 2, insert tubes; 3, container; 4, cooling device; 5, split-type furnace for combustion; 6, split-type furnace for heating silver; 7, split-type furnace for heating oxidizing zone. (From Ref. 64, courtesy Talanta.)

See Table 10.3 for the suggested filling and temperature of the quartz insert tube for the respective heteroelements. Use separate insert tubes for halogens and sulfur. If silvered pumice is used, one filling will serve for 80-100 determinations. An empty insert tube or one containing a filling which is not heated (e.g., gold, quartz wool) will last for at least 1 year. A quartz sample container can be used for 12-15 analyses; after each analysis, clean it with cotton wool, and then ignite or treat it with acid.

C. Procedure

Weighing the Sample

Weigh a solid sample in the quartz sample container, which can be hung horizontally on the hooks of the microbalance pan. For organic compounds that may decompose violently, distribute the sample on a layer of crushed quartz. Put a liquid sample into a capillary (preferably made of quartz), place it (open end downward) inside the quartz sample container, and weigh both the container and capillary together. Introduce a semisolid material to the quartz sample container by means of a thin quartz rod, which is then weighed together with the container.

TABLE 10.3. Conditions for Determination of Heteroelements from the Change in Weight of the Insert Tube.

| Elements to be determined | Quartz insert tube | |
	Temperature, °C	Filling
Cl, Br, I	575	Silver gauze or foil
Cl, Br, I	425	Silver-plated pumice
S	750	Silver gauze or foil
Rh	450	Silver gauze or foil
As, Mo, Sb, or Mn traces	Room temperature	None (or quartz wool)
Hg	0 (cool in ice)	Gold (or silver) gauze or foil

(From Ref. 64.)

Combustion

At the beginning of the day's work, burn a few milligrams of any standard substance to condition the apparatus. Then weigh the absorption tubes and quartz insert tubes. Weigh the sample into the quartz sample container and place the latter about 30 mm inside the quartz insert tube, which is subsequently pushed into the combustion tube (see positions in Fig. 10.16). Pass a current of oxygen through the combustion tube for 1-2 min and then connect the external absorption tubes and the aspirator. Heat (or cool) the narrow part of the quartz insert tube to the proper temperature (see Table 10.3). Adjust the position of the movable furnace so that the open end of the sample container is 10-15 mm inside it; then gradually advance the furnace at a rate dictated by the behavior of the sample. (The combustion is accompanied by the appearance of red-hot carbon particles moving toward the sample container outlet, or by flashes of flame, or by carbonization of the substance. The tempo of combustion can be controlled by moving the furnace backwards and forwards.) Keep the open end of the sample container in the heated zone all the time (to insure that the combustion takes place within the sample container). After combustion, if a carbonaceous residue remains in the sample container, cool it for 1-2 min and then heat it again until the carbon is all burned off. The total combustion time ranges from 15 to 30 min.

Determination

After the combustion is complete, disconnect the Pregl absorption tubes to be weighed (Sec. 10.5D, Note 1). Allow the sample container and quartz insert tube to cool in the combustion tube. Then withdraw them, remove any electrostatic charge (e.g., with a Tesla coil, vacuum detector), condition them on a metal block, and reweigh. Refer to Table 10.4 for the weighing forms of the respective heteroelements (Note 2).

TABLE 10.4. Heteroelements and Their Weighing Forms.

Element	Weighing form		
	In the container	In the insert tube	In the Pregl tube
Ru, Rh, Pd, Ag, Ir, Pt, Au	Ru*, Rh*, Pd, Ag, Ir*, Pt, Au		
Hg		Hg (gold amalgam)	
Tl	Tl_2O		
Be, Mg, Zn, Cd, Pb, Ni	BeO, MgO, ZnO, CdO, PbO, NiO		
B, Al, Cr, Fe, Ga, In, As, Eu, Er, Bi	B_2O_3, Al_2O_3, Cr_2O_3, Fe_2O_3, Ga_2O_3, In_2O_3, Eu_2O_3, Er_2O_3, Bi_2O_3	As_2O_3	
Mn	Mn_7O_{10}	Mn_7O_{10} (traces)	
Si, Ti, Ge, Zr, Sn, Hf, Th, Se, Te	SiO_2, TiO_2, GeO_2, ZrO_2, SnO_2, HfO_2, ThO_2, TeO_2	SeO_2, SeO_3 (Ag salt)	
Sb	Sb_2O_4	Sb_2O_4 (traces)	
P	P_2O_5		
Nb, Ta	Nb_2O_5, Ta_2O_5		
Mo, W	WO_3, MoO_3	MoO_3 (traces)	
S		SO_4 (Ag salt)	
Cl, Br, I		Cl, Br, I (Ag salt)	
Re		ReO_4 (Ag salt)	
Co	Co_3O_4		
Os			OsO_4[†] (Na salt)

*After reduction with hydrogen
[†]After freezing, evaporation, and absorption by Ascarite
(Information supplied by Professor N. E. Gel'man, October 1977.)

D. Notes

Note 1: To obtain correct results for carbon and hydrogen, empirical corrections are applied to the weights of water and carbon dioxide found. These corrections may be calculated from the results obtained for combustion of standard substances under similar conditions, and the values are usually less than 150 μg for water and 70 μg for carbon dioxide.

Note 2: If the heteroelement is determined as a sublimate collected on the cold walls of the quartz insert tube (e.g., As_2O_3, MoO_3), it is recommended to put some loose quartz wool in the insert tube outlet. For certain heteroelements (e.g., Sb, Mn) which are weighed as oxides retained in the sample container, the completeness of the retention should be checked by weighing the empty quartz insert tube. If the latter gains weight, this may be due to carry-over of finely dispersed particles or to volatilization of the residue in the sample container at high temperatures.

E. Comments

With two quartz insert tubes and the sample container to be weighed (see Fig. 10.16 for the arrangement of the quartz insert tubes), it is possible to determine four or even five elements using one weighed sample of the organic material. For example, in organomercury compounds it may be feasible to determine carbon, hydrogen, halogen or sulfur (in the Ag tube), mercury (in the Au tube), and one other element (e.g., P, B, Fe) which remains as residue in the sample container.

10.6 EXPERIMENTAL: SIMULTANEOUS DETERMINATION OF NITROGEN, PHOSPHORUS, AND BORON WITH ONE SAMPLE

A. Principle

The organic substance containing nitrogen, phosphorus, and boron (e.g., alkylaminophosphorodecaborane) is digested with concentrated sulfuric acid, hydrogen peroxide, selenium, and a 3:1 mixture of cupric sulfate-potassium sulfate in a quartz flask, whereupon nitrogen is converted to NH_4^+, phosphorus to PO_4^{3-}, and boron to BO_3^{3-}. The resulting solution is made up to a definite volume with distilled water, and suitable aliquots are then taken for the respective determinations. The nitrogen is determined by a Kjeldahl distillation and titration with 0.01 $\underline{N}$ HCl, the phosphorus is determined by spectrophotometry as the ammonium phosphovanadomolybdate complex, and the boron is determined by titration with 0.01 $\underline{N}$ NaOH in the presence of mannitol by the fixed pH method, using 6.30 as the critical pH.

B. Equipment

Quartz micro Kjeldahl flask, 10-ml capacity.
Digestion rack and distillation apparatus (see Sec. 3.3).
Spectrophotometer, Beckman Model B, or equivalent.
1.00-cm Corex cells.
Magnetic stirrer.

C. Reagents

$CuSO_4/K_2SO_4$ mixture (3:1).
Se powder.
H_2O_2, 30%.
Concentrated H_2SO_4.
Alundum boiling stones, 8-14 mesh (A. H. Thomas Co., Philadelphia).
NaOH solution, 30%.
H_3BO_3 solution, 2%.
Bromocresol green-methyl red indicator (see Sec. 3.3).
HCl standard solution, 0.01 $\underline{N}$.
Ammonium vanadate solution: Cautiously add to 500 ml of boiling distilled water first 2.35 g of NH_4VO_3 and then 100 ml of dilute H_2SO_4 (1:12). Cool, and make up to 1 liter.
Ammonium molybdate solution: Dissolve 122 g of $(NH_4)_6Mo_7O_{24} \cdot 4H_2O$ in 880 ml of distilled water.
Phosphate standard solution, 0.500 mg/ml: Dissolve 2.1315 g of pure $(NH_4)_2HPO_4$ or 2.196 g of pure KH_2PO_4 in a 1-liter volumetric flask and dilute to the mark with distilled water.
NaOH solution, 6 $\underline{N}$.
NaOH standard solution, 0.01 $\underline{N}$: Standardize this solution against pure boric acid.
Mannitol, CP.

D. Procedure

Decomposition

Accurately weigh the organic substance (5-10 mg) into a porcelain microboat and transfer to the 10-ml quartz micro Kjeldahl flask. Add

50 mg of the 3:1 $CuSO_4/K_2SO_4$ mixture and 10 mg of Se powder. Introduce 1 ml of concentrated H_2SO_4 and five drops (0.25 ml) of 30% H_2O_2. Add a few alundum boiling stones. Place the flask on the digestion rack and heat for about 45 min. During the course of the digestion, repeat the addition of five drops of 30% H_2O_2 several times; cool the flask quickly (with cold water) before each addition. At the completion of the digestion period, cool the flask again and add a few milliliters of distilled water. Boil the contents for several minutes to convert pyroacids (insoluble) to orthoacids (soluble). After cooling, transfer to a 50-ml volumetric flask and make up to volume. Withdraw aliquots of this solution for the determination of nitrogen, phosphorus, and boron, respectively.

Determination of Nitrogen

Withdraw an aliquot containing approximately 0.05 meq of N and transfer quantitatively into the micro Kjeldahl distillation apparatus. Add 15 ml of 30% NaOH and collect the distillate for 3 min in a 50-ml Erlenmeyer flask containing 5 ml of 2% boric acid and four drops of bormocresol green-methyl red indicator. Titrate the contents of the flask with the 0.01 $\underline{N}$ HCl standard solution to a pink endpoint.

Determination of Phosphorus

Withdraw a filtered aliquot containing 0.1-0.5 mg P and transfer to a 100-ml volumetric flask. Dilute the contents to about 65 ml with distilled water; add 1.5 ml of concentrated H_2SO_4, 10 ml of ammonium vanadate solution, 10 ml of ammonium molybdate solution, and make up to volume. After 30 min, read the absorbance at 410 nm against a blank. Prepare a calibration graph covering the range 0.1-0.5 mg P by use of a microburet and the phosphate standard solution.

Determination of Boron

Withdraw an aliquot containing 0.05-0.10 meq of B and transfer to a 150-ml beaker. Add distilled water to make a total volume of 50 ml. With a moderate rate of stirring, adjust the pH of the solution to 4-5 using 6 $\underline{N}$ NaOH, and then more carefully to exactly 6.30 using 0.01 $\underline{N}$ NaOH. Add excess mannitol (7-9 g) until the pH remains constant. Place the tip of the microburet containing the 0.01 $\underline{N}$ NaOH standard solution below the surface of the solution in the beaker, and record the volume of 0.01 $\underline{N}$ NaOH needed to bring the pH back to exactly 6.30.

E. Calculation

$$\%N = \frac{N_A \times V_A \times 14.008 \times 100}{W_N}$$

where N_A = normality of the 0.01 $\underline{N}$ HCl standard solution
V_A = volume of the 0.01 $\underline{N}$ HCl standard solution for titration of the distillate
W_N = weight (mg) of sample in aliquot for nitrogen determination

$$\%P = \frac{P \times 100}{W_P}$$

where P   = weight (mg) of P obtained from the calibration graph
W_P = weight (mg) of sample in aliquot for phosphorus determination

$$\%B = \frac{N_B \times V_B \times 10.82 \times 100}{W_B}$$

where N_B = normality (against boric acid) of the 0.01 $\underline{N}$ NaOH standard solution
V_B = volume of 0.01 $\underline{N}$ NaOH standard solution needed to return the pH to 6.30
W_B = weight (mg) of sample in aliquot for boron determination

F.   Comments

(1) The determination of only two of the above elements simultaneously is of course possible and in some instances the digestion procedure may be simplified. Thus, the simultaneous determination of boron and phosphorus is readily accomplished by digestion with concentrated H_2SO_4 and a few drops of concentrated HNO_3. In the simultaneous analysis of boron and nitrogen, the 30% H_2O_2 may be omitted. On the other hand, simultaneous determination of nitrogen and phosphorus requires all the reagents described in the above procedure.

(2) Only samples containing nitrogen in a form which can be determined by the micro Kjeldahl method may be analyzed for nitrogen by this procedure.

(3) The presence of H_2O_2 will interfere with the spectrophotometric determination of phosphorus. Therefore it is essential that the H_2O_2 be decomposed during the digestion.

REFERENCES

1.   F. Pregl, Die Quantitative Organische Mikroanalyse, 3rd Ed.,
     Springer, Berlin, 1930.
2.   J. B. Niederl and V. Niederl, Micromethods of Quantitative Organic
     Analysis, 2nd Ed., Wiley, New York, 1942.

3. H. Roth, Pregl-Roth Quantitative Organische Mikroanalyse, 7th Ed.,
    Springer, Wien, 1958.
4. G. Ingram, Methods of Organic Elemental Microanalysis, Reinhold,
    New York, 1961.
5. A Steyermark, Quantitative Organic Microanalysis, 2nd Ed., Aca-
    demic, New York, 1961.
6. R. Levy, Microanalyse Organique Elémentaire, Masson, Paris,
    1961.
7. T. S. Ma, unpublished work; see R. T. E. Schenck, Ph.D. Thesis,
    University of Chicago, 1939.
8. R. Belcher and C. E. Spooner, J. Chem. Soc. 1943:313.
9. G. Ingram, Mikrochim. Acta 1953:71.
10. M. O. Korshun (Obituary), Izv. Akad. Nauk SSSR, Otd. Khim. Nauk
    1959:759.
11. T. S. Ma, Anal. Chem. 30:762 (1958); 32:82R (1960).
12. T. S. Ma and M. Gutterson, Anal. Chem. 34:114R (1962); 36:155R
    (1964); 38:190R (1966); 40:150R (1968); 42:109R (1970); 44:451R
    (1972); 46:437R (1974).
13. J. A. Kuck (ed.), Methods in Microanalysis, Vol. 1, Simultaneous
    Rapid Combustion (translated by P. L. Bolton and K. Gingold),
    Gordon and Breach, New York, 1964.
14. J. A. Kuck (ed.), Methods in Microanalysis, Vol. 4, Determination
    of Carbon and Hydrogen in the Presence of Other Elements or Simul-
    taneously with Them (translated by K. Gingold), Gordon and Breach,
    New York, 1969.
15. W. Walisch, Chem. Ber. 94:2314 (1961).
16. C. Maresch, O. E. Sundberg, R. A. Hofstader, and G. E. Gerhard,
    Microchem. J., Symp. Ser. 2:387 (1962).
17. C. F. Nightingale and J. M. Walker, Anal. Chem. 34:1435 (1962).
18. J. T. Clerc, R. Dohner, W. Sauter, and W. Simon, Helv. Chim.
    Acta 46:2369 (1963).
19. C. D. Miller and J. D. Winefordner, Microchem. J. 8:334 (1964).
20. R. D. Condon, Microchem. J. 10:408 (1966).
20a. R. A. Nadkarni and G. H. Morrison, Radiochem. Radioanal. Lett.
    24:103 (1976).
20b. A. Gaudry, B. Maziere, D. Comar, and D. Nau, J. Radioanal.
    Chem. 29:77 (1976).
20c. M. H. Yang, S. J. Yeh, and S. F. Lai, Radioisotopes (Tokyo)
    22:118 (1973).
20d. R. L. Hutson, J. J. Reidy, K. Springer, H. Daniel, and H. B.
    Knowles, private communication, November 1976.
20e. S. J. Anderson, D. S. Brown, and A. H. Norbury, J. Organometal.
    Chem. 64:301 (1974).

21. M. O. Korshun and N. S. Sheveleva, Zh. Anal. Khim. 11:376 (1956).
22. M. O. Korshun, N. E. Gel'man and N. S. Sheveleva, Zh. Anal. Khim. 13:695 (1958).
23. V. A. Klimova and E. N. Merkulova, Izv. Akad. Nauk SSSR 1959:781.
24. O. Hadzija, Mikrochim. Acta 1970:970.
25. N. E. Gel'man and M. O. Korshun, Dokl. Akad. Nauk SSSR 86:685 (1953).
26. N. E. Gel'man, M. O. Korshun, and N. S. Sheveleva, Zh. Anal. Khim. 12:526 (1957).
27. M. O. Korshun, N. E. Gel'man, and K. I. Glazova, Dokl. Akad. Nauk SSSR 111:1255 (1956).
28. N. E. Gel'man, M. O. Korshun, and K. I. Novozhilova, Zh. Anal. Khim. 15:222, 342 (1960).
29. K. I. Novozhilova and N. E. Gel'man, Zh. Anal. Khim. 22:955 (1967).
30. T. S. Ma, in I. M. Kolthoff and P. J. Elving (eds.), Treatise on Analytical Chemistry, Part II, Vol. 12, Wiley, New York, 1965, p. 120.
31. N. E. Gel'man, M. O. Korshun, and K. I. Novozhilova, Zh. Anal. Khim. 15:628 (1960).
32. A. V. Zimin, S. V. Churmanteev, A. V. Gubanova, and A. D. Verina, Dokl. Akad. Nauk SSSR 126:784 (1959).
33. M. O. Korshun, Zh. Anal. Khim. 7:101 (1952).
34. M. O. Korshun and N. S. Sheveleva, Zh. Anal. Khim. 7:104 (1952).
35. N. V. Sokolova, Zh. Anal. Khim. 11:728 (1956).
36. E. I. Margolis and N. F. Egorova, Vestn. Moskv. Univ., Ser. Mat. Mekh. Astron. Fiz. Khim. 13:209 (1958).
37. V. A. Klimova, T. Z. Antipova, and G. K. Mukhina, Izv. Akad. Nauk SSSR Otd. Khim. Nauk 1962:19.
38. G. Ingram, Analyst 86:411 (1961); G. Ingram, in I. M. Koltshoff and P. J. Elving (eds.), Treatise on Analytical Chemistry, Part II, Vol. 11, Wiley, New York, 1965, p. 322.
39. G. Ingram, Microchem. J., Symp. Ser. 2:495 (1961).
40. N. P. Fedorovskaya and A. A. Zakharova, Tr. Inst. Goruchikh Iskopaemykh, Nauk SSSR 21:179 (1963).
41. A. A. Abramyan and A. A. Kocharyan, Izv. Akad. Nauk Arm. SSR, Khim. Nauk 17:301 (1964).
42. A. A. Abramyan, A. A. Kocharyan, and R. A. Megroyan, Izv. Akad. Nauk Arm. SSR, Khim. Nauk 20:25 (1967).
43. V. A. Klimova and G. K. Mukhina, Izv. Akad. Nauk. SSSR, Otd. Khim. Nauk 1959:2248.
44. A. A. Abramyan and S. M. Atashyan, Izv. Akad. Nauk Arm. SSR, Khim. Nauk 18:532 (1965).
45. H. L. Wang and H. H. Hsu, Hua Hsueh Tung Pao, 1962:50.

46. V. A. Klimova, M. O. Korshun, and E. G. Bereznitskaya, Dok. Akad. Nauk SSSR 96:287 (1954).
47. M. O. Korshun, E. A. Terent'eva, and V. A. Klimova, Zh. Anal. Khim. 9:275 (1954).
48. M. O. Korshun and E. A. Terent'eva, Dok. Akad. Nauk SSSR 100:707 (1955).
49. M. Papay and L. Mazor, Mikrochim. Acta 1967:269.
50. L. Mazor, Mikrochim. Acta 1957:113.
51. H. L. Wang, Acta Chim. Sinica 30:211 (1964).
52. V. A. Klimova, M. O. Korshun, and E. G. Bereznitskaya, Dok. Akad. Nauk SSSR 96:81 (1954).
53. V. A. Klimova, M. O. Korshun, and E. G. Bereznitskaya, Zh. Anal. Khim. 11:223 (1956).
54. V. A. Klimova, M. O. Korshun, and E. G. Bereznitskaya, Dok. Akad. Nauk SSSR 84:1175 (1952).
55. Y. N. Platonov, Tr. Komis. Anal. Khim. Akad. Nauk SSSR 13:15 (1963).
56. T. Aranyi and A. Erdey, Magy. Kem. Lapja 20:164 (1965).
57. V. A. Klimova and E. G. Bereznitskaya, Zh. Anal. Khim. 12:424 (1957).
58. V. A. Klimova and E. G. Bereznitskaya, Zh. Anal. Khim. 11:292 (1956).
59. R. C. Rittner and R. Culmo, Anal. Chem. 34:673 (1962).
60. H. Trutnovsky, Mikrochim. Acta 1968:371.
61. A. I. Lebedeva and K. S. Kramer, Izv. Akad. Nauk SSSR, Otd. Khim. Nauk 1962:1309.
62. V. Pechanec, Collect. Czech. Chem. Commun. 27:2009 (1962).
63. M. O. Korshun, N. S. Sheveleva, and N. E. Gel'man, Zh. Anal. Khim. 15:99 (1960).
64. N. E. Gel'man, Talanta 14:1423 (1967); private communication, January 1975.
65. A. S. Zabrodina and V. P. Miroshina, Vestn. Mosk. Univ. 1958:187.
66. M. K. Dzhamaletdinova, Zavod. Lab. 29:1175 (1963).
67. M. K. Dzhamaletdinova and L. V. Ivanova, Izv. Akad. Nauk Kazakh. SSR, Ser. Khim. 1962:112.
68. A. I. Lebedeva, N. A. Mikolaeva, and E. V. Shikhman, Zh. Anal. Khim. 20:832 (1965).
69. N. E. Gel'man and I. I. Bryushkova, Zh. Anal. Khim. 19:369 (1964).
70. E. C. Olson, in I. M. Kolthoff and P. J. Elving (eds.), Treatise on Analytical Chemistry, Part II, Vol. 14, Wiley, New York, 1971, p. 19.
71. F. Cassani, Chim. Ind. (Milan) 51:167 (1969).
72. M. O. Korshun and M. N. Chumachenko, Dok. Akad. Nauk SSSR 99:769 (1954).

73. S. S. M. Hassan and M. B. Elsayes, Mikrochim. Acta 1972:115.
74. H. A. Laitinen and K. W. Boyer, Anal. Chem. 44:920 (1972).
75. V. J. Shiner, Jr. and M. L. Smith, Anal. Chem. 28:1043 (1956).
76. R. C. Rittner and T. S. Ma, Mikrochim. Acta 1976(I):243.
77. V. Nuti and P. L. Ferrarini, Farmaco, Ed. Sci. 24:930 (1969).
78. M. O. Korshun and E. A. Bondarevskaya, Dokl. Akad. Nauk SSSR 110:220 (1956).
79. N. E. Gel'man, M. O. Korshun, M. N. Chumachenko, and N. I. Larina, Dokl. Akad. Nauk SSSR 123:468 (1958).
80. M. N. Chumachenko, M. O. Korshun, V. P. Burlaka, and V. N. Simonova, Dokl. Akad. Nauk SSSR 133:138 (1960).
81. P. N. Fedoseev and N. P. Ivashova, Zh. Anal. Khim. 7:116 (1952).
82. T. S. Ma and W. G. Zoellner, Mikrochim. Acta 1972:333.
83. N. T. Faithful, Lab. Prac. 20:41 (1969).
84. R. C. Rittner and R. Culmo, Anal. Chem. 34:673 (1962).
85. S. K. Ng and P. T. Lai, Appl. Spectrosc. 26:369 (1972).
86. S. K. Ng, R. C. Hsia, and P. T. Lai, Appl. Spectrosc. 24:583 (1970).
87. J. F. Kopp and R. C. Kroner, Appl. Spectrosc. 19:155 (1965).
87a. E. A. Terent'eva, N. N. Smirnova, and I. M. Pruslina, Zavod. Lab. 40:143 (1974).
87b. F. Mikolanda, Chem. Prum. 23:308 (1973).
88. H. Trutnovsky, Z. Klim. Chem. Klin. Biochem. 9:341 (1971).
88a. M. J. Pedrosa and J. Paul, Microchem. J. 19:314 (1974).
89. G. Dugan and V. A. Aluise, Anal. Chem. 41:495 (1969).
90. V. A. Aluise, H. K. Alber, H. S. Conway, C. C. Harris, W. H. Jones, and W. H. Smith, Anal. Chem. 23:530 (1951).
91. G. Dugan and V. A. Aluise, Hercules Chemists 60:7 (1970); U. S. Patent 3,838,969 (1974); private communication, November 1974.
91a. G. Dugan, Anal. Lett. 10:639 (1977); private communication, October 1977.
92. R. Culmo, Mikrochim. Acta 1968:811.
93. NUMEC-Carlo Erba Model 1100 Elemental Analyzer, NUMEC Instrument Department, Apollo, Pa.; Carlo Erba Scientific Instruments Division, Rodano (Milano), Italy.
94. W. C. Heraeus G. M. B. H., Hanau, Germany.
95. R. F. Culmo, Microchem. J. 17:499 (1972).
96. V. A. Klimova and K. S. Zabrodina, Akad. Nauk SSSR 1959:582.
97. E. I. Margolis and A. G. Shevkoplyas, Vestnk. Mosk. Univ. 1960:73.
98. I. Monar, Mikrochim. Acta 1965:208.
99. L. Foissac, Chim. Anal. 48:354 (1966).

100. M. N. Chumachenko and I. E. Pakhomova, Izv. Akad. Nauk SSSR, Ser. Khim. 1968:235.
101. J. W. Frazer, Mikrochim. Acta 1962:993.
102. J. W. Frazer and R. K. Stump, Mikrochim. Acta 1968:1326.
103. M. Marzadro and A. Mazzeo-Farina, Ann. Ist. Super Sanita 4:608 (1968).
104. E. Wachberger, A. Discherl, and K. Pulver, Microchem. J. 16:318 (1971).
105. C. Maresh, O. E. Sundberg, R. A. Hofstader, and G. E. Gerhardt, Microchem. J., Symp. Ser. 2:390 (1962).
105a. V. R. Larionov, M. I. Al'yanov, Y. M. Khlyupin, V. F. Borodkin, Y. I. Yashin, and R. P. Smirnov, Tr. Ivanovsk. Khim. Tekhnol. Inst. 1973(15):71.
106. C. F. Nightingale and J. M. Walker, Anal. Chem. 34:1435 (1962).
107. J. T. Clerc, Promotionsarb. (Zurich) 1964:3496.
108. W. Walisch, Trans. N.Y. Acad. Sci. 25:693 (1963).
109. H. Weitkamp and F. Korte, Chem.-Ing.-Tech. 35:429 (1963).
110. R. Donner, Z. Chem. 5:466 (1965).
111. E. Kozlowski and E. Sienkowska-Zyskowska, Bull. Acad. Polon. Sci., Ser. Sci. Chim. 13:323 (1965).
112. K. Hofmann, Naturwissenschaften 52:428 (1965).
113. B. Lewandowska, Chem. Anal. (Warsaw) 10:1353 (1965).
114. M. N. Chumachenko and I. E. Pakhomova, Dokl. Akad. Nauk SSSR 170:125 (1966).
115. G. Kainz and E. Wachberger, Mikrochim. Acta 1968:395.
116. M. N. Chumachenko, I. E. Pakhomova, and R. A. Ivanchikova, Izv. Akad. Nauk SSSR, Ser. Khim. 1970:1219.
117. M. N. Chumachenko and N. B. Levina, Dokl. Akad. Nauk SSSR 180:894 (1968).
118. V. Rezl, Microchem. J. 15:381 (1970).
119. A. S. Zabrodina and N. F. Egorova, Vestn. Mosk. Univ., Ser. Khim. 15(4):66 (1960).
119a. Y. Tomida, M. Okano, and T. Ando, Japan Analyst 25:161 (1976).
120. A. A. Abramyan and S. M. Atashyan, Izv. Akad. Nauk Arm. SSR, Khim. Nauk 14:401 (1961).
120a. V. N. Tkacheva and V. K. Bukina, Dokl. Akad. Nauk Uzb. SSR 1975(5):31.
121. G. Gutbier and G. Rockstroh, Mikrochim. Acta 1962:686.
122. A. I. Lebedeva, N. A. Nikolaeva, and V. A. Orestova, Zh. Anal. Khim. 17:993 (1962).
123. P. B. Olson and R. E. Kolb, Microchem. J. 12:117 (1967).
124. A. A. Abramyan and R. A. Megroyan, Arm. Khim. Zh. 21:115 (1968).

125. O. Hadzija, Mikrochim. Acta 1969:59.
126. M. Marzadro and J. Zavattiero, Mikrochim. Acta 1971:67.
126a. V. S. Bostoganashvili, D. G. Turabelidze, and D. G. Narchemashvili, Tr. Inst. Farm.-Khim., Akad. Nauk Gruz. SSR, Ser. 1 1973(12):19.
127. H. Rotzsche and H. V. Jurczyk, Z. Chem. 8:263 (1968).
127a. M. A. Volodina, A. A. Barysheva, and V. I. Garnova, Zh. Anal. Khim. 28:977 (1973).
128. H. Malissa, Mikrochim. Acta 1960:127.
128a. A. A. Abramyan and A. A. Kocharyan, Arm. Khim. Zh. 27:745 (1974).
129. E. Pell and H. Malissa, Talanta 9:1056 (1962).
130. H. Malissa and W. Schmidt, Microchem. J. 8:180 (1964).
131. R. C. Rittner and R. Culmo, Microchem. J. 11:269 (1966).
132. A. A. Abramyan and R. A. Megroyan, Arm. Khim. Zh. 21:111 (1968).
133. O. Hadzija, Mikrochim. Acta 1968:619.
134. D. R. Beuerman and C. E. Meloan, Anal. Chem. 34:1671 (1962).
135. S. I. Obtemperanskayz and F. I. Mullayanov, Vestnk. Mosk. Gos. Univ., Ser. Khim. 11:118 (1970).
136. A. P. Terent'ev, M. A. Volodina, and E. G. Fursova, Dokl. Akad. Nauk SSSR 169:851 (1966).
137. A. P. Terent'ev, M. A. Volodina, E. G. Fursova, and G. A. Martynova, Zh. Anal. Khim. 23:953 (1968).
137a. L. Futekov, O. Kütschukov, and H. Specker, Z. Anal. Chem. 284:197 (1977).
138. N. E. Gel'man, N. S. Sheveleva, and N. I. Shakhova, Zh. Anal. Khim. 23:1067 (1968).
138a. N. E. Gel'man, N. S. Sheveleva, and A. G. Gorova, Usp. Anal. Khim. 1974:297.
139. T. Onoe, C. Furukawa, and H. Otsuka, Ann. Rep. Takamina Lab. 11:100 (1959).
140. A. A. Abramyan, A. A. Kocharyan, and R. A. Megroyan, Izv. Akad. Nauk Arm. SSR, Khim. Nauk 20:29 (1967).
141. O. Hadzija, Mikrochim. Acta 1969:1114.
142. D. R. Beuerman and C. E. Meloan, Anal. Lett. 1:195 (1967).
143. V. S. Bazalitskaya and G. M. L'dokova, Izv. Akad. Nauk Kaz. SSR, Ser. Khim. 1968:71.
143a. N. E. Gel'man, V. I. Skorobogtova, Y. M. Faershtein, and I. M. Korotaeva, Zh. Anal. Khim. 28:611 (1973).
144. S. I. Obtemperanskayz and I. V. Dudova, Zh. Anal. Khim. 24:1241 (1969).
145. S. I. Obtemperanskaya and I. V. Dudova, Vestnk. Mosk. Gos. Univ., Ser. Khim. 11:461 (1970).

146. A. I. Lebedeva and E. F. Fedorova, Zh. Anal. Khim. 16:87 (1961).
147. V. Pechanec, Collect. Czech. Chem. Commun. 27:2976 (1962).
148. G. Kainz and F. Scheidl, Mikrochim. Acta 1964:998.
149. M. K. Papay and L. Mazor, Magy. Kem. Lapja 24:621 (1969).
150. M. Bigois, R. Levy, and M. Marzin, Bull. Soc. Chim. Fr. 1970:388.
151. M. A. Volodina, S. Z. Ivin, and M. V. Pal'yanova, Vestn. Mosk. Gos. Univ., Ser. Khim. 1970:632.
152. Y. A. Gawargious, G. M. Habashy, and B. N. Faltaoos, Indian J. Chem. 7:610 (1969).
153. A. Radecki, Chem. Anal. (Warsaw) 8:607 (1963).
153a. L. S. Ignatenko, V. V. Mashchenko, and T. Y. Fedoseeva, Izv. Vyssh. Ucheb. Zaved., Khim. Khim. Tekhnol. 16:1850 (1973).
154. J. C. Mamaril and C. E. Meloan, J. Chromatogr. 17:23 (1965).
155. P. N. Fedoseev and T. E. Chernysheva, Izv. Vyssh. Ucheb. Zaved., Khim. Khim. Tekhnol. 10:1024 (1967).
156. A. Campiglio, Mikrochim. Acta 1968:106.
157. J. G. Gagnon and P. B. Olson, Anal. Chem. 40:1856 (1968).
157a. A. A. Abramyan, A. S. Tevosyan, and R. A. Megroyan, Arm. Khim. Zh. 28:614 (1975).
158. P. N. Fedoseev and V. D. Osadchii, Izv. Vyssh. Ucheb. Zaved., Tekhnol. Legk. Prom. 1969:57.
158a. V. M. Vladimirova, P. N. Fedoseev, and V. D. Osadchii, Izv. Vyssh. Ucheb. Zaved., Tekhnol. Legk. Prom. 1972(5):51.
159. M. N. Chumachenko and V. P. Miroshina, Zavod. Lab. 26:1084 (1960).
159a. A. Mazzeo-Farina and P. Mazzeo, Microchem. J. 21:198 (1976).
160. G. Giesselmann and I. Hagedorn, Mikrochim. Acta 1960:390.
161. E. Pella, Mikrochim. Acta 1961:472.
162. R. Belcher and J. E. Fildes, Anal. Chim. Acta 26:155 (1962).
163. D. C. White, Mikrochim. Acta 1962:807.
164. K. Yoshikawa and T. Mitsui, Japan Analyst 10:723 (1961).
165. K. Hozumi and K. Mizuno, Japan Analyst 10:383 (1961).
166. V. M. Gorokhovskii and A. M. Kazymova, Zh. Anal. Khim. 19:499 (1964).
167. L. Heinrich and H. Diedrich, Z. Anal. Chem. 197:360 (1963).
168. J. Horacek and V. Pechanec, Collect. Czech. Chem. Commun. 31:4268 (1966).
169. S. Utsumi, W. Machida, and S. Ito, Japan Analyst 16:674 (1967).
170. D. Pitre and M. Grandi, Mikrochim. Acta 1967:347.
171. V. D. Osadchii and P. N. Fedoseev, Izv. Vyssh. Ucheb. Zaved., Tekhnol. Legk. Prom. 1969:76.
171a. O. Hadzija and Z. Korzarac, Z. Anal. Chem. 277:191 (1975).

172. E. A. Bondarevskaya, V. M. Kuznetsova, and S. V. Syavtsillo, Zh. Anal. Khim. 16:472 (1961).

173. M. A. Volodina and T. A. Gorshkova, Zh. Anal. Khim. 24:1437 (1969).

173a. M. A. Volodina and M. D. Khamed, Zh. Anal. Khim. 27:1828 (1972).

174. M. Marzadro and J. Zavattiero, Mikrochim. Acta 1969:1262.

174a. K. Imaeda, K. Ohsawa, and K. Ohgi, Japan Analyst 22:1568 (1973).

175. N. E. Gel'man and I. I. Bryushkova, Zh. Anal. Khim. 18:1100 (1963).

175a. M. A. Volodina, T. A. Gorshkova, O. P. Morozova, and A. A. Barysheva, Zh. Anal. Khim. 28:1833 (1973).

176. C. Eger and J. Lipke, Anal. Chim. Acta 20:548 (1959).

176a. M. Gachon, A. Gehenot, and G. Maire, Bull. Soc. Chim. Fr. 1975(I):2442.

177. R. Roncucci, G. Lambelin, M. J. Simon, and W. Soudyn, Anal. Biochem. 26:118 (1968).

177a. V. G. Shah, S. S. Ramdasi, R. B. Malvankar, S. Y. Kulkarni, and V. S. Pansare, Indian J. Chem. 12:419 (1974).

178. V. A. Klimova and M. D. Vitalina, Zh. Anal. Khim. 22:406 (1967).

179. M. P. Strukova and A. A. Lapshova, Zh. Anal. Khim. 24:1577 (1969).

180. W. Z. Leithe, Z. Anal. Chem. 195:93 (1963).

181. K. I. Nakamura, K. Ono, and K. Kawada, Microchem. J. 15:364 (1970).

182. A. P. Terent'ev, A. M. Turkel'taub, E. A. Bondarevskaya, and L. A. Domochkina, Dokl. Akad. Nauk SSR 148:1316 (1963).

182a. K. Ubik, Microchem. J. 17:556 (1972).

182b. A. A. Abramyan, A. S. Tevosyan, and R. A. Megroyan, Zh. Anal. Khim. 30:817 (1975).

183. D. S. Galanos and V. M. Kapoulas, Anal. Chim. Acta 34:360 (1966).

183a. J. W. B. Stewart and J. Ruzicka, Anal. Chim. Acta 82:137 (1976).

184. T. R. F. W. Fennell and J. R. Webb, Talanta 2:389 (1959).

185. W. Radmacher and A. Hoverath, Z. Anal. Chem. 167:336 (1959).

186. M. N. Chumachenko and V. P. Burlaka, Izv. Akad. Nauk SSSR, Otd. Khim. Nauk 1963:5.

187. A. J. Christopher, T. R. F. W. Fennell, and J. R. Webb, Talanta 11:1323 (1964).

187a. N. A. Gradskova, E. A. Bondarevskaya, and A. P. Terent'ev, Zh. Anal. Khim. 28:1846 (1973).

188. B. M. Luskina, A. P. Terent'ev, and N. A. Gradskova, Zh. Anal. Khim. 20:990 (1965).

188a. Z. Sliepcevic, M. S. Siroki, and Z. Stefanac, Mikrochim. Acta 1973:945.

189. K. Umemoto, S. Hirose, K. Sakamoto, T. Kouri, and K. Hozumi, Japan Analyst 19:191 (1970).

190. D. Filippo, G. Peyronel, and C. Preti, Ann. Chim. (Rome) 53:1552 (1963).

191. V. D. Osadchii and P. N. Fedoseev, Izv. Vyssh. Ucheb. Zaved., Tekhnol. Legk. Prom. 1969:47.

192. A. P. Terent'ev, B. M. Luskina, and S. V. Syavtsillo, Zh. Anal. Khim. 16:635 (1961).

193. B. M. Luskina, A. P. Terent'ev, and S. V. Syavtsillo, Tr. Komis. Anal. Khim. Akad Nauk SSSR 13:3 (1963).

194. A. P. Kreshkov, L. V. Myshlyaeva, E. A. Kuchkarev, and T. G. Shetunova, Lakokrasochnyi Materialy Prum. 1966:60.

195. A. P. Terent'ev and B. M. Luskina, Zh. Anal. Khim. 14:112 (1959).

195a. A. D. Horton, W. D. Shults, A. S. Meyer, and D. R. Mathews, Environ. Sci. Technol. 7:449 (1973).

196. S. A. Liebman, D. H. Ahlstron, T. C. Creighton, G. D. Pruder, R. Averitt, and E. J. Levy, Anal. Chem. 44:1411 (1972).

196a. Elemental Analyzer and Peak Identifier (brochure), Chemical Data Systems, Inc., Oxford, Pa., 1976.

197. American Microchemical Society, Round Table Discussion on Techniques with Automated Elemental Analyzers, East Brunswick, N.J., April 1973.

198. International Symposium on Microchemical Techniques, Automated Elemental Analyzers--Ten Years Later, University Park, Pa., August 1973.

199. E. Wachberger, A. Dirscherl, and K. Pulver, Microchem. J. 16:318 (1971).

200. B. Cousin, Bull. Soc. Chim. Fr. 1971:361.

201. K. Ono, A. Saito, K. Kawada, K. Nakamura, and M. Yamamuro, Ann. Sankyo Res. Lab. 19:62 (1967); 21:28 (1969); 22:34, 41, 49 (1970).

202. M. Shimizu and K. Hozumi, Japan Analyst 19:1041 (1970).

203. R. Stoffel, Mikrochim. Acta 1972:242.

204. K. Hozumi, O. Tsuji, and H. Kushima, Microchem. J. 15:481 (1970).

Ultramicro Analysis

11.1 GENERAL CONSIDERATIONS

A. Introductory Remarks

In this chapter we discuss the methods for determining the elements in organic materials using less than 1 mg of the sample. Various names have been advocated to designate this kind of analysis. Thus, in order to differentiate these analytical procedures from the conventional ones employing milligram quantities of organic material, the terms "submicro-method" and "ultramicromethod" are used interchangeably. Some workers prefer to indicate the amounts of sample taken for analysis and hence classify the methods into "decimilligram," "centimilligram," and "microgram" procedures, respectively. It should be noted that these procedures stipulate that the initial sample size is in the specified range. If the final product to be measured is in the microgram region while more than milligram amounts of the organic material are required to perform the determination, the analytical technique is known as trace analysis (see Chap. 12). Understandably, procedures of ultramicro analysis and trace analysis may use an identical mode of finish when the final product for measurement is obtained at the same level, say 5 μg.

The techniques of quantitative ultramicro organic analysis were first investigated in the biochemical laboratories. During the 1930s Lindestrøm-Lang and co-workers [1, 1a] developed many useful methods in Copenhagen, Denmark; these were followed up by Kirk [2] in Berkeley, California. It is understandable that these workers were not concerned with the determination of carbon and hydrogen, and that the precision and accuracy of their methods did not need to meet the requirements of the conventional

organic elemental analysis. In 1951 Belcher [3] and co-workers initiated
a project in Birmingham, England, to scale down the Pregl micromethods
about 100-fold, using sample weights approximating to the 30-50 μg range.
About the same time, Kirsten [4] studied methods in Uppsala, Sweden, to
determine the elements with decimilligram amounts of organic material;
he was later joined by Hozumi [5] who continued the investigation in Kyoto,
Japan. Subsequently Tölg [6] carried out determinations in Mainz, Germany, using a few micrograms of the analytical sample. Thus, the development of ultramicro analysis has been an international endeavor.

B. Applications

The chief advantage of ultramicro analysis over the conventional procedures is the conservation of the material submitted for analysis. This ex-
plains the general acceptance of ultramicro analysis in biochemical and
clinical tests [7, 8]. In the case of new or unknown organic compounds,
ultramicro procedures are not intended to replace the micromethods using
milligram amounts of the organic substance. Ultramicro analysis is indispensable, however, when less than 1 mg of the compound can be isolated for investigation; such is the case with some metabolic compounds or
minor products of certain organic reactions. Recently, thanks to the development of new micro separation techniques, it is feasible to purify and
recover microgram quantities of organic compounds, for instance, by
means of high-speed liquid chromatography [9]. Furthermore, since it is
now possible to carry out organic synthesis at the microgram level [10],
the applications of ultramicro analytical methods become apparent.

C. Approaches

An obvious approach to scaling down the sample size while maintaining the
desired precision and accuracy is to refine the equipment and techniques
used in microanalysis. Unfortunately, attempts in this direction have
frequently ended in despair. For instance, it is impossible to determine
carbon and hydrogen below the milligram region by weighing the carbon
dioxide and water when they are obtained in microgram amounts, since
there is a limit to which the dimensions of the absorption tubes and the
quantities of absorbents can be reduced. Gravimetric procedures which
involve precipitation and filtration usually fail owing to the loss of the precipitate in transfer, or the difficulty in removing the contaminants. Titrimetry using visual indicators becomes less accurate as the concentration of reacting species in the solution diminishes. Griepink et al. [10a]
have discussed the general problems in lowering the limits of determination; Römer [10c] has studied the minimization of C, H analysis.

In order to circumvent the above-mentioned difficulties, electrochemical determinations are favored in ultramicro analysis [11-12a], taking advantage of the fact that electrical measurements are precise and extremely sensitive. For similar reasons, spectrophotometric finishes have been advocated; these methods are characterized by high sensitivity although they are sometimes of insufficient accuracy. Gas chromatography has been extensively exploited for ultramicro separation of combustion products, though satisfactory quantitative determinations of such products in the microgram region still await investigation.

Another approach is to convert the combustion product into a substance which has a very large conversion factor (e.g., determination of phosphate [13]), or into a suitable derivative with high spectroscopic absorptivity. Still another approach involves the utilization of "amplification" or "multiplication" reactions in which the normal equivalence is altered in some way so that a more favorable measurement can be made [14]. One example is the iodimetric titrimetry described for the determination of iodine which amplifies the original equivalence sixfold (see Sec. 5.5). It is also possible to amplify the equivalence 24-fold by the following sequence of reactions [15]:

$$2I^- + IO_4^- + 2H^+ \longrightarrow I_2 + IO_3^- + H_2O$$
$$I_2 + 5IO_4^- + H_2O \longrightarrow 7IO_3^- + 2H^+$$

Combining,

$$2I^- + 6IO_4^- \longrightarrow 8IO_3^-$$

Finally,

$$8IO_3^- + 40I^- + 48H^+ \longrightarrow 24I_2 + 24H_2O$$

An interesting scheme for the determination of carbon which can be multiplied indefinitely has been proposed by Schöniger [16]. It consists of reducing the carbon dioxide obtained from the organic sample to carbon monoxide and reoxidizing the latter to carbon dioxide, thus doubling the original quantity of carbon dioxide in each step, as depicted below:

$$\text{Organic compound} \xrightarrow{O} CO_2 \xrightarrow{C} 2CO \xrightarrow{O} 2CO_2 \xrightarrow{C} 4CO \xrightarrow{O} 4CO_2$$

A similar scheme for the determination of hydrogen was suggested by Malissa [14]. It was based on an indirect method to obtain a multiple yield of carbon dioxide from water via acetylene according to the reaction

$$H_2O + CaC_2 \longrightarrow C_2H_2 + CaO$$

The flow diagram showing a twofold increase of carbon dioxide is as follows:

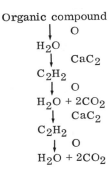

Unfortunately, however, the experimental results revealed that quantitative amplification could not be accomplished in the above manner [14], probably due to side reactions such as

$$CaC_2 + 2H_2O \longrightarrow Ca(OH)_2 + C_2H_2$$

Other examples of amplification reactions will be discussed in connection with trace analysis (see Sec. 12.2B). It should be noted that amplification multiplies the yield of a product but does not increase the accuracy of the analysis.

11.2 CURRENT PRACTICE

A. Weighing Devices For Ultramicro Analysis

One of the essential pieces of equipment for quantitative ultramicro analysis is a balance which provides adequate precision at and below the microgram region. The reader is referred to the comprehensive survey by Tölg [6] on such balances. Weighing devices which are sensitive to the nanogram (10^{-9} g) level are constructed from quartz fiber based on the principle of torsion restoration. A schematic diagram is shown in Fig. 11.1. The torsion member is fixed at one end to an adjusting screw and at the other end to the axis of a rotating dial. The quartz fiber which serves as the horizontal beam is mounted centrally and at a right angle to the torsion member. A pan is attached at one end of the beam and a

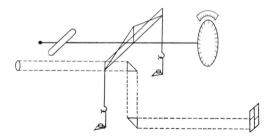

FIG. 11.1.   Principle of quartz fiber torsion balance.   (From Ref. 28, courtesy Japan Analyst.)

counterpoise at the other.   When a load is placed on the pan, the amount of torsion applied to the torsion member is determined by rotating the dial until the beam is restored to its original position, as observed by the optical attachment.   Since the tension of the torsion member is constant and the amount that the dial has to be rotated is proportional to the load on the pan, the weight of the load can be evaluated.   It should be noted that each balance requires careful calibration to obtain a conversion factor.   A detailed discussion of working with quartz fiber has been written by Benedetti-Pichler [17].   El Badry and Wilson [18] have constructed a simple balance by joining the various parts together with cement, thus avoiding the specialized skill necessary to obtain fused quartz joints.   Several models of quartz fiber torsion balances are commercially available [19-22].

Ultramicro beam balances [23-25] can be used for weighing samples in the 0.5-2 mg range.   The electric microbalances [26, 27] in which the change of equilibrium position of the beam produced by a load is electromagnetically compensated are intended for weighing in the 100 μg region. Hozumi [28] has undertaken a study of three commercial ultramicro-balances (Mettler UM-7, Rodder Model E, and Oertling Model Q_{01}) with respect to their sensitivities and the reproducibities of their rest points.

B.   Determination in the Decimilligram Region

Between 0.5 and 1 mg of Sample

Merz and Pfab [29] have reported on their experience in performing organic elemental analysis using samples of approximately 0.5-1 mg on a routine basis for over 5 years.   The coefficients of variation have been found to range from 0.1% to 0.3% absolute, showing that the decimilligram methods are comparable to the micromethods in precision and accuracy. An ultramicrobalance with precision of ±0.3 μg is employed. Carbon, hydrogen, and nitrogen, respectively, are determined by measurements of the gaseous products in the apparatus depicted in Fig. 11.2. Oxidation of the organic material is carried out in the Pregl-Lieb combustion tube (see Sec. 2.6) in the presence of copper oxide. Helium mixed with 3% oxygen is used as the carrier gas. The excess oxygen is removed by passing the gas through metallic copper which also converts the nitrogen oxides to elementary nitrogen. The gas mixture is then driven into the vacuum line where water is condensed at -78°C and carbon dioxide at -196°C, in their respective vessels, while nitrogen is trapped in activated charcoal at -196°C. These three products are subsequently measured in the manometers [30,31] filled with mercury and di-2-ethylhexyladipate.

When nitrogen alone is determined, Merz and Pfab [29] use the micro Dumas method (see Sec. 3.4) with an azotometer of 0.5 ml capacity. The combustion tube is packed with metallic copper, copper oxide, and cobaltic oxide. A supplementary burner maintained at 450°C is placed near the azotometer in order to decompose methane and other hydrocarbons if formed.

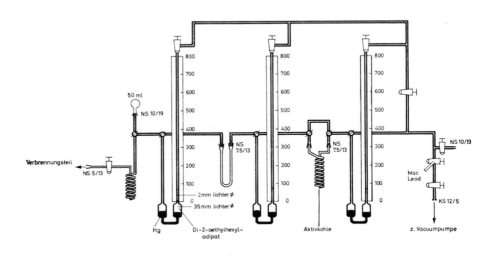

FIG. 11.2. Decimilligram determination of carbon, hydrogen, and nitrogen by measurement of gas volume. (From Ref. 29, courtesy Mikrochim. Acta.)

Determination of oxygen is carried out by scaling down the pyrolytic procedure of Schütze [32] and Zimmerman [33]. The resulting carbon monoxide
is oxidized by means of iodine pentoxide to carbon dioxide which is conducted into the absorption vessel [34] containing barium hydroxide solution
(pH 11) connected to a pH meter. Precipitation of barium carbonate causes
the pH to decrease; then standardized barium hydroxide solution is added
into the absorption vessel to restore the pH, and the amount of carbon dioxide can be calculated. Analysis of sulfur is performed by fusion with
metallic potassium [35], followed by distillation of hydrogen sulfide into a
special receiver containing cadmium acetate solution. The resulting cadmium sulfide is treated with a known amount of standardized potassium
iodate solution and the excess is determined iodimetrically by titration
with 0.01 $\underline{N}$ sodium thiosulfate.

Merz and Pfab [29] employ sodium peroxide fusion in a metal bomb to
decompose organic compounds containing fluorine, chlorine, or bromine.
Subsequently, fluorine is determined potentiometrically [36] after steam
distillation of hydrogen fluoride from sulfuric acid solution into the titration cell shown in Fig. 11.3. The cell consists of two compartments separated by sintered glass. Both compartments have been filled with a 1:1
mixture of Ce(III):Ce(IV) solution and connected to the platinum electrodes.
When fluoride ions enter the vertical compartment, formation of the cerium fluoride complex causes a potential difference. Addition of standardized sodium fluoride solution to restore the null point gives the amount of
fluorine in the original sample. Chloride and bromide are determined
titrimetrically with 0.01 $\underline{N}$ AgNO$_3$ using calomel-silver indicating electrodes in aqueous acetone solution. Acidification of the peroxide fusion
product should be carried out below 5°C, otherwise low bromine values are
obtained. Since the presence of peroxide interferes with the determination
of iodine, closed-flask combustion is employed for iodo compounds, followed by iodimetric titration with 0.01 $\underline{N}$ Na$_2$S$_2$O$_3$ (see Sec. 5.5). Decomposition of phosphorus-containing materials is performed either by
closed-flask combustion or by wet digestion; the former technique tends to
give high blanks. The resulting phosphate ions are determined by extraction [37] of the phosphomolybdic acid with isobutyl alcohol at pH 2.8, reduction with monomethyl-p-aminophenol sulfate, and measurement of
absorbance at 710 nm. Silicon analysis is carried out by sodium peroxide
fusion in a metal bomb, followed by extraction [37] of silicomolybdic acid
with isobutyl alcohol at a pH below 2, reduction with stannous chloride,
and measurement of the absorbance at 578 nm.

Determination of nitrogen by the Kjeldahl method (see Sec. 3.3) can
be performed with high precision and accuracy in the 0.5-1 mg range.
After sulfuric acid digestion, ammonia is steam-distilled into 2% boric
acid solution in a suitable apparatus [38]. The ammonium solution is then
titrated with standardized 0.01 $\underline{N}$ acid using the indicator described by Ma
and Zuazaga [39].

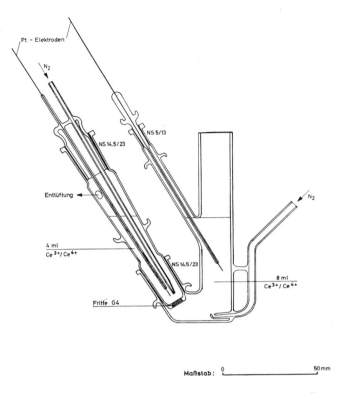

FIG. 11.3.   Titration cell for fluoride determination, after Merz and Pfab. (From Ref. 29, courtesy Mikrochim. Acta.)

Between 0.1 and 0.5 mg of Sample

After if was found that scaling down the Pregl gravimetric method for carbon and hydrogen analysis suffered from insurmountable difficulties [40], several groups of workers utilized different approaches and successively solved the problem. As mentioned in the previous section, organic material weighing slightly less than 1 mg can be oxidized in the Pregl type combustion tube and the amounts of carbon dioxide and water vapor obtained are then determined separately in a vacuum line. For the decomposition of samples below 0.5 mg, Kirsten [41] employs the sealed-tube combustion technique which has since been improved [42]. The organic substance is sealed in a quartz tube of 3.5-4.5 mm i.d. containing a copper gauze roll, Ag/Co_2O_3 catalyst [43], and pure oxygen. After combustion at 500°C in an electric furnace [44], the tip of the sealed tube is cut open (see Fig. 11.4b) in the assembly (Fig. 11.4a) which has been evacuated. Water vapor and carbon dioxide, respectively, are condensed in the freezing traps while nitrogen, if present, is swept out of the system through the vacuum pump. Subsequently, the system is closed and the carbon dioxide is evaporated into the U-tube manometer to be measured, followed by the water. This method can be used for organic materials containing nitrogen, halogens, sulfur, and phosphorus.

A coulometric method to determine water and carbon dioxide is used by Nakamura and co-workers [45]. As depicted in Fig. 11.5, the organic material (mixed with Co_3O_4) is decomposed in a vertical quartz combustion tube with a rapid stream of nitrogen passing through. The upper section of the combustion tube (heated at 900°C) has no packing while the lower section (heated at 700°C) contains platinum gauze, copper oxide, and Ag/Co_3O_4 mixture. The combustion gases then pass through metallic copper at 500°C into the Pt/P_2O_5 hygrometer [46] where water is absorbed and determined by electrolysis. Carbon dioxide proceeds to the pulse coulometric titrator [47] after being diluted with air. The diluted carbon dioxide is absorbed in 5% $Ba(ClO_4)_2$ solution which has been electrolyzed to pH 10. When the pH decreases with the absorption of carbon dioxide, the gate circuit of the pulse generator is activated and pulse electrolysis generation of OH^- ions takes place to neutralize the carbon dioxide. Alternatively [48], water vapor and carbon dioxide can be separated in a gas chromatograph; the carbon dioxide is then converted to water by passing through a tube containing LiOH at 210°C and determined by means of the same Pt/P_2O_5 electrolytic cell which later also receives the water that represents the hydrogen content of the organic sample.

Determination of nitrogen in 0.2- to 0.5-mg samples can be carried out in sealed tubes in the presence of copper gauze and pure oxygen [42]. After being heated at 700°C for 2 h, the sealed tube is transferred to the gas expansion device (Fig. 11.6) designed by Hozumi and Umemoto [49]. Several tubes can be affixed to the holding plate with their tips submerged in the 50% KOH solution. The tips are then carefully broken. Water and carbon dioxide in the tubes are absorbed in the liquid phase, leaving nitrogen gas on top to be measured.

The Kjeldahl method for determining nitrogen in 0.1- to 0.5-mg samples can be performed in various ways. The sample can be digested in an open test tube or in a sealed tube. In the latter case, the reaction temperature should be kept below 300°C to prevent oxidation of ammonia. The ammonia can be liberated from the digestion mixture by aeration [50]

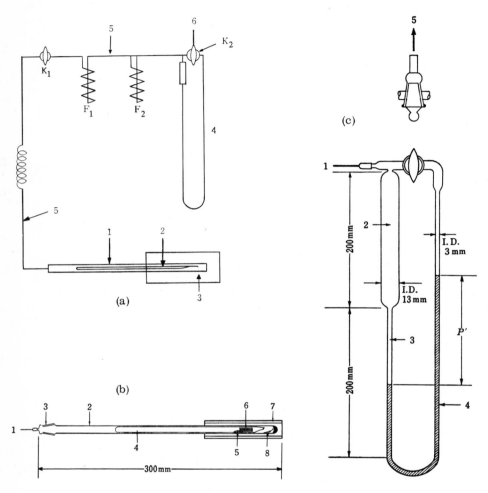

FIG. 11.4. Determination of carbon and hydrogen using 0.1-0.5 mg of sample. (a) Schematic diagram of the assembly. 1, quartz tube; 2, combustion tube; 3, heating furnace; 4, glass U-tube; 5, stainless steel capillary, 1.5 mm o.d.; 6, to vacuum pump: K_1, K_2, stopcocks; (F_1) H_2O trap, (F_2) CO_2 trap. (b) Breaking the tip of the sealed tube after combustion. 1, to freezing system; 2, quartz tube, 9 mm o.d.; 3, cap; 4, combustion tube; 5, sulfix (Ag + Co_2O_2); 6, Cu gauze; 7, heating furnace; 8, broken tip. (c) U-tube manometer. 1, to freezing trap; 2, diffusion chamber; 3, U-tube; 4, liquid paraffin; 5, to vacuum pump. (From Ref. 44, courtesy Japan Analyst.)

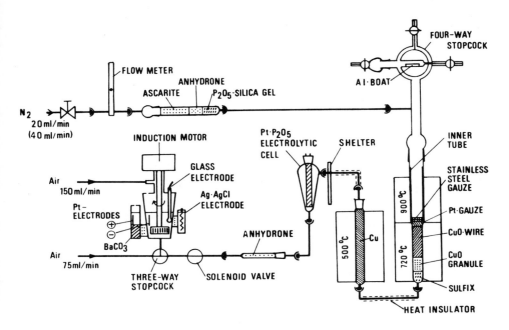

FIG. 11.5. Coulometric determination of carbon and hydrogen. (From Ref. 45, courtesy Microchem. J.)

(see Sec. 11.6), and determined by titrimetry or colorimetry. Griepink and Terlouw [51] percolate the ammonium sulfate in sulfuric acid over an anion exchanger in OH⁻ form, and determine the ammonium hydroxide collected in the effluent by titration with 3×10^{-3} M perchloric acid potentiometrically.

It has been pointed out in Chaps. 2 and 3 that the automated analyzers for carbon, hydrogen, and nitrogen which are based on gas chromatography prescribe sample sizes in the region of 0.3-0.5 mg. The reasons for such restriction are (1) the organic material should be totally and instantaneously mineralized so that all combustion gases enter the chromatographic column simultaneously, and (2) precise separation and quantitation of gas chromatograms incur many problems above the decimilligram level. Berezkin [52] has reviewed the papers published between 1960 and 1965 concerning organic elemental analysis for which the gas chromatograph is utilized. Besides carbon, hydrogen, and nitrogen, determinations of oxygen and sulfur, respectively, have been reported. Recently Liebman and

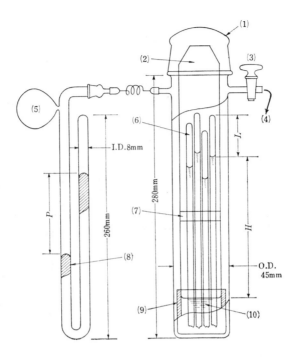

FIG. 11.6. Gas expansion arrangement for nitrogen determination.
1, cap; 2, holding plate; 3, stopcock; 4, vacuum; 5, bulb; 6, N_2; 7, scotch
tape; 8, Hg; 9, reservoir; 10, 50% KOH solution. (From Ref. 49, courtesy Japan Analyst.)

co-workers [53] have described a system for on-line elemental analysis of
gas chromatographic effluents. The apparatus [54] can determine the
atomic ratios of carbon, hydrogen, and nitrogen or oxygen using samples
between 10 μg and 1 mg. The reactor for carbon, hydrogen, and nitrogen
consists of a two-stage oxidation/reduction furnace, while that for oxygen
is a single stage using 5% platinum on charcoal.

Based on the principle of closed-flask combustion (see Sec. 5.1B),
Kirsten [55] has constructed a quartz vessel of 14 mm i.d. and 100 mm
overall length for the determination of halogens in less than 0.5 mg of
organic substance. One procedure, illustrated in Fig. 11.7, is called the
hot-flask combustion method. The combustion vessel A is placed in a
furnace B which is maintained at 850°C. (The commercial furnace shown

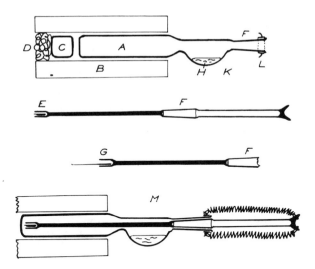

FIG. 11.7. Hot-flask combustion method, after Kirsten. A, combustion vessel of quartz, i.d. about 14 mm, wall thickness about 1.5 mm, length of part in furnace about 60 mm; B, tube furnace, temperature 850°C; C, quartz bulb filling out dead space in furnace; D, asbestos; E, sample holder; F, joint B 7; G, sample holder with capillary of Supremax glass containing volatile liquid sample; H, absorbent; K, bulb comtaining absorbent; L, metal band with hooks fixed around joint for holding of spring; M, figure showing position of vessel and sample holder at beginning of combustion. (From Ref. 55, courtesy Microchem. J.)

in Fig. 9.1 is suitable for this purpose.) An appropriate absorption solution H is pipetted into the bulb K. The sample, weighed in a platinum microboat, is placed into the sample holder E. The widest part of the ground joint F is lubricated with silicon grease. After oxygen is blown into vessel A for 30 s, the holder EF with the sample is quickly inserted into the vessel and the wire springs are affixed. The electric furnace is immediately tilted so that the joint F points downward and the stopper is covered with the absorption liquid. The apparatus is heated for 1 min and is then removed from the furnace. Upon cooling, the vessel is shaken well and allowed to stand for 5 min to insure complete absorption of the combustion gases. The sample holder is then taken out from the vessel and rinsed. The halogen to be determined can be titrated in the vessel. Volatile liquids are weighed in capillaries G which are broken when the sample holder F hits the wall of the vessel A upon insertion. A slightly modified sample holder is used for the analysis of sulfur [56]. After combustion and absorption by means of bromine water, the contents of the vessel are transferred into a reduction/distillation apparatus for the subsequent determination of the sulfur by the sensitive methylene blue - sulfide color reaction. An alternative procedure [55], called the hot-flask combustion - diffusive absorption method, employs a vertical furnace and an absorption flask separate from the combustion vessel. The improved version [57] of this procedure is given in detail in Sec. 11.4.

An apparatus has been patented [58] for halogen or sulfur analysis in which the organic material (100-200 μg) is subjected automatically to combustion in oxygen; the absorbent solution then circulates in the system and is subsequently analyzed by a titrimetric or colorimetric method.

C. Determination in the Centimilligram Region

Between 50 and 100 μg of Sample

For elemental analysis using approximately 50-μg samples, the reader is referred to the manual published by Belcher [3]. It describes in detail the apparatus and experimental procedures for the determination of nitrogen, carbon, and hydrogen, halogens, sulfur, phosphorus, and arsenic by the methods developed at the University of Birmingham, England. Except for the first three elements mentioned, closed-flask combustion is employed as the technique of mineralization. A special flask (Fig. 11.8) of 25-ml total capacity has been designed to allow ignition with a sample wrapper and fuse. Polyethylene sheet is the most satisfactory wrapper since it is practically free from impurities. Linen thread, pretreated with 50% ethanol and dried at 110°C, serves as the fuse for ignition. Griepink and Krijgsman [58a] wrap the sample and fuse in the following manner. A layer of collodion is formed on a porcelain dish from a drop of 1.5% collodion solution in diethyl ether. The sample and fuse are placed on the collodion layer and covered with a further drop of collodion. The resulting packet is then stripped from the dish and inserted into the platinum basket.

Gas chromatography can be used for carbon, hydrogen, nitrogen, and oxygen analysis in this range. Belcher and co-workers [59] perform simultaneous determination of carbon, hydrogen, and nitrogen using 40-80 μg of sample. The organic material is decomposed in a stream of helium at 850°C in the presence of Co_3O_4 in a combustion tube containing CuO and electrolytic silver, and the H_2O formed is converted into C_2H_2 by passage over CaC_2. The N_2, CO_2, and C_2H_2 are then separated on a column (50 cm x 2 mm) of silica gel at 50-70°C. The C_2H_2 is subsequently converted to H_2O and CO_2 by passage through CuO at 930°C. The H_2O is removed on a second silica gel column at 80°C, and the peaks of N_2 and of the two

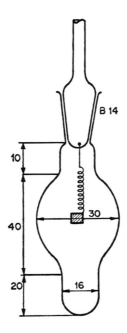

FIG. 11.8. Combustion flask for decimilligram samples, after Belcher. Dimensions in millimeters. (From Ref. 3, courtesy Elsevier Scientific Publishing Co.)

discrete volumes of CO_2 are measured with a GowMac thermister micro-cell. Organic oxygen is converted to CO by pyrolysis through platinized graphite wool; after passing through a column (50 cm x 1.5 mm) of molecular sieve which retains N_2 and hydrocarbons, CO is determined by thermal conductivity [60].

Hozumi and Umemoto [61] determine 30–80 μg of organic material for nitrogen by sealed tube combustion and measurement of the nitrogen gas collected over 50% KOH solution under reduced pressure (see Fig. 11.6). The volume of water delivered from a piston microburet to replace the volume of nitrogen is corrected by measuring the difference between the two volumes which is related to the inside diameter of the combustion tube. The natural capillary rise of the KOH solution in the combustion tube and the volume adhering to the walls are precisely estimated in order to obtain the actual volume of nitrogen. Miyahara and Takaoka [61a] have described suitable techniques for handling hygroscopic samples and volatile liquids or solids.

Less Than 50 μg of Sample

Tölg [6] has compiled the ultramicro procedures which appeared in the literature up to about 1967 for the determination of carbon, hydrogen, oxygen, nitrogen, sulfur, halogens, phosphorus, and arsenic using sample weights in the microgram region. In this section, we discuss recently published methods as well as microgram-level analytical techniques of general interest. The reader is also referred to manuals [62, 63] on microchemical manipulations for the equipment and operations in handling minute amounts of solid material and volumes of liquid in the microliter region.

Römer and co-workers [64] determine carbon by combustion in the presence of pure oxygen in a quartz tube (10 mm i. d.) packed with a 60-mm length of the thermally decomposed products of silver permanganate [65] at 525°C. The CO_2 formed is absorbed in 0.1 $\underline{M}$ barium perchlorate (pH 9.5) in a mixed solvent containing 10% t-butyl alcohol and 90% deionized water, and is determined by titration with 0.03 $\underline{M}$ sodium hydroxide dissolved in the same solvent until the pH returns to the original value. Ballschmiter and Tölg [66] also determine carbon by titrimetry using sample weights of 10-30 μg. Sulfur-free substances are decomposed at 1000°C in pure oxygen; sulfur-containing materials are mineralized at 550°C in the presence of an $Ag/Mn/PbCrO_4$ catalyst. The combustion is carried out with an electrically heated platinum spiral in a slow stream of oxygen. The carbon dioxide is absorbed in 0.01 $\underline{M}$ barium hydroxide solution (400 ± 0.1 μl) using a special stirrer. The excess of Ba^{2+} ions is photometrically back-titrated with 0.01 $\underline{M}$ EDTA solution and phthalein purple as indicator. The complete assembly is shown in Fig. 11.9.

For the determination of 0.5-3 μg of hydrogen in 10-40 μg of organic compounds, Tölg and Ballschmiter [67] burn the sample in a combustion tube in the presence of oxygen using a heating spiral which is connected to a variable autotransformer and can be heated up to above 1000°C in less than a second. The H_2O generated is converted to H_2S by reaction with CS_2 in a stream of nitrogen at 500°C with aluminum oxide as catalyst. The H_2S is argentimetrically titrated by a potentiometric method.

Tölg [68] utilizes the ter Meulen hydrogenation principle [69, 69a] to determine nitrogen in samples of 5-20 μg. The organic substance is pyrolyzed in a stream of hydrogen at a temperature above 1000°C; subsequently the nitrogen is converted to ammonia by an iron catalyst at 400°C, and the ammonia is titrated iodimetrically with biamperometrical indication. Krijgsman and co-workers [70] convert amino nitrogen to ammonium ion by digestion with perchloric acid. Ammonia is liberated from the digestion tube (0.5-cm diameter, 10-cm length) by aeration after the addition of alkali, and subsequently determined by titration with 0.03 $\underline{M}$ perchloric acid or coulometrically [71]. For the liberation of microgram amounts of ammonia, the Conway diffusion [72] or the aeration technique is recommended (see experimental details in Sec. 11.6).

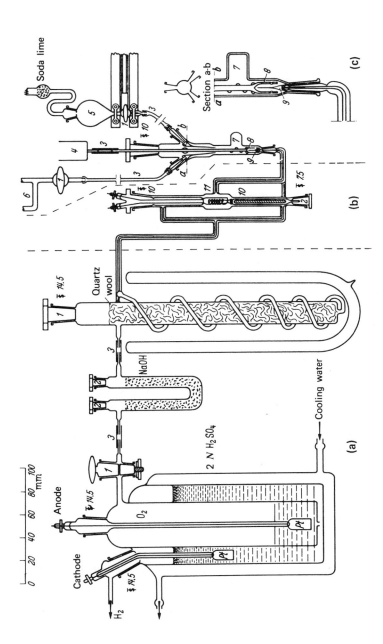

FIG. 11.9. Apparatus for ultramicro carbon determination, after Ballschmiter and Tölg. a, oxygen generation and purification section; b, decomposition section; c, carbon dioxide absorption and determination section; 1, Teflon plug, 2, Teflon stopper; 3, polyethylene tubing connection; 4, stirrer motor; 5, micro piston buret; 6, 1-ml semimicroburet; 7, cuvet; 8, bell-shaped stirrer; 9, polyethylene capillary; 10, combustion tube and holder; 11, platinum heating coil. (From Ref. 66, courtesy Z. Anal. Chem.)

Schwab and Tölg [73] determine chlorine in 1-10 μg of organic substance by combustion in a stream of oxygen at 1000°C. The hydrogen
halide formed is absorbed in glacial acetic acid and determined by argentimetric titration with bipotentiometric endpoint indication (see Sec. 11.5).
A similar procedure is employed for the analysis of bromine [74] in samples
of 1-5 μg, the organic material being pyrolyzed in a hydrogen stream at
900°C. Krijgsman and co-workers [75, 76] determine 0.01-2 μg chlorine
by burning the organic sample in oxygen mixed with nitrogen. The combustion gases are conducted into a coulometric cell [77] with a darkened
surface, and the chlorine ions are determined by automatic titration with a
solution containing 80% acetic acid and 1% sodium perchlorate. A silvered
platinum strip is used as the sensor electrode.

Taking advantage of the fact that hydrogen sulfide, unlike sulfur dioxide, is not adsorbed on quartz surfaces, Tölg and Grünert [78] employ hydrogenation to determine sulfur in less than 2 μg of organic substance.
The pyrolysis is carried out at 800°C, and the hydrogen sulfide formed is
determined either argentimetrically by means of bipotentiometric endpoint
indication or fluorimetrically by quenching of fluorescein-mercuric acetate.

For the determination of selenium, Engler and Tölg [79] employ closed-
flask combustion with 3 x 10^{-3} N NaOH as absorbent. After mineralization,
hypophosphoric acid is added to convert all selenium to hydrogen selenide,
which is determined by titration bipotentiometrically with 10^{-4} M lead
acetate solution.

Phosphorus in organic materials can be determined at the microgram
level utilizing the intense blue color of the phosphomolybdate complex [80].
Wet digestion in an open tube is the preferred technique of decomposition.
Some phosphoric acid esters, however, are resistant to mineralization by
heating with oxidizing acids and hence require closed-flask combustion [6].
It should be mentioned that some of the phosphorus may remain in the platinum basket. Thus Römer and Griepink [81] recommend putting the platinum gauze in the absorption liquid after the combustion. For the determination of arsenic in organic substances by closed-flask combustion, a
quartz coil is employed in place of platinum, which alloys with arsenic.

11.3 REVIEW OF LITERATURE

A. Summary of Recent Publications

White [82] has compiled the literature references on ultramicromethods
up to 1962. Therefore only publications which appeared since about that
time are summarized in this section. The methods dealing with carbon,
hydrogen, oxygen, and nitrogen are listed in Table 11.1, while those dealing with the other elements are given in Table 11.2. Certain salient points

TABLE 11.1. Summary of Recent Publications on Ultramicro Analysis of Carbon, Hydrogen, Oxygen, and Nitrogen.

Element	Decomposition method	Product determined	Finish technique	Reference
C	Combustion tube	CO_2	Titrimetric	66–66b
	Empty tube	CO_2	Titrimetric	82
	Empty tube	CO_2	Infrared spectrometry	83
H	Combustion tube	$H_2O \rightarrow H_2S$	Titrimetric	67
	Sealed tube	H_2	Gravimetric	84
	Sealed tube	H_2	Diffusion	85
O	Combustion tube	$CO \rightarrow CO_2$	Gas chromatography	86
	Combustion tube	CO	Titrimetric	87
	Combustion tube	CO	Gas chromatography	60
	Combustion tube	$CO \rightarrow H_2O + CH_4$	Conductivity	88
	Sealed tube	$CO + CO_2$	Gasometric	89
	Neutron activation		Radiometric	90
N	Combustion tube	N_2	Gasometric	91–92
	Combustion tube	NH_3	Titrimetric	68
	Combustion tube	NO_2	Spectrophotometric	93
	Sealed tube	N_2	Gasometric	49, 61, 94–100a
	Sealed tube	NH_4^+	Turbidimetric	101

TABLE 11.1.   (Continued)

Element	Decomposition method	Product determined	Finish technique	Reference
N	Sealed tube	NH_4^+	Titrimetric	101a
	Kjeldahl	NH_4^+	Titrimetric	51, 102, 103
	Kjeldahl	NH_4^+	Spectrophoto-metric	104–106
C and H	Sealed tube	CO_2, H_2O	Gasometric	44, 92
	Combustion tube	CO_2, H_2O	Gasometric	107, 108
	Combustion tube	CO_2, H_2O	Titrimetric	108a
	Combustion tube	CO_2, H_2O	Gas chroma-tography	109–113
C, H, and N	Combustion tube	CO_2, H_2O, N_2	Gas chroma-tography	59, 114–119
	Combustion tube	CO_2, H_2O, N_2	Gasometric	29, 120
	Sealed tube	CO_2, H_2O, N_2	Gasometric	121

in ultramicro analysis, as compared with the conventional methods using milligram amounts of the organic sample, can be seen from these two tables.   For instance, when the sample falls below the milligram region, separate determinations of carbon and hydrogen are sometimes advisable. The sealed-tube decomposition technique can be employed with advantage in ultramicro analysis of carbon, hydrogen, nitrogen, and oxygen.   This technique utilizes a minimal quantity of reagents, thus reducing the blank errors; it also confines the gaseous combustion products in a small volume, thus preventing losses.   The closed-flask decomposition technique has been extensively investigated for the ultramicro determination of halogens, sulfur, phosphorus, and arsenic in organic substances.   When a wrapper is used to hold the sample, it may be difficult to find a suitable material which is free from the element to be determined.   Similarly, the mineral acids for wet digestion and the sodium peroxide for fusion in a metal bomb should be meticulously tested for possible interfering impurities.   Needless to say, the finish step calls for measuring devices commensurate with the precision and accuracy required in the determination.

TABLE 11.2. Summary of Recent Publications on Ultramicro Analysis of Halogens, Sulfur, Phosphorus, Metalloids, and Metals.

Element	Decomposition method	Product determined	Finish technique	Reference
Cl, Br, I	Closed flask	Cl^-, Br^-, I^-	Titrimetric	57, 58, 122–126a
	Closed flask	I_2O_5	Spectrophotometric	127
	Gas chromatographic column	HCl, HBr, HI	Coulometric	127a
	Combustion tube	Cl^-, Br^-	Titrimetric	74, 128
	Combustion tube	Cl^-	Spectrophotometric	55, 129, 130
	Combustion tube	Cl^-	Coulometric	131
	Sodium biphenyl	Cl^-, I^-	Titrimetric	132, 133
	Electrolysis	I^-	Coulometric	134
F	Closed flask	F^-	Spectrophotometric	135–138
	Closed flask	F^-	Colorimetric	139–142
	Ashing	HF	Titrimetric	143
S	Closed flask	$SO_4^{2-} \to H_2S$	Titrimetric	135
	Closed flask	SO_4^{2-}	Titrimetric	144, 145
	Closed flask	SO_4^{2-}	Spectrophotometric	146
	Empty tube	SO_4^{2-}	Titrimetric	129, 147, 148
	Empty tube	SO_2	Coulometric	149, 149a
	Empty tube	H_2SO_4	Conductimetric	150
	Empty tube	$SO_4^{2-} \to H_2S$	Colorimetric	151
	Empty tube	SO_3	Colorimetric	152
	Hydrogenation	H_2S	Titrimetric	100, 153, 154

TABLE 11.2.   (Continued)

Element	Decomposition method	Product determined	Finish technique	Reference
S	Hydrogenation	H_2S	Colorimetric	155
	HNO_3 digestion	SO_4^{2-}	Spectrophoto-metric	156
	Flame	S	Photometric	157
Se	Closed flask	H_2Se	Titrimetric	79
	X-ray fluor-escence	Se	Spectrometry	158
P	Closed flask	PO_4^{3-}	Spectrophoto-metric	135, 159
	Closed flask	PO_4^{3-}	Titrimetric	159
	Wet digestion	PO_4^{3-}	Spectrophoto-metric	160–165
	Flame	P	Gas chroma-tography	166
As	Closed flask	AsO_4^{3-}	Spectrophoto-metric	58, 159
	Closed flask	AsO_4^{3-}	Titrimetric	167, 168
Si	Na_2O_2 bomb	SiO_3^{2-}	Spectrophoto-metric	29
Hg	Combustion tube	Hg	Photometric	169
	Closed flask	Hg^{2+}	Spectrophoto-metric	170
	Wet digestion	Hg^{2+}	Spectrophoto-metric	171, 172
	X-ray fluor-escence	Hg	Spectroscopy	158
Metals	Wet digestion	Metal ions	Colorimetric	173, 174
	Wet digestion	Metal ions	Spectrophoto-metric	175, 176
	Wet digestion	Metal ions	Phototitrimetric	177

B. Comparison of Methods

Since ultramicro elemental analysis is a recent development, a method developed and used by one group of workers may not have yet found its way to other laboratories. For this reason, hardly any published reports have appeared on the evaluation and comparison of the methods mentioned in Tables 11.1 and 11.2. At the International Symposium on Microchemical Techniques, 1973, a session [178] was devoted to the evaluation of several automated C,H,N analyzers which use samples around the 0.5 mg region. It is expected that the other methods and procedures for the determination of other elements will be under scrutiny in the future.

The ideal method for ultramicro determination is one which maintains the same level of precision and accuracy within a wide range of sample weights. Thus Griepink and Sandwijk [124] have described a method for determining 1-200 μg of iodine using 12-500 μg of several iodo compounds with standard deviations of 0.2-0.5% absolute. Kirsten [119] has reported that by modifying the Technicon C,H,N analyzer acceptable results (within 0.2% absolute) can be obtained using 200 μg, 100 μg, or 50 μg of various organic samples.

As a rule, the microchemist is not surprised to find that the precision and accuracy of the analytical data deteriorate as the experimental procedure is scaled downward. In general, the effect of "blank errors" becomes magnified exponentially in reverse proportion to the sample weights. The reagents are the chief culprit, since they are not reduced to the same extent as the sample size. In a study employing the Hewlett-Packard C,H,N analyzer, Scheidl [111] has described an oxygen donor which gave C, H, and N results within ±0.3% for sample weights of 0.5-0.8 mg, whereas even an improved oxygen donor with very low blanks produced results deviating as much as 1.5% absolute for sample weights of 50-150 μg. Naturally, the purest reagents should be used in ultramicro analysis. It is not possible, however, always to procure a reagent which is completely free from the element in question. Thus, in the titrimetric determination of chlorine using a medium of ethanol-water, Belcher [3] cautions that even spectroscopically pure ethanol contains up to 0.3 μg/ml of chlorine.

11.4 EXPERIMENTAL: DETERMINATION OF CHLORINE USING
      DECIMILLIGRAM SAMPLES

A. Principle

Decimilligram amounts of organic material containing chlorine can be decomposed in the presence of phosphoric acid in a vertical combustion vessel (the hot flask, see Sec. 11.2B) heated at 850°C. The chlorine and hydrogen chloride produced are quantitatively diffused into a dilute sodium hydroxide solution. The chloride ions are then determined by differential electrolytic potentiometric titration with 0.005 $\underline{N}$ silver nitrate standard solution [57]. The procedure described below is based on the information supplied by Prof. W. J. Kirsten, Department of Chemistry, Royal Agricultural College of Sweden, Uppsala, Sweden. The experimental details were verified by Prof. Kirsten [57a] in October 1974.

B. Apparatus

Combustion apparatus (see Fig. 11.10).

Titration apparatus (see diagram shown in Fig. 11.11): The electrodes are made of silver wire, 0.6 mm in diameter, 100-200 mm long. One of them is inserted into a polyethylene tube which fits snugly. The two wires are then twisted and introduced into a wider polyethylene tube. Liquid polyethylene is then allowed to drop upon their lower ends from a heated polyethylene rod and is shaped to a tight seal. After cooling, the end is shaped with a razor blade and a fine, sharp file so that the clean, flat ends of the silver wire lie in the same plane as the surrounding polyethylene. Between the silver and the polyethylene there should be no gaps into which the solution might enter, causing potential shifts.

Magnetic stirrer with stirring bars made of iron wire (0.6 mm thick, 4-5 mm long) sealed into snugly fitting thin-walled glass tubing.

An Agla syringe-buret mounted with a motor drive for 10 constant speeds and a Sargent SRL recorder connected to a Metrohm pH meter E300 B or equivalent arrangement for automatic recording of potentiometric microtitration. The buret is adjusted to deliver 50 μl/min and the recorder to a chart speed of 1 in./min.

C. Reagents

Orthophosphoric acid, AR, specific gravity 1.7.

Sodium hydroxide solution, AR, 0.2 $\underline{M}$ (see Sec. 11.4F).

Glacial acetic acid, AR.

Silver nitrate standard solution, AR, 0.005 $\underline{N}$, accurately standardized against pure NaCl.

Absolute alcohol, AR.

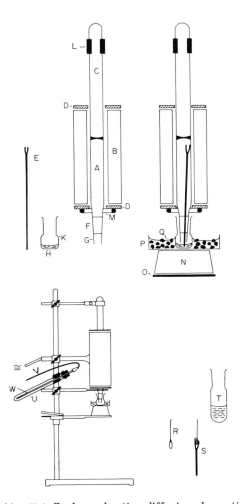

FIG. 11.10. Hot-flask combustion diffusion absorption method, after Kirsten et al. A, combustion vessel of quartz, i.d. approximately 14 mm, wall thickness about 1.5 mm, length of part in furnace about 60 mm; B, tube furnace, temperature 850°C; C, handle of opaque quartz tubing sealed upon combustion vessel A; D, asbestos sheet; E, sample holder (quartz tubing with i.d. about 3 mm is sealed upon quartz rod with diameter of about 2 mm. Length of container about 6 mm. Total length about 95 mm); F, joint B 12 with extension G, which ends about 2 mm above absorbing solution H, when sample has been inserted; K, vessel with joint B 12 (diameter like A, length of part below joint about 18 mm); I, support clamp which holds combustion vessel and handle C; M, ring support which holds D; O, adjustable support table; N, soft rubber stopper securing steady press upon O from below; P, metal vessel; Q, ice and ice water; R, sample of volatile liquid weighed out in capillary of Supremax glass; S, holder E with capillary R inserted; T, glass tube containing sodium hydroxide; U, test tube clamped to support; V, stainless steel capillary tubing which connects quartz capillary W to oxygen tank. (From Ref. 57, courtesy Microchem. J.)

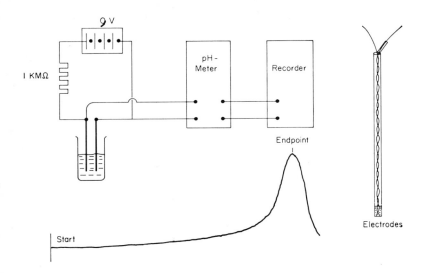

FIG. 11.11. Layout of titration setup, electrodes, and titration graph.
(From Ref. 57, courtesy Microchem. J.)

D. Procedure

Combustion

Weigh the sample (not more than 300 μg) containing up to 75 μg of
chlorine into a platinum microboat (Sec. 11.4E, Note 1). Add 1 μl of
phosphoric acid (with a syringe-buret) to wet the sample. Place the boat
in the holder E. Pipet 0.3 ml of sodium hydroxide solution (or the mixed
absorbent, see Sec. 11.4F) into vessel K. Dry joint F with filter paper
and lubricate its upper part with a small amount of silicone grease. Blow
oxygen into vessel A with capillary W at the rate of 100 ml/min. Put the
capillary back into test tube U, place rod E into vessel K and introduce it
quickly into the hot flask A, tightening the joint immediately. Hold vessel
K firmly to the joint with the left hand and raise table O with stopper N
and vessel P with the right hand and press them firmly against vessel K.
Fill vessel P with ice water. Allow to stand for 20 min for the diffusion.

After 20 min, lower table O and remove vessel K with rod E standing in it for titration. Blow oxygen once more into the hot flask and introduce the next sample (Note 2).

Titration

Wash the lower end of rod E and the joint of the titration vessel K with 1 ml of glacial acetic acid. Add a magnetic stirring bar and turn on the stirrer. Introduce the electrodes and buret-tip, and begin the titration. (Vigorous and steady stirring is important.) After the titration, lower K and wash the electrode and buret-tip with water and replace K with a beaker containing distilled water, in which the end of the electrode is immersed.

E. Notes

Note 1: Volatile liquids are weighed in Supremax capillaries R (see Fig. 11.10), which are closed and placed into holder E as shown in S. Phosphoric acid is added at the side of the capillary. The thin handle of the capillary is then crushed when it meets the upper wall of flask A.

Note 2: When the apparatus is not in use, joint F is closed with vessel T which is partially filled with soda lime. No alkali should touch G or joint F. Microboats and rod E are cleaned with hot sulfuric acid and distilled water.

F. Comments

(1) According to Kirsten [57a], a better absorbent consists of a mixture of tris(hydroxymethyl)-aminomethane and hydrazinium hydroxide, prepared by mixing 120 ml of a 0.3 $\underline{M}$ aqueous solution of the former and 20 ml of a 2% aqueous solution of the latter. This absorbent solution keeps for 1 week.

(2) Using the above absorbent solution, the same procedure can be applied to the determinations of bromine or iodine at the decimilligram region.

(3) The mixed absorbent and acetic acid are kept in respective Erlenmeyer flasks; they are delivered by means of Grunbaum micropipets [62] which are provided with stoppers and kept in the flasks so that the pipet tips are immersed at least 2 cm in the liquid.

11.5 EXPERIMENTAL: DETERMINATION OF CHLORINE USING CENTIMILLIGRAM SAMPLES

A. Principle

The organic sample containing chlorine is weighed directly or measured out by solution partition. Decomposition is by combustion in a current of oxygen. The hydrogen chloride formed is absorbed in glacial acetic acid and subsequently determined by argentometric titration with bipotentiometric endpoint indication. The procedure given below is based on information supplied by Prof. Günther Tölg [6], Max-Planck-Institut für Metallforschung, Schwäbisch Gmünd, Germany. The experimental details were verified by Prof. Tölg [6a] in October 1974.

B. Apparatus

The apparatus is illustrated in Fig. 11.12A. It is connected to an oxygen generator (similar to the electrolysis cell shown in Fig. 11.12B) through a bubble-counter filled with water, a U-tube with sodium hydroxide pellets, and a cold trap with Dry Ice-methanol. The path between the decomposition vessel (Fig. 11.12A, B_1) and the absorption vessel C_1 has a very small surface area. The combustion chamber is heated by the P/Rh coil B_3. The absorption vessel is provided with a cooling jacket, and has four $\mathbf{B}$ inlet joints (see C_3). One joint is for admission of the polyethylene titration capillary which is connected via a capillary joint and polyethylene tube to a 500-μl glass piston buret with micrometer screw (see Fig. 11.13). The glass body of the buret is lacquered black. The second joint is for introduction of glacial acetic acid from a 2-ml buret. The third opening is for drawing off the solution after titration and for rinsing. The fourth opening is for introduction of the chloride standard solution by means of a 100-μl displacement buret.

For the electric circuit for bipotentiometric titration [179] (Fig. 11.14), a 24-V anode battery, a vacuum tube voltmeter (0-500 mV) with input resistance over 100 MΩ and zero suppression or a corresponding millivolt recorder, high ohmic resistance of 10^9, 10^{10}, and 10^{11} Ω, and shield cable are needed.

C. Reagents

Chloride standard solution, 0.001 $\underline{M}$: Accurately weigh 58.45 mg of pure sodium chloride (previously dried for several hours at 180-200°C and stored over P_2O_5) into a 1-liter quartz volumetric flask

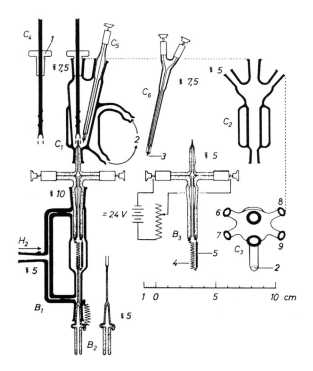

FIG. 11.12A. Apparatus for ultramicro determination of chlorine, after Tölg. B_1, quartz decomposition vessel; B_2, quartz sample holder; B_3, Pyrex heating coil holder; C_1, absorption and titration vessel, side view; C_2, absorption and titration vessel, side view (rotated 90° about its long axis with respect to C_1); C_3, absorption and titration vessel, view from above; C_4, bell stirrer; C_5 and C_6, electrode rod; 1, Teflon stopper; 2, water connection to and from thermostat; 3, platinum electrodes; 4, heating coil of 0.3 mm Pt-Rh wire; 5, quartz insulating tube; 6, $\overline{S}$ 5 joint for connection to 500-μl piston buret; 7, $\overline{S}$ 5 joint for connection to a 2-ml semimicroburet; 8, $\overline{S}$ 5 joint for connection to a rinsing system; 9, $\overline{S}$ 5 joint for introduction of standard solutions. (From Ref. 6, courtesy J. Wiley & Sons, Inc.)

(previously steamed for 30 min) and make up to volume with chlorine-free water.

Glacial acetic acid with a chloride content less than 0.01 μg/ml: Distil glacial acetic acid, AR, with some silver powder added, through a column. Store in a previously steamed brown borosilicate glass bottle.

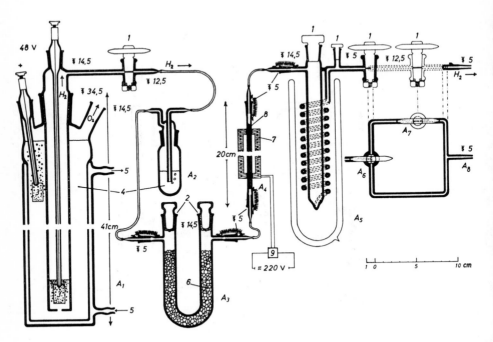

FIG. 11.12B. Hydrogen generation and purification setup. A_1, hydrogen generator; A_2, bubble counter with 2 $\underline{N}$ NaOH; A_3, U-tube with NaOH pellets; A_4, catalyst tube filled with palladium-asbestos, heated to 300–400°C; A_5, cold trap in a Dewar filled with Dry Ice-methanol; A_6, three-way stopcock for dual path; A_7, stopcock; A_8, joint for connection to the pyrolysis section; 1, Teflon parts; 2, polyethylene stopper; 3, platinum foil electrodes (30 x 10 x 0.1 mm); 4, 2 $\underline{N}$ potassium hydroxide solution; 5, cooling water; 6, sodium hydroxide pellets; 7, electric furnace (two 200-W heating elements); 8, palladium-asbestos; 9, autotransformer. (From Ref. 6, courtesy J. Wiley & Sons, Inc.)

0.0002 $\underline{N}$ Silver nitrate-glacial acetic acid solution: Dilute 5.00 ml of aqueous 0.02 $\underline{N}$ silver nitrate solution (weakly acidified with acetic acid) to volume with chloride-free glacial acetic acid in a 500-ml quartz volumetric flask. Store in the black lacquered reservoir of the 500-μl piston buret (Fig. 11.13). Standardize daily against the 0.001 $\underline{M}$ chloride standard solution.

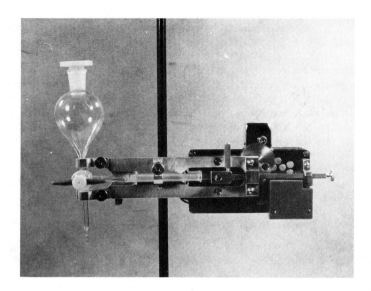

FIG. 11.13. A 500-μl piston buret with reservoir and 10-stage gearing for automated titration. (Courtesy Prof. G. Tölg.)

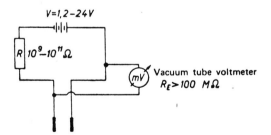

FIG. 11.14. Circuit for bipotentiometric titration, after Bishop and Dhaneshwar. (From Ref. 179, courtesy Analyst.)

Methanol, ethanol, and acetone, with chloride content less than 0.01
ppm: Distil these solvents (used for rinsing and solution partition),
after addition of some solid silver nitrate, through an efficient
column. Store in steamed brown borosilicate bottles.

Bath solution for preparation of silver electrodes: Prepare silver
cyanide by mixing equimolar amounts of dissolved silver nitrate
and potassium cyanide, collecting the precipitate on a filter, washing free of nitrate, and drying at 120°C for 1 h. Then dissolve 2.7
g of this silver cyanide and 2.6 g of potassium cyanide in 100 ml of
double-distilled water.

D.   Procedure

Apparatus Preparation

Clean the two Pt wires in the electrode holder (Fig. 11.12A, C_6) with
30% nitric acid and ignite. Then alternately connect both electrodes anodi-
cally and cathodically for several minutes in 2 $\underline{N}$ sulfuric acid. After
thorough rinsing with double-distilled water, put the electrodes into the
silver bath in the electrolysis vessel shown in Fig. 11.15 and electrolyze
for 1-2 h with stirring. (Connect the Pt electrodes as cathodes against a
Pt wire auxillary electrode 3. With an emf of 6 V, the electrolysis current
should be 350-400 μA.) Thoroughly rinse the silvered electrodes with
water and chlorinate them by connecting them as anodes against a Pt wire
electrode for 10 min in 0.5 $\underline{N}$ hydrochloric acid (400-500 μA electrolysis
current). After rinsing free of chloride, store the electrodes in glacial
acetic acid and protect them from direct sunlight.

Before use, etch a new piston buret for a few minutes with 2% hydrofluoric acid solution, thoroughly rinse with water, and steam for 30 min.
Then fill the buret with the 0.0002 $\underline{N}$ silver nitrate titrating solution. Clean
the 2-ml buret for measuring glacial acetic acid with concentrated nitric
acid (not with chromosulfuric acid); rinse with water and glacial acetic
acid before filling.

Decomposition and Titration

Flush the apparatus (Fig. 11.12) for 5 min with oxygen (20-50 ml/min,
300 mm water pressure). With the absorption vessel C_1 empty, place the
quartz combustion vessel containing the sample about two-thirds into the
Pt/Rh heating coil. Rinse the absorption vessel several times and introduce 1 ml of glacial acetic acid from the 2-ml buret. Turn on the bell
stirrer and combust the sample by rapid heating of the Pt/Rh coil to over
1100°C. Maintain this temperature for 30 s and then adjust it to about
900°C (Sec. 11.5E, Note 1).

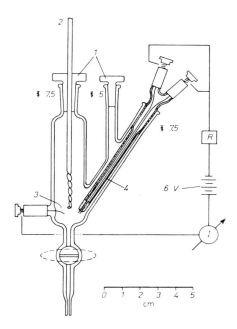

FIG. 11.15. Electrolysis vessel for preparation of Ag or AgCl electrodes, after Tölg. 1, Teflon stopper; 2, spiral stirrer; 3, platinum
electrode; 4, electrode rod with platinum electrodes to be silvered. (From
Ref. 6, courtesy J. Wiley & Sons, Inc.)

Keep the stirrer and oxygen going throughout the operation. When the
absorbent solution has attained the temperature of the coolant in the jacket,
introduce the 0.0002 $\underline{N}$ silver nitrate solution continuously (2-3 μl/s) from
the piston buret. When a potential change is noted on the vacuum tube
voltmeter, add 0.5-μl portions and observe the change of direction of the
pointer deflection (Note 2). In order to test the completeness of the HCl
transfer, perform a checking titration after a few minutes.

E. Notes

Note 1: The time required for complete transfer of chlorine must be determined for the apparatus by calibration. The time and temperature of
heating are then kept constant for all analyses.

Note 2: In using a motor-driven buret, it is recommended to titrate with a flow rate of 25 μl/min and a recorder paper speed of 1.5 m/h. The paper distance between the beginning of titration and the curve maximum is evaluated, and from this, after calibration with known amounts of chloride, $AgNO_3$ solution consumption is obtained.

11.6 EXPERIMENTAL: ULTRAMICRO KJELDAHL DETERMINATION OF NITROGEN USING THE AERATION TECHNIQUE

A. Principle

After the digestion of the aminoid material in a test tube provided with two side arms, the tube is closed with a syringe cap, alkali solution is added, and ammonia is driven into 2% boric acid solution in the receiving tube by means of a current of air. The ammonia trapped in the receiving tube is then determined by titration with standardized acid. The procedure described below is adapted from the method proposed by Sobel and co-workers [50].

B. Equipment

Digestion-aeration tubes, made of 15-mm Pyrex tubing, 120 mm long (see Fig. 11.16), with side arms and a bubbling tube sealed on about 35 mm from the top of the tube. The bubbling tube, kept closed to the wall, is 3-4 mm in outside diameter and is constricted to a 1-mm bore at the tip. The neck of the digestion-aeration tube is fitted with a syringe cap (rubber stopper of sleeve type).
Digestion rack (see Sec. 3.3), or metal heating block [62].
Capillary microburet.

C. Reagents

Digestion mixture: Dissolve 30 g of potassium sulfate and 5 g of cupric sulfate in 500 ml of distilled water, add 0.5 g of selenium powder in 20 ml of concentrated sulfuric acid, and finally 480 ml of concentrated sulfuric acid.
Hydrogen peroxide, 30%.
Sodium hydroxide solution, 30%.

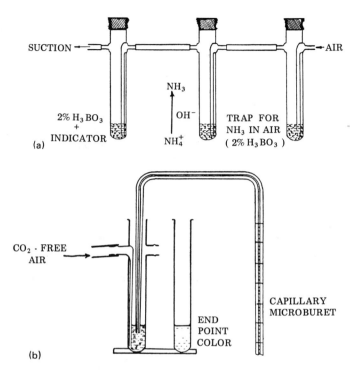

FIG. 11.16. Ultramicro Kjeldahl determination of nitrogen. (a) Schematic diagram of microaeration apparatus for 10-100 μg of nitrogen. (b) Schematic diagram of titration assembly. (Courtesy Anal. Chem.)

Bromocresol green-methyl red mixed indicator (see Sec. 3.3).

2% Boric acid with indicator: Add 1 ml of the above indicator to 100 ml of 2% boric acid solution.

Ammonium sulfate standard solution: Accurately weigh 4.7186 g of ammonium sulfate, AR (dried at 105°C), into a 100-ml volumetric flask and fill to volume with distilled water (1 ml = 10 mg N). For use, transfer 1.00 ml of this stock solution to a 100-ml volumetric flask and fill to the mark with distilled water (1 ml = 100 μg N).

Standardized acid (HCl, H_2SO_4, or $KHIO_3$), 0.01 N (Sec. 11.6E, Note 1).

D. Procedure

Digestion

Accurately weigh 50-300 μg of solid sample (or measure 10-100 μl of
solution) containing 10-150 μg N into the digestion-aeration tube. Add 0.2
ml of the digestion mixture and heat the contents of the tube for several
minutes. If the solution does not clarify, cool the tube, carefully introduce
one drop of 30% hydrogen peroxide, and reheat. Upon cooling, wash the
walls of the tube with 1.0 ml of distilled water. Then connect the tube between two other tubes as illustrated in Fig. 11.16a.

Aeration

Place 1.5 ml of 2% boric acid with indicator in the air-inlet tube and
the receiving tube, respectively. Close the tubes with syringe caps as
shown in Fig. 11.16a. Start the suction so that a steady stream of air is
bubbled through the digestion mixture now containing ammonium ions.
Using a plastic syringe, introduce 0.5 ml of 30% sodium hydroxide solution
to the middle tube. Continue the aeration for 20-30 min to drive all the
ammonia into the receiving tube. Then remove the caps, turn off the suction, and disconnect the receiving tube.

Titration

Bring the receiving tube, now containing ammonia in boric acid solution, to the titration stand as depicted in Fig. 11.16b. Pass a slow current
of pure nitrogen gas through the solution and submerge the tip of the capillary microburet below the surface. Using a dropper, wash the walls of the
tube with distilled water. Gradually deliver the standardized acid from the
microburet until the color of the solution in the receiving tube matches the
original color of the 2% boric acid with indicator (Note 2).

E. Notes

Note 1: It is recommended to standardize the acid for titration by aerating
known quantities (in the range of 10-150 μg N) of ammonium sulfate
standard solution under similar conditions and determining the volumes of
acid required for neutralization.

Note 2: If preferred, the titration may be carried to the gray transition
point, keeping the original color of the 2% boric acid with indicator as
reference.

F. Comments

(1) For accurate and precise analyses, blank determinations should be performed in order to correct for reagent and operational errors.

(2) The aeration process in multiple analyses can be conveniently carried out simultaneously by joining the tubes either in series or in parallel to one suction line.

(3) Certain aminoid compounds can be hydrolyzed by enzymes to yield NH_4^+ under mild conditions. For instance, urea can be treated with urease [180]; D-α-amino acids are deaminated by oxidase [181]. These reactions can be carried out in the digestion-aeration tubes.

(4) For the determination of 1-10 μg N, Sobel and co-workers [50] have described a similar procedure employing 5-ml conical centrifuge tubes and 0.1 ml of the 2% boric acid in the receiving tube.

(5) As an alternative to the titrimetric finish, the ammonia can be determined spectrophotometrically by applying the indophenol blue method [182].

REFERENCES

1. K. Lindestrøm-Lang and H. Holter, Z. Physiol. Chem. 220:5 (1933).
1a. K. Lindestrøm-Lang, D. Glick, and H. Holter, Compt. Rend. Trav. Lab. (Carlsberg), Ser. Chim. 22:300 (1937); 24:334, 400 (1943).
2. P. L. Kirk, Quantitative Ultramicro-analysis, Wiley, New York, 1950.
3. R. Belcher, Submicro Methods of Organic Analysis, Elsevier, Amsterdam, 1966.
4. W. J. Kirsten, Mikrochem. 40:170 (1953); Mikrochim. Acta 1955:1086.
5. W. J. Kirsten and K. Hozumi, Mikrochim. Acta 1962:777; Microchem. J. 6:591 (1962).
6. G. Tölg, Ultramicro Elemental Analysis, Wiley, New York, 1970; see also G. Tölg, Elemental Analysis with Minute Samples, in G. Svehla (ed.), Wilson and Wilson's Comprehensive Analytical Chemistry, Vol. 3, Elsevier, Amsterdam, 1975, p. 1.
6a. G. Tölg, private communication, October 1974.
7. S. Natelson, Microtechniques of Clinical Chemistry, 2nd Ed., Thomas, Springfield, Ill., 1961.
8. H. Mattenheimer, Micromethods for the Clinical and Biochemical Laboratory, Ann Arbor Science Publishers, Ann Arbor, Mich., 1970.
9. R. Stillman and T. S. Ma, Mikrochim. Acta 1973:491.
10. T. S. Ma, A. Fono, A. Sapse, B. J. Marcus, J. Polesuk, and S. Wassef, Mikrochim. Acta 1965:1098; 1967:960; 1968:436; 1969:352, 815; 1970:677; 1971:267, 622; 1972:313; 1975:649.

10a. B. F. A. Griepink and W. Krijgsman, Z. Anal. Chem. 265:241
     (1973).
10b. B. Griepink, F. G. Römer, and W. J. van Oort, Microchem. J.
     20:33 (1975).
10c. F. G. Römer, Ph. D. Thesis, University of Utrecht, Holland, 1975.
11. I. P. Alimarin, M. N. Petrikova, and T. A. Kokina, Mikrochim.
     Acta 1971:494.
12. I. M. Korenman, Quantitative Ultramicroanalysis, Academic, New
     York, 1965.
12a. M. Ozcimder, W. Krijgsman, and B. Griepink, Mikrochim. Acta
     1974:219.
13. S. S. M. Hassan and S. A. I. Thoria, Mikrochim. Acta 1974:561.
14. R. Belcher, Talanta 15:357 (1968).
15. J. L. Lambert, in McGraw-Hill Yearbook of Science, McGraw-Hill,
     New York, 1973, p. 361.
16. W. Schöniger, Microchem. J. 11:469 (1966).
17. A. A. Benedetti-Pichler, in F. Hecht and M. K. Zacherl (eds.),
     Handbuch der Mikrochemischen Methoden, Band I, Teil 2, Springer-
     Verlag, Wien, 1959, p. 70.
18. H. M. El Badry and C. L. Wilson, in Symposium on Microbalance,
     Royal Institute of Chemistry, London, 1950, Report No. 4, p. 23.
19. Microchemical Specialities Co., Berkeley, Calif.
20. Microtech Services Co., Los Angeles, Calif.
21. L. Oertling, Ltd., Kent, England.
22. Vortex Co., Claremont, Calif.
23. Type 25 UM, Bunge Co., Hamberg, Germany.
24. Type UM-7, Mettler Co., Zürich, Switzerland.
25. Electronic Microbalance, Sartorius Division, Brinkmann Instruments,
     Westbury, N.Y.
26. Electrobalances, Cahn Instrument Co., Paramount, Calif.
27. Autobalances, Model AM-1 and Model AD-1, Perkin-Elmer Co.,
     Norwalk, Conn.
28. K. Hozumi, Japan Analyst 19:163 (1970).
29. W. Merz and W. Pfab, Mikrochim. Acta 1969:905.
30. H. Simon and G. Mülhofer, Z. Anal. Chem. 181:85 (1961).
31. W. Merz and W. Pfab, Microchem. J. 10:346 (1966).
32. M. Schütze, Z. Anal. Chem. 118:241 (1939).
33. W. Zimmermann, Z. Anal. Chem. 118:258 (1939).
34. F. Ehrenberger, S. Gorbach, and W. Mann, Z. Anal. Chem.
     198:242 (1963).
35. W. Zimmermann, Mikrochem. 40:162 (1952).
36. T. A. O'Donnell and D. F. Stewart, Anal. Chem. 34:1347 (1962).
37. E. Ruf, Z. Anal. Chem. 161:1 (1958).

38. S. Natelson, Microtechniques of Clinical Chemistry, 2nd Ed., Thomas, Springfield, Ill., 1961, p. 311.

39. T. S. Ma and G. Zuazaga, Anal. Chem. 14:280 (1942).

40. J. A. Kuck, P. L. Altieri, and A. K. Towne, Mikrochim. Acta 1954:1.

41. W. J. Kirsten, Anal. Chem. 26:1091 (1954); Mikrochim. Acta 1956:836; Chim. Anal. 40:253 (1958); Z. Anal. Chem. 181:1 (1961); 191:161 (1962).

42. K. Hozumi and W. J. Kirsten, Anal. Chem. 34:434 (1962); 35:666, 1522 (1963); 38:641 (1966); Mikrochim. Acta 1962:777; Mikrochem. J. 6:591 (1962); 12:512 (1967).

43. J. Körbl, Mikrochim. Acta 1956:1705.

44. K. Umemoto and K. Hozumi, Japan Analyst 17:1302 (1968).

45. K. Nakamura, K. Ono, and K. Kawada, Microchem. J. 17:338 (1972).

46. K. Nakamura, K. Ono, and K. Kawada, Microchem. J. 15:364 (1970).

47. K. Nakamura and K. Kawada, Ann. Sankyo Res. Lab. 22:62 (1970).

48. K. Nakamura, K. Kuboyama, K. Ono, and K. Kawada, Mikrochim. Acta 1972:353.

49. K. Hozumi and K. Umemoto, Japan Analyst 16:800 (1967).

50. A. E. Sobel, A. Hirschman, and J. Besman, Anal. Chem. 19:927 (1947).

51. B. Griepink and J. K. Terlouw, Mikrochim. Acta 1968:624.

52. V. G. Berezkin, Analytical Reaction Gas Chromatography (translated from Russian by L. S. Ettre), Plenum, New York, 1968, p. 135.

53. S. A. Liebman, D. H. Ahlstrom, T. C. Creighton, G. D. Pruder, R. A. Averitt, and E. J. Levy, Anal. Chem. 44:1411 (1972).

54. CDS Bulletin 900-0372, Chemical Data Systems, Inc., Oxford, Pa.

55. W. J. Kirsten, Microchem. J. 7:34 (1963).

56. J. E. Fildes and W. J. Kirsten, Microchem. J. 9:411 (1965).

57. W. J. Kirsten, B. Danielson, and E. Ohren, Microchem. J. 12:177 (1967).

57a. W. J. Kirsten, private communication, October 1974; Mikrochim. Acta 1976(II):300.

58. M. Pont, British Patent 1,334,925 (April 19, 1971); French Patent 16.2.70, 4.1.71.

58a. B. Griepink and W. Krijgsman, Mikrochim. Acta 1968:330.

59. R. Belcher, G. Dryhurst, A. M. G. Macdonald, J. R. Majer, and G. J. Roberts, Anal. Chim. Acta 43:441 (1968).

60. R. Belcher, G. Dryhurst, and A. M. G. Macdonald, Anal. Lett. 1:807 (1968).

61. K. Hozumi and K. Umemoto, Microchem. J. 12:512 (1967).

61a. K. Miyahara and T. Takaoka, Microchem. J. 22:7, 210, 216 (1977).

62. T. S. Ma and V. Horak, Microscale Manipulations in Chemistry, Wiley, New York, 1976.

63. A. A. Benedetti-Pichler, Identification of Materials, Springer-Verlag, New York, 1964.

64. F. G. Römer, G. W. S. van Osch, and B. Griepink, Mikrochim. Acta 1971:772.

65. J. Körbl, Mikrochim. Acta 1956:1705.

66. K. H. Ballschmiter and G. Tölg, Z. Anal. Chem. 203:20 (1964).

66a. F. G. Römer, P. H. van Rossum, and B. Griepink, Mikrochim. Acta 1975:337, 345.

66b. E. Buijsman, F. G. Römer, and B. Griepink, Microchem. J. 22:328 (1977).

67. G. Tölg and K. H. Ballschmiter, Microchem. J. 9:257 (1965).

68. G. Tölg, Z. Anal. Chem. 205:40 (1964).

69. H. ter Meulen and J. Heslinga, Nieuwe Methoden voor Elementair-analyse, Meinema, Delft, 1925.

69a. C. J. van Niewenburg and J. W. L. van Ligten, Quantitative Chemical Micro-Analysis (translated from Dutch by C. G. Verver), Elsevier, Amsterdam, 1963, p. 152.

70. W. Krijgsman, J. G. Simons, B. Griepink, and D. M. Verduyn, Z. Anal. Chem. 259:274 (1972).

71. W. Krijgsman, W. P. van Bennekom, and B. Griepink, Mikrochim. Acta 1972:42.

72. K. J. Conway, Micro-diffusion Analysis and Volumetric Error, Lockwood, London, 1940.

73. G. Schwab and G. Tölg, Z. Anal. Chem. 205:29 (1964).

74. W. H. List and G. Tölg, Z. Anal. Chem. 226:127 (1967).

75. W. Krijgsman, G. de Groot, W. P. van Bennekom, and B. Griepink, Mikrochim. Acta 1972:364.

76. W. Krijgman, Ph. D. Thesis, University of Utrecht, Holland, 1972.

77. D. M. Coulson and L. A. Cavanagh, Anal. Chem. 52:1245 (1960).

78. G. Tölg, personal communication; A. Grünert and G. Tölg, Talanta 18:881 (1971).

79. R. Engler and G. Tölg, Z. Anal. Chem. 235:151 (1968).

80. B. L. Horeker, T. S. Ma, and E. Haas, J. Biol. Chem. 136:775 (1940).

81. F. G. Römer and B. Griepink, Mikrochim. Acta 1970:867.

82. D. C. White, Talanta 10:727 (1963).

83. C. E. van Hall, J. Safranko, and V. A. Stenger, Anal. Chem. 35:315 (1963).

84. E. Maly, Mikrochim. Acta 1963:1046.

85. E. Maly and J. Krsek, Mikrochim. Acta 1964:778.

86. W. Walisch and W. Marks, Mikrochim. Acta 1967:1051.
87. A. Campiglio, Mikrochim. Acta 1964:114.
88. I. Klesment, Mikrochim. Acta 1969:1237.
89. K. Yoshikawa and T. Mitsui, Microchem. J. 9:52 (1965).
90. H. J. Born and N. Riehl, Angew. Chem. 72:559 (1960).
91. G. Gutbier and M. Boetius, Mikrochim. Acta 1960:636.
92. W. J. Kirsten, K. Hozumi, and L. Kirk, Z. Anal. Chem. 191:161 (1962).
92a. K. Miyahara and T. Takaoka, Microchem. J. 21:325 (1976).
93. T. A. Norris and J. E. Flynn, Anal. Chem. 37:152 (1965).
94. W. J. Kirsten, and K. Hozumi, Mikrochim. Acta 1962:777.
95. W. J. Kirsten and K. Hozumi, Microchem. J. 6:591 (1962).
96. K. Hozumi, Anal. Chem. 35:666 (1963).
97. K. Hozumi and W. J. Kirsten, Anal. Chem. 34:434 (1962).
98. A. Pietrogrande, Mikrochim. Acta 1964:1106.
99. K. Hozumi, Anal. Chem. 38:641 (1966).
100. W. Merz, Z. Anal. Chem. 207:424 (1965).
100a. K. Miyahara, Microchem. J. 19:416, 423 (1974); 20:453 (1975).
101. R. A. Shah and N. Bhatty, Mikrochim. Acta 1967:81.
101a. H. J. Yu, Chem. Bull. (Peking) 1976:178.
102. R. Belcher, A. D. Campbell, and P. Gouverneur, J. Chem. Soc. 1963:531.
103. H. Roth, Mikrochim. Acta 1960:663.
104. A. A. Kanchukh, Biokhimiya 26:393 (1961).
105. C. J. F. Bottcher, C. M. van Gent, and C. Pries, Rec. Trav. Chim. Pays-Bas 80:1157 (1961).
106. S. E. Dixon and R. W. Shuel, Chemist-Analyst 51:84 (1962).
107. W. Pfab and W. Merz, Z. Anal. Chem. 200:385 (1964).
108. P. Gouveneur, H. C. E. van Leuven, R. Belcher, and A. M. G. Macdonald, Anal. Chim. Acta 33:360 (1965).
108a. F. G. Römer, J. W. van Schaik, K. Brunt, and B. Griepink, Z. Anal. Chem. 273:109 (1975).
109. F. Cacace, R. Cipollini, and G. Perez, Science 132:1253 (1960).
110. R. Belcher and B. Fleet, Anal. Lett. 1:525 (1968).
111. F. Scheidl, Z. Anal. Chem. 245:30 (1969).
112. I. A. Revel'skii, R. I. Borodulina, V. G. Klimova, and T. M. Sovakova, Neftekhimiya 1965:417.
113. C. D. Miller, Microchem. J. 11:366 (1966).
114. W. Walisch, Trans. N.Y. Acad. Sci., Ser. II 25:693 (1963).
115. Technicon Instruments Corp., British Patent 1,010,330 (1963).
116. W. Walisch, G. Scheuerbrandt, and W. Marks, Microchem. J. 11:315 (1966).
117. J. Binkowski and R. Levy, Bull. Soc. Chim. Fr. 1968:4289.

118. C. D. Miller and J. D. Winefordner, Microchem. J. 8:334 (1964).

119. W. J. Kirsten, Microchem. J. 16:610 (1971).

120. C. W. Koch and E. E. Jones, Mikrochim. Acta 1963:734.

121. K. Hozumi and W. J. Kirsten, Anal. Chem. 35:1522 (1963).

122. R. Belcher, P. Gouverneur, and A. M. G. Macdonald, J. Chem. Soc. 1962:1938.

123. R. Belcher, Y. A. Gawargious, P. Gouveneur, and A. M. G. Macdonald, J. Chem. Soc. 1964:3560.

124. B. Griepink and A. van Sandwijk, Mikrochim. Acta 1969:1014.

125. Y. Imai, K. Yamauchi, S. Terabe, R. Konaka, and J. Sigita, J. Chromatogr. 36:345 (1968).

125a. A. Nara, N. Kobayashi, K. Honba, and S. Baba, Microchem. J. 20:200 (1975).

126. B. F. A. Griepink and A. van Sandwijk, Mikrochim. Acta 1969:1246.

126a. W. J. Kirsten, Mikrochim. Acta 1976(II):299.

127. H. Tamura and Y. Kondo, J. Biochem. Soc. Japan 35:278 (1963).

127a. M. Ozcimder, G. F. Ernst, and B. Griepink, Microchem. J. 20:227 (1975).

128. W. Walisch and O. Jaenicke, Mikrochim. Acta 1967:1147.

129. A. Discherl and F. Erne, Mikrochim. Acta 1963:242.

130. R. D. Rowe, Anal. Chem. 37:368 (1965).

131. R. A. Hofstader, Microchem. J. 11:87 (1966).

132. R. C. Blinn, Anal. Chem. 32:292 (1960).

133. B. Zak and E. S. Baginski, Anal. Chem. 34:257 (1962).

134. J. Kis and C. Shejtanow, Period. Polytech. (Budapest) 4:163 (1960).

135. G. Tölg, Z. Anal. Chem. 194:20 (1963).

136. B. Lewandowska, Chem. Anal. (Warsaw) 10:1353 (1965).

137. T. Onoe and H. Shimada, Sankyo Res. Lab. 17:82 (1965).

138. M. E. Fernandopulle and A. M. G. Macdonald, Microchem. J. 11:41 (1966).

139. R. Belcher, M. A. Leonard, and T. S. West, J. Chem. Soc. 1959:3577.

140. W. Oelschlager, Z. Anal. Chem. 191:408 (1962).

141. H. Soep, J. Chromatogr. 6:122 (1961).

142. H. Soep, Nature 192:67 (1961).

143. R. J. Hall, Analyst 85:560 (1960).

144. R. Belcher, A. D. Campbell, P. Gouveneur, and A. M. G. Macdonald, J. Chem. Soc. 1962:3033.

145. G. Gorbach and E. Regula, Mikrochim. Acta 1966:615.

146. W. Machida, T. Okutani, and S. Utsuimi, Japan Analyst 17:436 (1968).

147. J. P. Dixon, Chem. Ind. (London) 1959:156.

148. W. Walisch and G. Humme, Mikrochim. Acta 1968:748.

149. D. Fraisse and R. Levy, C. R. Acad. Sci., Paris, Ser. C 271:49
      (1970).
149a. G. de Groot, P. A. Greve, and R. A. A. Maes, Anal. Chim. Acta
      79:279 (1975).
150. A. Campiglio and E. Pell, Mikrochim. Acta 1969:467.
151. L. Gustafsson, Talanta 4:227 (1960).
152. J. Dokladalova, Mikrochim. Acta 1965:345.
153. L. Granatdli, Anal. Chem. 31:434 (1959).
154. B. Griepink, J. Slanina, and J. Schoonman, Mikrochim. Acta
      1967:984.
155. S. D. Iordanov and C. P. Ivanov, C. R. Acad. Bulg. Sci. 21:451
      (1968).
156. P. Stoffyn and W. Keane, Anal. Chem. 36:397 (1964).
157. G. A. Robinson, Can. J. Biochem. Physiol. 38:643 (1960).
158. E. C. Olson and J. W. Shell, Anal. Chim. Acta 23:219 (1960).
159. R. Belcher, A. M. G. Macdonald, S. E. Phang, and T. S. West,
      J. Chem. Soc. 1965:2044.
160. C. J. Bottcher, C. M. van Gent, and C. Pries, Anal. Chim. Acta
      24:203 (1961).
161. A. C. Bhattacharyya, B. Bhaduri, and P. Banerjee, Analyst 86:195
      (1961).
162. P. M. Saliman, Anal. Chem. 36:112 (1964).
163. T. Salvage and J. P. Dixon, Analyst 90:24 (1965).
164. A. J. Christopher and T. R. F. W. Fennell, Microchem. J. 12:593
      (1967).
165. W. J. Kirsten, Microchem. J. 12:307 (1967).
166. A. J. Karmen, J. Gas Chromatogr. 3:336 (1965).
167. B. F. A. Griepink, W. Krijgsman, A. J. M. E. Leenaers-Smeets,
      J. Slanina, and H. Cuijpers, Mikrochim. Acta 1969:1018.
168. B. F. A. Griepink and W. Krijgsman, Mikrochim. Acta 1968:1003.
169. J. A. Wenniger, J. Ass. Offic. Agric. Chem. 48:826 (1965).
170. W. H. Gutenmann and D. J. Lisk, J. Agr. Food Chem. 8:306 (1960).
171. K. Ozawa and S. Egashira, Japan Analyst 11:506 (1962).
172. Committee Report, Analyst 90:515 (1965).
173. T. R. F. W. Fennell and J. R. Webb, Talanta 9:795 (1962).
174. G. Westoo, Analyst 88:287 (1963).
175. R. Belcher, B. Crossland, and T. R. F. W. Fennell, Talanta
      17:639 (1970).
176. R. Belcher, B. Crossland, and T. R. F. W. Fennell, Talanta
      17:112 (1970).
177. S. Kotrly and J. Vrestal, Collect. Czech. Chem. Commun. 25:1148
      (1960).

178. D. Ketchem, Automated Elemental Analyzers, reported at the International Symposium on Microtechniques, University Park, Pa., August 1973.
179. E. Bishop and R. G. Dhaneshwar, Analyst 87:845 (1962).
180. A. E. Sobel, A. M. Moyer, and S. P. Gottfried, J. Biol. Chem. 156:355 (1944).
181. N. D. Cheronis and T. S. Ma, Organic Functional Group Analysis, Wiley, New York, 1964, p. 245.
182. H. K. Boo and T. S. Ma, Mikrochim. Acta 1976(II):515.

Trace Analysis

12.1 GENERAL REMARKS

A. Importance of Trace Analysis

The importance of trace analysis was fully recognized only during the present decade when the general public became concerned about the environment and aware of the serious effects of air and water pollution. Because of the public clamor, both federal and local governments have promulgated laws and regulations with regard to the tolerance limits of contaminants in the air, water, foods, etc. As a result, the demands for trace analysis have increased by leaps and bounds during the past several years. Concurrently there has been intensive activity in the investigation of methods of trace analysis. Thus, it is significant that in 1971 the Division of Analytical Chemistry of the American Chemical Society sponsored a Summer Symposium entitled "Analytical Chemistry: Key to Progress on National Problems" which focused on six areas (agriculture, air pollution, clinical chemistry and biomedicine, oceanography, solid state research, and water pollution), all requiring trace analysis techniques [1], and that the Summer Symposium in 1973 dealt entirely with organic pollutant analysis [2]. In 1974, the National Bureau of Standards sponsored a symposium on accuracy in trace analysis [2a]. Reports [2b-e] from the trace metals project of the petroleum industry began to appear in 1975.

It should be noted, however, that trace analysis is not a new field. The senior author (Ma) was involved in the early 1940s in research projects which utilized methods for the determination of minute amounts of phosphorus in proteins, sulfur in petroleum products, chlorine in a synthetic polymer, and carbonaceous material in uranium metal. A symposium on

trace analysis [3] was held at the New York Academy of Medicine in 1955,
which discussed the analysis of trace constituents in industrial, agricultural, biological, and medical fields. The implications of trace elements
in human and animal nutrition have been studied for many years [4]. The
importance of trace elements in some industries has long been recognized.
For example, in the petroleum industry [5], halogen and sulfur may cause
corrosion of refining equipment. Tarry oxygenated compounds and inorganic materials may be deposited in equipment and necessitate costly shut-
down and replacements. Certain metallic elements may deposit on hydrocarbon conversion catalysts and adversely affect yields or product distribution. On the other hand, minute amounts of certain additives are
incorporated into petroleum products to obtain beneficial effects, such as
lead in gasoline and metal salts of fatty acids in lubricating oil. The significance of trace elements in animal and plant systems has been established for some time [5a]; bibliographies have been compiled for a number
of elements in soil and plants [5b].

The recent need for extremely pure chemicals calls for analysis of
trace impurities [6]. Extra pure solvents for use in spectroscopy are
common examples. In the manufacture of phosphoric acid, the concentrated sulfuric acid employed is analyzed for carbon (see Sec. 12. 8) because the carbon content which is indicative of the amount of organic impurities present affects the yield of phosphoric acid.

B. Scope of This Chapter

In the present chapter we confine our discussions to the methods of analysis
for the elements in organic materials at the level of less than 1% down to
about 0.0001% (= 1 part per million, abbreviated as ppm). The fact that the
element in question exists in the sample in such a low percentage may be
due to the fact that (1) this element is a small fraction of a large molecule
(e.g., terminal chlorine of a polymer), or (2) the organic compound containing this element is a minute constituent of the bulk material (e.g., insecticide residue on foods). In the latter case, the ideal answer is to determine the specific compound using analytical methods that can character-
ize its molecular structure, such as spectroscopic analysis [7]. Frequently,
however, this approach is neither feasible nor necessary. Determination
of the compound may entail elaborate processes of separation and isolation.
For instance, if the compound contains arsenic or fluorine which is not
present in the rest of the mixture, elemental analysis is much simpler than
the analysis of the compound by gas chromatography-mass spectrometry.
In the example of determination of carbon in sulfuric acid (see Sec. 12. 8),

it is not pertinent to identify the nature of the carbonaceous impurities. This also applies to the analysis of fuel for sulfur which is the culprit responsible for sulfur dioxide in the air.

Two distinct features will be noted in trace analysis when compared with the analytical techniques discussed in the previous chapters. First, the size of the sample taken for trace analysis is much larger than that for the corresponding procedure described for regular analysis. Secondly, the precision and accuracy required in trace analysis are considerably less demanding. The large sample size imposed on trace analysis is to obtain a suitable quantity of the product to be measured. For example, in order to determine phosphorus in coke [8], the minimum sample size is about 30 mg; for the analysis of carbon-13 and tritium in biological materials, 250-mg samples are prescribed [9]. On the other hand, precision and accuracy are sacrificed in trace analysis because it is difficult to control the experimental conditions. The increase in sample size and complexity of the material to be analyzed call for modifications of the regular methods as well as different approaches in order to accomplish the purpose of trace analysis.

12.2 APPROACHES FOR TRACE ANALYSIS OF THE ELEMENTS

A. Decomposition Techniques

While the same principles can be utilized to decompose pure organic compounds as well as mixtures, application of the decomposition technique for trace analysis is dependent on the nature of the sample. For instance, wet digestion in micro Kjeldahl flasks was recommended by Horeker and co-workers [10] for the determination of trace amounts of phosphorus in natural polymers. Recently Majowska and co-workers [11] have compared the results obtained by using the closed-flask, open-crucible, and wet-oxidation decomposition techniques, respectively, for determining phosphorus in synthetic fibers and confirmed the superiority of the last procedure. This gave the most accurate results over the widest concentration range and was found to be suitable for all types of fibers. The precision of determination at the 0.01% level was ±0.001%.

For trace element analysis in petroleum products [5], the decomposition techniques employed include ashing, wet oxidation, closed-flask combustion, high-speed burning, and extraction decomposition. It will be recalled that the last-mentioned technique is not used for the analysis of pure organic compounds. For trace metal determination by atomic absorption, Gaylor [11a] has recommended a procedure which involves absorbing the liquid sample on magnesium sulfonate, adding H_2SO_4, and ashing at 650°C. The sulfonated ash is then dissolved in dilute acid to provide the aspiration solution.

A decomposition technique specific for trace analysis is known as the oxygen bomb method and involves combustion of the sample in oxygen under pressure in a metal bomb. The capacities of the metal bombs [12-14] range from 200 to 500 ml. This method has been used to determine practically all nonmetals and metalloids, and also many metallic elements in organic materials, employing sample weights up to 1 g. Figure 12.1 shows the oxygen bomb recommended by Fujiwara and Narasaki [15]. Constructed

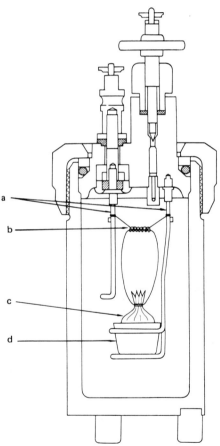

FIG. 12.1. Schematic diagram of the oxygen bomb for trace analysis. a, electrodes; b, coil; c, sample wrapped with rice paper; d, combustion capsule. (From Ref. 15, courtesy Anal. Chem.)

of stainless steel, it has a capacity of 300 ml and can accommodate 1 g of the sample. The combustion cup and electrodes are made of platinum; the interior of the bomb is platinum plated. The sample is tied up in a sheet of rice paper by means of cotton thread and placed in the cup. Then the remaining parts of the thread are passed through a coil of platinum wire between the electrodes. Oxygen pressure in the bomb is brought up to 25 kg/cm^2, and the sample is ignited by passing a small alternating current through the platinum wire coil under a potential difference of 10 V.

Generally speaking, wet oxidation (also known as wet ashing) and dry ashing are the most commonly employed techniques for the destruction of organic matter for trace element determinations. Mixtures of HNO_3/H_2SO_4/$HClO_4$ [15a] are frequently used, and hydrogen peroxide [15b] is sometimes added. Thompson and Blanchflower [15c] have described an improved wet-ashing apparatus for biological materials. Feldman [15d] has designed a temperature-programming procedure for perchloric acid digestion of up to 5 g of sample. Bessman [15e] has constructed an automated dry-ashing apparatus that permits quantitative aliquot measurement of the sample; it consists of a turntable fitted with 40 silica tubes in which the samples are slowly rotated through the various stages of the ashing, washing, and extracting system. Van Raaphorst et al. [15f] have used ^{60}Co and ^{65}Zn to investigate the loss of cobalt and zinc during dry ashing of biological material in porcelain crucibles for 20 h. No significant losses were observed at temperatures up to 1000°C. After ashing at 1000°C, however, as much as 20% of the total cobalt content was not extracted by leaching with HCl, part of the melt being difficult to remove from the crucible.

Recently the use of a low-temperature oxygen plasma ashing technique has been proposed, e.g., for the determination of cadmium in tobacco [15g] and of lead in fish [15h]. The apparatus is commercially available [15i], and a monograph [15j] has been published in Japan. According to Hozumi [15j], there are still problems in the quantitative recovery of the elements when they are plasma ashed. While one group of workers claimed excellent recoveries, other investigators reported very unreliable results for the same elements in similar organic materials.

B. Modes of Finish

Owing to the fact that the quantity of the product obtained from the element to be determined in trace analysis is usually extremely small, the first requirement of its finishing mode is high sensitivity. Secondly, the method should be reliable and the experimental conditions easily controllable so as to obtain reproducible results. For example, the molybdenum blue method for the determination of phosphorus (see Sec. 7.5) is sensitive at the microgram level and can be performed by means of a simple colorimeter [10]. The color intensity, however, is temperature dependent; hence the reaction is carried out in a boiling water bath instead of at ambient temperature. Since many physicochemical measurements are inherently very sensitive and precise, they are favorable modes of finish in trace analysis. On the other hand, some finishing techniques which are not employed in regular determinations using milligram amounts of pure organic compounds are singularly suited for trace analysis, for example, the amplification and catalytic methods.

Amplification Methods

Amplification methods are those in which the end product from the element in question is caused to increase considerably in order to facilitate measurement. Typical examples of such methods are based on multiplication reactions [16]. For instance, Emich [17] suggested in 1933 to multiply the carbon dioxide in the combustion gases by the following scheme.

The traces of carbon dioxide are conducted into a tube in which a carbon rod and a copper oxide filament are mounted. The tube is closed. By heating the tube electrically, first at the carbon rod and then at the copper oxide filament, the original amount of carbon dioxide is doubled. Repeating these operations successively, the carbon dioxide increases in a geometric progression.

Recently Schöniger [18] has reported the application of this method to the determination of oxygen by carbon reduction (see Sec. 4.1B) and obtained a multiplication factor of 16. The sequence of chemical reactions can be depicted as follows:

$$CO \xrightarrow[500°C]{CuO} CO_2 \xrightarrow[900°C]{Pt/C} 2CO \xrightarrow[500°C]{CuO} 2CO_2$$

$$\xrightarrow[900°C]{Pt/C} 4CO \xrightarrow[500°C]{CuO} 4CO_2 \xrightarrow[900°C]{Pt/C} 8CO \xrightarrow[500°C]{CuO} 8CO_2 \ldots$$

The schematic diagram of the apparatus is shown in Fig. 12.2. The carbon monoxide from the oxygen combustion train is introduced through the three-way stopcock and freezing trap (liquid nitrogen) into a short tube containing copper oxide at 500°C. The Anhydrone tube serves to remove water if hydrogen is present. In the next tube, which is filled with platinized carbon at 900°C, the carbon dioxide is reduced to carbon monoxide. The following tube is again filled with copper oxide at 500°C and the process is repeated. Finally the carbon dioxide is driven into a titration cell [19] to be determined.

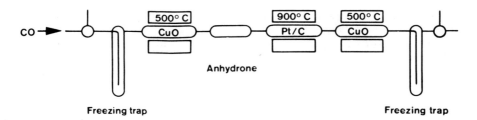

FIG. 12.2. Multiplication unit for oxygen determination. (From Ref. 18, courtesy Microchem. J.)

It is obvious that the above assembly can be used for the trace analysis of carbon. Application of this principle to the determination of hydrogen through the water obtained from combustion is indicated in the following series of reactions:

$$\text{Organic H} \xrightarrow{(O)} H_2O \xrightarrow[900°C]{Pt/C} H_2 + CO \xrightarrow[500°C]{CuO} H_2O + CO_2 \xrightarrow[900°C]{Pt/C} 2CO$$

$$\xrightarrow[500°C]{CuO} 2CO_2 \cdots$$

Weisz and Fritsche [20] have suggested a number of possibilities for carrying out cyclic multiplication reactions. The use of ion-exchange, complex formation, and liquid-liquid extraction are included. Experimental proofs of the practicability of these processes have not been reported.

The amplification method most frequently utilized is the iodide-iodate reaction which permits a sixfold increase of equivalence (see Sec. 5.5). Besides the determination of iodine, this reaction can be employed for the indirect amplification analysis of other elements. Thus, Belcher and Coulden [21] have determined chloride by reacting it with silver or mercury iodate; an equivalent amount of soluble iodate is released, which, after separation, is titrated iodimetrically. Awad and co-workers [22] have used the sparingly soluble calcium iodate to determine fluoride according to the following equation:

$$Ca(IO_3)_2 + F^- \longrightarrow IO_3^- + CaF_2 \downarrow$$

Isopropyl alcohol is added to the solution so that the alcohol concentration is 65%; the solution is filtered, and the iodate ions in the filtrate are

determined. Hassan and Thoria [23] have also used calcium iodate to amplify phosphate analysis. Hashmi [24] has described similar reactions for
determining Ag(I) and Au(III). A scheme for 24-fold amplification of iodide
[25] is given in Sec. 11.1C.

Gravimetry is rarely used in trace analysis. In such cases [5], it is
necessary to select a precipitant that has a large conversion factor, e.g.,
oxine (8-hydroxyquinoline, molecular weight 145), for trace metal analysis.

Catalytic Methods

Catalytic methods are based on the observation that the rates of certain
reactions are related to trace amounts of some specific element present
in the system. A well-known example is the catalytic effect of iodide on the
reduction of cerium(IV) by arsenious acid [26], and on the reduction of
ferric thiocyanate [27]. In 1937 Sandell and Kolthoff [28] described a procedure for iodide determination by measuring the time required to reduce
cerium(IV) as indicated by a visual colorimeter. Subsequently, many
variations of this method have been reported.

The basic reaction catalyzed by iodide ions in the reduction of ceric
ions by arsenite ions in an acid medium can be expressed in a cyclic form
as follows [29]:

$$
\begin{array}{ccc}
Ce(IV) & I^- & 1/2\ As(V) \\
\left.\right) & \left(\ \right) & \left(\ \right. \\
Ce(III) & I^0 & 1/2\ As(III)
\end{array}
$$

Ceric ions are yellow while cerous ions are colorless, and the change
can be determined spectrophotometrically [30]. Aleksiev et al. [30a] have
determined silver in human saliva by the catalytic effect of Ag^+ on the
oxidation of sulfanilic acid by $K_2S_2O_8$. Ke et al. [31] have developed a
method to determine iodine in serum by measuring the difference in reaction rate due to catalysis by iodide. Hadjiioannou [31a] has proposed a
spectrophotometric kinetic method for determining copper in urine utilizing the catalytic effect of Cu^{2+} on the periodate/thiosulfate reaction.

Physicochemical Methods

Although the physical methods of finish entail more expensive equipment and instrumentation than titrimetry, they are frequently employed in
trace analysis [32]. Their chief advantage is high sensitivity; some are
also rapid and selective. In general, when the bulk material is a solid,
emission spectrometry, neutron activation, spark source mass spectrometry, and X-ray fluorescence [32a,b] are suggested. When the sample is

in the form of a liquid, flame and atomic absorption spectrometry, absorptiometry in the visual and ultraviolet regions, fluorimetry, and polarography [32c] are utilized. Techniques like atomic absorption and emission spectrometry [32d] are favored because they can determine several elements simultaneously [33, 33a]. In contrast, colorimetry has the drawback of being a single-element method, besides requiring time to prepare the solution and develop the color.

Continuing efforts are being made to lower the limit of trace element analysis. For instance, Layman and Hieftje [33b] have described a device in which a microwave-excited plasma is combined with a microarc sample atomizer under computer control to achieve a stable atomic emission source. This combination is shown to possess many desirable features, including sensitivity (down to picograms for many elements), precision (±5%), stability, low background, and minimal interference effects. Belcher and co-workers [33c] have reinvestigated the capabilities of candoluminescence which is the ultraviolet-visible radiation, in excess of black body radiation, stimulated at the surface of a solid matrix placed in a hydrogen-based flame. The phenomenon has been recognized for more than a century and its application to analytical chemistry was proposed by Donau [33d] in 1913. It has been established that the presence of traces of certain metal ions (activators) is essential for luminescence, and that the wavelengths of the emissions are characteristic of the activating ions. By using the candoluminescence spectrophotometry and molecular emission cavity analysis (MECA) techniques, a wide range of metals and nonmetals, e.g., sulfur, phosphorus, halogens, can be determined below the nanogram level. Chambers and McClellan [33e] have studied the enhancement of atomic absorption sensitivity by means of solvent extraction. By extracting 200 ml of aqueous solution with 2 ml of organic solvent and aspirating the organic phase, it is possible to detect copper at 0.01 parts per billion (ppb), cadmium at 0.1 ppb, and antimony at 10 ppb, based upon the original aqueous phase concentration.

C. Difficulties and Precautions

In trace analysis we should bear in mind that the substance to be determined is only a very minute portion of the bulk material. Hence special effort is needed in the process of sampling and sample preparation [34]. Besides, during the process of decomposition, some compounds containing the element in question may be lost before the bulk is decomposed, while other compounds may be left behind unchanged; mercaptans in petroleum are examples of the former and refractory nitrogen-heterocyclics of the latter. Finally, since the product to be collected for measurement in the finish step is mixed with large amounts of extraneous matter, there is always the chance of its escaping without being noticed. For these reasons, it is advisable to perform multiplicate determinations and also to add known amounts of the substance to be determined in order to check the recovery. It should be noted that satisfactory recovery of added inorganic salts does not necessarily prove that the recovery of the particular element in the organic material is also complete. For example, in the determination of chromium in refined and unrefined sugars, Wolf et al. [34a] have found that there are losses of organic chromium when sugar samples are introduced directly into the graphite furnace or ashed at 450°C prior to analysis, while inorganic chromium is not lost. In determining lead in blood, Kopito et al. [34b] have reported that recovery of added inorganic lead may not accurately reflect the fate of endogenous lead in some of the procedures.

The reliability of analytical data frequently depends on the experimental conditions and the nature of the samples. Thus, Lachica and Robles [34c] have studied the influence of the method of mineralization of samples in the analysis of plant materials. Lofberg and Levri [34d] have shown that visual criteria are inadequate for the analysis of copper and zinc in hemolyzed serum samples. For the determination of potassium in serum by flame photometry, Anand et al. [34e] have found that errors of the order of ± 0.2 μmol/ml can be caused by the presence of sodium. Olthuis et al. [34f] have reported on the interference of fatty acids with the fluorimetric titration of serum calcium. In the direct-current arc emission methods, Boumans and Maessen [34g] have found that the spectral line intensities are affected by the arc operating conditions and the thermochemical conditions in the electrode cavity.

Needless to say, contamination presents a serious problem in trace analysis [35]. Special considerations should be directed to the purity of the laboratory, instruments, reaction vessels, and reagents. Every effort should be made to eliminate or minimize the effects of impurities and contamination [36].

12.3 CURRENT PRACTICE

A. Trace Analysis of Nonmetallic Elements

Carbon, Hydrogen

Trace analysis of organic carbon in water has attracted attention because of increased concern with ecology [37]. Historically, the concentration of organic matter in water and wastewater has been evaluated by determining the biochemical oxygen demand (BOD). Besides being subject to many interferences [38], the BOD method takes 5 days to obtain results. Chemical oxygen demand (COD) measurements also have been used as an estimate of organic loading in water. This method requires 2 h and is not specific to organic compounds since it responds to any substance that can be oxidized by dichromate in acidic media [39]. It has been demonstrated that determination of the total organic carbon (TOC) content of water is an effective method for estimating organic loading [40]. A number of instruments have been developed to measure TOC, which can be determined accurately and rapidly. Van Hall and co-workers [41] measure the carbon dioxide in the combustion gases from the water sample with a nondispersive infrared detector. Lysyj and co-workers [42] use high-temperature reductive pyrolysis to obtain methane. The pyrolytic products are led through a glass-bead column without separation and methane is measured as a single peak by a flame ionization detector. The procedure is calibrated with solutions similar in composition to the waters being analyzed.

Oda and co-workers [43] have constructed a compact combustion system for the determination of carbon in aqueous solutions containing only a few percent of organic material. The assembly is shown in Fig. 12.3. The combustion products pass through the absorption cell where carbon dioxide is retained and measured with an electrophotometer.

Pennington and Brown [44] perform carbon and hydrogen analysis of the components of a mixture utilizing separation-combustion gas chromatography. The method involves separating the components on a gas chromatographic column and splitting the effluent so that a portion is led to a flame ionization detector. The remainder passes to a stainless-steel combustion tube packed with copper oxide. The water and carbon dioxide formed are separated in a polytetrafluoroethylene column and analyzed by thermal conductivity. The analysis can be carried out on less than 10 μg of a component. Theer [45] has employed a similar method to determine trace amounts of hydrocarbons in air and in oxygen. Titrimetric [45a], gasometric [45b], gravimetric [45c], and coulometric [45d] determinations of CO_2 for trace analysis have been advocated.

Ober and co-workers [9] use closed-flask combustion for the analysis of ^{14}C and ^{3}H in biological samples. Since 250 mg of the bulk material is decomposed, the combustion flask is made from a 1-liter round-bottom flask. Ignition is carried out by means of a lamp shining on a cotton wick. The wick is made by applying India ink to cotton, air drying the blackened cotton, and placing a 10-mg piece in the wire basket along with the ladle containing the sample. According to these workers, this type of wick gives more rapid ignition, less color, and hence higher and more uniform counting efficiencies than do paper wicks. The combustion flask is fitted with an absorption tube at the base which is cooled in Dry Ice during ignition. The [14]CO_2 and [3]HOH are absorbed by a phenethylamine-methanol mixture and absolute ethanol, respectively, and the ^{14}C and ^{3}H are counted with liquid-scintillation techniques.

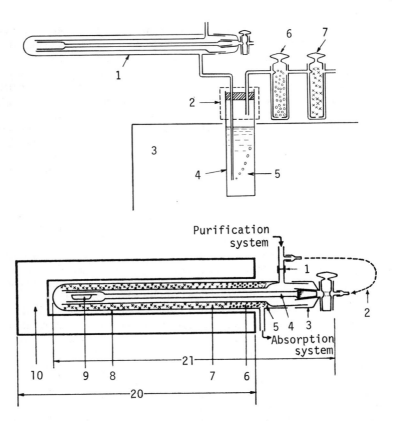

FIG. 12.3. Assembly for trace analysis of carbon in aqueous solutions.
(top) Complete assembly. 1, combustion tube; 2, black paper; 3, colorimeter; 4, absorption cell (1 in.); 5, absorption solution (alizarin yellow-
alkaline solution); 6, silica gel; 7, soda asbestos. (bottom) Details of the
combustion unit. 1, vinyl tube; 2, vinyl tube (50 cm); 3, cap; 4, internal
tube (i.d. 0.7 cm); 5, quartz wool; 6, cerium dioxide; 7, copper oxide;
8, combustion tube (o.d. 2.2 cm); 9, sample boat; 10, combustion furnace.
(From Ref. 43, courtesy Japan Analyst.)

Oxygen

Because petroleum contains small amounts of oxygenated compounds of
interest, shortly after the micromethod for direct determination of oxygen
was established, the American Petroleum Institute undertook a study to
test the procedures [46]. It was found that the pyrolytic hydrogen caused

high results when the analyses were performed by the regular iodimetric procedure for pure organic compounds (see Sec. 4. 4). Fortunately, only slight modifications are necessary in order to obtain correct results for oxygen in the range below 1%. Four modifications are equally satisfactory:

1. Conversion of the CO to CO_2 by means of CuO at 300°C, and measurement of the CO_2 manometrically.
2. Conversion of the CO to CO_2 by means of I_2O_5, absorption of the CO_2 in standardized NaOH solution, precipitation of the carbonate with $BaCl_2$, followed by back-titration of the excess alkali.
3. Conversion of the CO to CO_2 by means of I_2O_5, absorption of the liberated I_2 with $Na_2S_2O_3$ crystals, and determination of the CO_2 gravimetrically.
4. Removal of the pyrolytic H_2 by diffusion through metallic Pd at 350°C, conversion of the CO to CO_2 by means of I_2O_5, and titration of the liberated I_2 with standardized $Na_2S_2O_3$ solution.

Recently Smith and co-workers [47] have used the Perkin-Elmer Elemental Analyzer (see Sec. 4. 6) to determine oxygen in petroleum products and have obtained results comparable to those from the modified iodimetric method mentioned above.

Gouverneur and Bruijn [48] have constructed an apparatus for analyzing samples containing traces of oxygen down to 5 ppm. The schematic diagram of the assembly is shown in Fig. 12. 4, and the detailed description of the combustion tube is given in Fig. 12. 5. The determination is finished manometrically by measuring the carbon dioxide obtained by oxidation of the carbon monoxide by means of iodine pentoxide and sulfuric acid on silica gel.

Klesment [49] has described a method to determine oxygen by gas chromatography and twofold conversion. The sample, in a stream of argon and hydrogen, is pyrolyzed in the presence of activated carbon, and the CO, after separation on molecular sieve 5A, is hydrogenated over a nickel catalyst giving CH_4 and H_2O; the H_2O is then removed by means of molecular sieve 5A. The inlet and outlet argon-hydrogen mixtures pass through the differential thermal conductivity detector, and changes in hydrogen concentration are thus indicated. Because six atoms of hydrogen react with one molecule of CO, the sensitivity of determination is enhanced correspondingly. The method is applicable to contents of oxygen down to 0. 1% or less.

Liebman and co-workers [50] have recently designed an on-line elemental-reaction analyzer which can determine oxygen and other nonmetallic elements at trace levels (10^{-6} mole/liter). It is a realistic combination of gas chromatography and reaction microchemistry. A stop-flow arrangement in the separating gas chromatography unit allows such detailed information to be obtained on individual species in multicomponent mixtures which may be transferred as desired to the elemental-reaction analyzer for any or all of the denoted analysis.

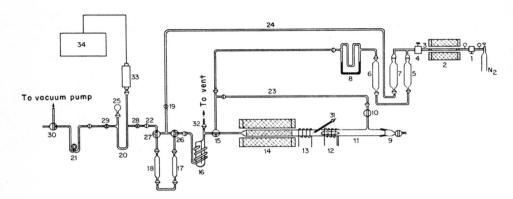

FIG. 12.4. Schematic diagram of the assembly for trace analysis of oxygen. 1, reducing valve; 2, furnace, 300°C; 3, purification tube filled with copper; 4, pressure regulator; 5, purification tube filled with BTS catalyst; 6, purification tube filled with soda asbestos, $Mg(ClO_4)_2$ and P_2O_5 on pumice; 7, purification tube filled with BTS catalyst, soda asbestos, $Mg(ClO_4)_2$ and P_2O_5 on pumice; 8, flowmeter; 9, cap with stopcock; 10, stopcock; 11, reaction tube filled with platinized carbon; 12, primary heating coil with quartz core; 13, sample heating coil; 14, electric furnace, 950°C; 15, 26, 27, 30, three-way stopcocks; 16, freezing trap for hydrocarbons and sulfur compounds; 17, oxidation tube filled with "Schütze contact" (80 mm); 18, guard tube, filled with copper (40 mm) and magnesium perchlorate (40 mm); 19, 28, 29, Teflon membrane valves; 20, carbon dioxide trap; 21, guard trap; 22, restricted capillary; 23, reverse flushing tube; 24, by-pass tube; 25, expansion bulb; 31, spherical inlet for syringe; 32, stopcock to vent; 33, pressure transducer; 34, digital voltmeter. (From Ref. 48, courtesy Talanta.)

Lauff and co-workers [51] have developed a differential method for the analysis of oxygen in the presence of fluorine by the complementary use of a californium-252 neutron source and a 14-MeV neutron generator; the sensitivity is 0.04 mg oxygen in a 10-g sample.

Nitrogen

Petroleum products containing 0.05-0.1% nitrogen can be analyzed by the micro Kjeldahl method or the automated C,H,N analyzer [47]; when the latter instrument is used, the carbon and hydrogen signals will be off scale

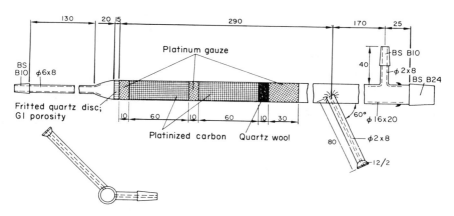

FIG. 12.5. Combustion tube for trace analysis of oxygen. Top view (upper figure) and end view (lower figure). Dimensions in millimeters. (From Ref. 48, courtesy Talanta.)

and they should therefore be ignored. For the determination of nitrogen in coal, Prasad [52] recommends decomposition by means of 70% perchloric acid followed by colorimetric determination of the ammonia. For feeding-stuff, Law and Norton [53] use sulfuric acid digestion and indophenol blue colorimetry.

Several methods have been published for the determination of ^{15}N. Lundeen and co-workers [54] use acid digestion and isobutane chemical ionization mass spectrometry. Proksch [55] and Fielder and Proksch [56] decompose plant materials mixed with copper oxide, copper, and calcium oxide in an evacuated tube at 600°C. The gas is released into an evacuated chamber of known volume and the nitrogen pressure is used to determine the nitrogen content of the sample. The gas is then admitted into the mass spectrometer. This method has been automated [57].

Halogens

Trace amounts of chlorine in organic materials have been determined by oxyhydrogen flame combustion, followed either by potentiometric titration with silver nitrate or by a spectrophotometrical finish [58]. Because this method is time-consuming, two groups of analysts have advocated the coulometric finish. Lädrach and co-workers [59] determine traces of chlorine in liquid samples as follows. The sample (1–100 μl) is vaporized in a stream of helium and then fed to an oxygen atmosphere where it undergoes combustion. The HCl formed passes into a silver coulometer cell and the chloride ion is measured by the Dohrmann microcoulometric titration technique. A precision (repeatability) of 0.05 ppm of chlorine was reported. The apparatus is shown in Fig. 12.6. Krijgsman and co-workers [60,61] have constructed a simple automatic coulometer system for the determination of chlorine in organic compounds. Sulfur, phosphorus, and arsenic do not interfere even if present in large amounts. The interference caused by the presence of large amounts of nitrogen can be eliminated by means of sulfamic acid.

Neutron activation is often used for the determination of bromine. This technique has been employed by Lunde [62] on biological materials, and by Bunus and Murgalescu [63] to determine trace bromine in vinyl chloride.

Trace analyses of fluorine in plant materials [64-66] and biological fluids [67,68] have been carried out by chemical methods (see Sec. 12.5), neutron activation [51], and mass spectrometry. The last-mentioned technique can be employed to determine more than one halogen in the same sample. For instance, Gregory [69] determines fluorine, chlorine, and bromine in blood by direct-current mass spectrometry; Tong and co-workers [70] determine fluorine and bromine in halogenated herbicide residues by spark source mass spectrometry.

Sulfur, Selenium, Tellurium

There is a wide range of methods for trace sulfur analysis. Slanina and co-workers [71,72] determine sulfur at ppm levels by hydrogenation of the organic substance at 1050°C over quartz. The resulting H_2S is absorbed in a solution 1 $\underline{M}$ in KOH and 1.5 $\underline{M}$ in hydroxylamine and titrated with 2 x 10^{-4} $\underline{M}$ Pb(II) automatically using an ion-selective electrode. Halides, cyanide, nitrate, and phosphate do not interfere. Stephen [73] employs closed-flask combustion and nephelometry to determine ppm amounts of sulfur, using 2-aminopyrimidine hydrochloride to precipitate the sulfate; Mendes-Bezerra and Uden [74] recommend 4-chloro-4'-aminobiphenyl as the precipitation reagent. Dokladalova and Nekovarova [75] decompose the sample in an empty quartz tube to obtain sulfur dioxide which is determined spectrophotometrically. Radiometric determination using $Cd^{14}CO_3$ as reagent [76], mass spectrometry [69], and X-ray fluorescence [77] have also found applications in trace sulfur analysis.

The determination of sulfur in petroleum has been performed in several ways. Heistand and Blake [78] decompose the sample in an oxyhydrogen burner, collect the combustion products in a solution containing sodium nitrite, and after removal of excess nitrite and adding 1,4-dioxane, titrate potentiometrically with 0.0025 $\underline{M}$ lead perchlorate using the lead ion-selective electrode. Killer and Underhill [79] incorporate a Dohrmann microcoulometer to the combustion unit (similar to the assembly shown in

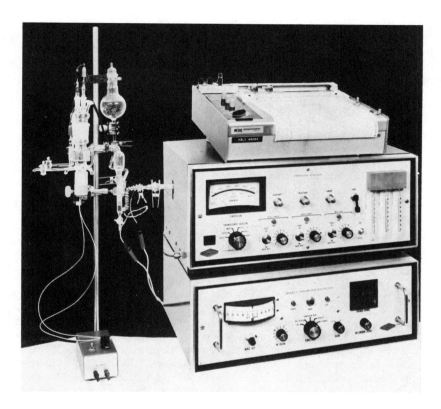

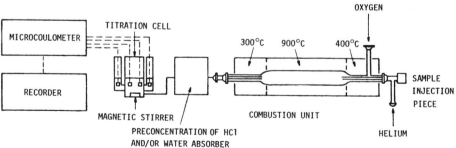

FIG. 12.6. Photograph and schematic diagram of the apparatus for microcoulometric chlorine determination. (From Ref. 59, courtesy Anal. Chim. Acta and Koninklijke/Shell-Laboratorium.)

Fig. 12.6) and finish the analysis coulometrically. Kellner and Pell [80] use the Wickbold apparatus, precipitate the sulfate ion with barium acetate solution in excess, and observe the change of conductivity with an oscillo-titrator. A commercial sulfur titrator [81] utilizes iodimetric titration. Iodimetric determination of sulfur by amplification reaction has been investigated by Gawargious and Farag [82]. Veillon and Park [83] determine sulfur in petroleum fractions by the emission of S_2 in a low-temperature hydrogen flame. Kirkbright and co-workers [84] determine sulfur in oils by atomic absorption spectrometry using an inert gas shielded nitrous oxide-acetylene flame.

Trace amounts of selenium [85-88] in biological materials and soils are usually determined by fluorimetry. Trace tellurium [89] analysis is carried out by atomic absorption spectrometry.

Arsenic, Antimony

For the determination of trace quantities of arsenic from herbicide residues in plant materials, Sachs and co-workers [90] recommend digestion of the sample with HNO_3 first, to avoid loss of arsenic as volatile chloride during charring, and then addition of H_2SO_4 to complete the mineralization. The resulting As(V) is determined with silver diethyldithio-carbamate.

Decomposition by oxidation with $HClO_4/HIO_4$ is employed by Spielholtz and co-workers [91] for trace arsenic analysis of coal and insecticide, followed by atomic absorption spectrometry. Coprecipitation and X-ray fluorescence are utilized as follows by Reymont and Dubois [92] to determine traces of arsenic in air, water, soil, or vegetation. After treatment of the sample with bromine in CCl_4 and with HNO_3, the solution is evaporated to fumes with H_2SO_4, and diluted with water. Arsenic is precipitated as sulfide with thioacetamide from a perchloric acid solution containing an aliquot of the test solution and added molybdenum for coprecipitation. The precipitate is collected on a Millipore filter. The intensities of the As K_{α} line and of the background are measured at $2\theta = 34.00°$ and at $2\theta = 35.00°$. Down to 1 μg of arsenic can be determined.

Neutron activation analysis has been applied to the determination of arsenic and of antimony in biological materials at submicrogram levels. Byrne [93] irradiates the samples (0.1-1 g) and standards (As_2O_3 in HNO_3; Sb_2O_3 in HCl) in a neutron flux of about 1.8×10^{12} neutrons per cm^2 per s for 20 h. After processing, ^{76}As is counted at the 0.559-MeV peak and ^{122}Sb at 0.564 MeV. Differential pulse polarography has been reported by Myers and Osteryoung [94] to be useful for the determination of As(III) at the ppb level.

Phosphorus

The most favored technique for mineralization of organic materials for trace phosphorus analysis is wet oxidation in acidic media. Thus, sulfuric acid, alone or in combination with other acid, is used for the decomposition of feeding stuffs [53], organophosphonates in biological samples [95], and complex lipids [96]; perchloric acid is used in the analysis of coal [52] and proteins [97]. Closed-flask combustion is employed by Kirk and Wilkinson [8] to decompose coke, while dry ashing without fixatives is recommended by Tusl [98] for biological samples. As expected, the extremely sensitive molybdenum blue reaction [10] remains to be the method of choice for colorimetric or spectrophotometric determination of trace amounts of phosphorus [53, 95-98]. Nevertheless, the yellow molybdophosphoric acid [99] has been used by Ghimicescu and Dorneanu [100] for the determination of phosphorus in blood, serum, and urine, measurement of absorption being made at the ultraviolet region in a water-acetone solution.

Flame emission spectrometry [33, 83, 101] is suitable for trace phosphorus analysis. Atomic absorption spectrometry [102] has been used to determine phosphorus in biological materials. Neutron activation [63] based on β counting has been employed to determine phosphorus impurity in vinyl chloride, while mass spectrometry [69] has been applied to the analysis of phosphorus in ashed blood samples. Smirnov and Khislavski [103] determine phosphorus in fabrics by X-ray absorption; Fieldes and Furkert [77] use X-ray fluorescence to analyze plant materials for a number of trace elements including phosphorus.

B. Trace Analysis of Metallic Elements

General Techniques

It should be noted that trace metals in the organic matrix are in general, not present as ionizable cations which can be extracted into aqueous solutions and determined directly by chemical or physical methods. Most metals are in complexed forms with miscellaneous organic ligands [104, 105]. Therefore the first step involves the destruction of the organic matter. This is usually done by digestion with oxidizing acids and occasionally by dry ashing (see Sec. 12.2A). The latter technique consists of heating the material in a muffle furnace at 400-700°C; it is time-consuming and some elements can be lost. A number of variations of the acid digestion procedure have been proposed; the British Society of Analytical Chemistry recommend four different methods [106].

    The acid most frequently employed for digestion is concentrated nitric
acid, taking advantage of the fact that all metal nitrates are soluble in
aqueous medium. However, since nitric acid is a strong solvent for metals,
the danger always exists that the digestion reagent may be contaminated
with the element to be determined. Thomas and Smythe [107] have described a procedure for vapor phase oxidation with nitric acid which eliminates this difficulty and is considerably more rapid than the conventional wet
digestion. The apparatus is illustrated in Fig. 12.7. Nitric acid vapor is
generated by dropping concentrated nitric acid (specific gravity 1.4) into a
reservoir of a nitric-sulfuric acid mixture (1:1) at 152°C. The nitric acid
vapor distils at 122°C; this corresponds to a vapor consisting of 67% HNO_3.
In the 250-ml round-bottom flask is placed 100 ml of 1:1 nitric-sulfuric
acid mixture. The flask is fitted with the dropping funnel containing the
concentrated nitric acid and the splash head. A "gas finger" is fitted to the
other end of the splash head and this conducts the nitric acid vapor onto the
organic material, which is contained in the reaction tube (14 cm long, 4.5
mm o.d.) with a side-arm. During the entire digestion period the reaction
tube is heated by an infrared lamp; this prevents excessive condensation of

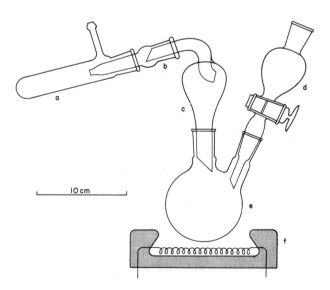

FIG. 12.7. Apparatus for destruction of organic matter by vapor
phase oxidation. a, reaction tube; b, gas finger; c, splash-head; d, dropping funnel; e, acid reservoir; f, heating mantle. (From Ref. 107, courtesy Talanta.)

the nitric acid vapor before it reaches the sample to be oxidized. For 0.5 g of dry plant material, about 90% of the organic matter is destroyed in 5-6 min. Then the reaction tube is removed, 2 ml of perchloric acid (70-72%) is added and the sample is heated in a furnace at 280°C to perchloric acid fumes, indicating that oxidation of the organic matter is completed. According to these workers, the presence of nitric acid during the final perchloric acid oxidation eliminates any explosion hazard. Nevertheless, it is prudent to perform the experiment behind a protective shield.

The reader is referred to the comprehensive manuals on the determination of trace elements [108,108a] for the selection of the most suitable method for the metal in the particular sample to be analyzed. Some general techniques applicable to multielement analysis are briefly discussed below.

Atomic absorption spectrometry is a popular and relatively inexpensive tool for trace metal analysis. This technique can be used directly on coal samples [109] and crude oils [109a]. Glenn et al. [110] determine iron in serum without any sample pretreatment. Puyo and Thiel [111] use atomic absorption to determine nickel and vanadium in fuel oil; Morie and Morrisett [111a] use it for transition metals in cigarette smoke; Henn [111b] uses it for metals in polymers. Varghese et al. [112] have described an improved method to determine major cations in biological extracts. Falchuk et al. [112a] have constructed a multichannel instrument for simultaneous analysis of Zn, Cu, and Cd in biological materials. Tse et al. [112b] put forward a "stream combinator" that mechanically combines a stream of aqueous standard solution with the organic sample stream.

Trace analysis by mass spectrometry [113] will become a general technique when the cost of the instrument reaches a moderate level. Guidoboni [114] employs spark source mass spectrometry to determine several heavy metals in coal. Although the amounts found are at the ppm level, they are significant because of their possible effects on the environment in view of the enormous tonnages of coal burned yearly. The data obtained by mass spectrometry check roughly with those obtained by atomic absorption spectrometry. On the other hand, comparison between results from wet digestion and those from dry ashing shows that the latter decomposition technique tends to give low values. McHugh and Stevens [115] use ion microprobe mass spectrometry to analyze metals in oil soot particles which are air pollutants. Analysis of single particles in the 2-8 μm range showed high contents of vanadium, calcium, potassium, and sodium. Moore and Machlan [116] have described a technique for the determination of 100 μg/g calcium in blood serum with an accuracy of 0.2%, using isotope dilution mass spectrometry.

If facilities are available, neutron activation analysis [117,117a] is the most generally applicable technique for the determination of trace elements. This method is sensitive to the nanogram level and can be used for the analysis of some 70 elements [116]. It has been employed extensively in

many fields. For example, Garrec [119] utilizes 14-MeV neutrons to
analyze plant material for aluminum, calcium, potassium, magnesium,
and also phosphorus, silicon, and chlorine. The dried, powdered sample
is placed in a polyethylene tube and irradiated with neutrons from an electrostatic accelerator for 15 min, with oxides, nitrates, or carbonates of
the metals as standards. After a decay period of 5 min, the γ-ray spectrum is recorded by means of a NaI(Tl) crystal and a 200-channel analyzer,
and is then analyzed by a computer. There is no interference from carbon, hydrogen, oxygen, or nitrogen. Haller and co-workers [120] determine 15 trace elements, including arsenic, mercury, and selenium, in
plant tissues. The samples are freeze-dried, neutron-activated, and
analyzed by γ-ray spectrometry with a Ge(Li) detector. Fritze and
Robertson [121] have put forward a procedure for the determination of
protein-bound trace metals in human serum by neutron activation analysis.

Quantitative fluorimetry is a useful technique for trace metal determination. Thus Zweidinger and co-workers [122] determine amounts of gallium
as low as 15 ng/g of biological materials using wet digestion, ion-exchange,
and fluorimetry. Chelate gas chromatography is employed by Kaiser and
co-workers [123] to determine trace beryllium in blood, foods, and sewage.
Flame atomic emission is utilized by Chuang et al. [124] to analyze alkali
and alkaline earth metals in blood serum. Seitz and Hercules [125] use
chemiluminescence to determine trace amounts of iron(II). Levit [126]
determines traces of thallium in urine by anodic stripping alternating-
current voltammetry.

Trace Analysis of Mercury

The toxic effect of mercury has been known for centuries. The seri-
ousness of mercury poisoning, however, was established only in recent
years. It has been confirmed that elemental mercury and inorganic mercury compounds in water end up in fish tissues, and humans who have consumed fishes containing traces of mercury show signs of brain damage.
Since fish is an important source of protein and is the staple food for certain people, analysis of mercury in fish and foodstuffs has received much
attention. It should be recognized that accurate determination of mercury
in fish is difficult because (1) the element is organically bound and hence
requires mineralization, and (2) mercury and many of its compounds are
volatile. Uthe and co-workers [127] use acid digestion and permanganate
oxidation. Mercury is then released by chemical reduction and determined
photometrically by ultraviolet light absorption at 253.7 nm. Thomas and
co-workers [128] have found that this method gave low and erratic results
on fishes containing fatty tissue, and have developed a dynamic approach
by which the total mercury in fish is pyrolytically converted to the elemental form and quantitated photometrically. The sample is burned in a
stream of air at 900°C and combustion gases are conducted over copper

oxide at 850°C. The air stream containing the elemental mercury is then passed through an ultraviolet photometer. Electronic integration of the detector signal is employed to improve analysis statistics. The detailed experimental procedure is given in Sec. 12.6.

Closed-flask combustion is employed by Pappus and Rosenberg [129] to decompose biological materials, including fish; the products are concentrated on cadmium sulfide pads, which are subsequently burned to release mercury vapor. A similar procedure is used by Hatch and Ott [130]. The 500-ml combustion flask, now commercially available [131], is shown in Fig. 12.8. The sample, up to 50 mg, is burned in oxygen over a nitric acid absorption liquid. Reagents are added to the closed flask through an addition funnel and the injection port with a serum stopper. A built-in bubbler tube and inlet and outlet stopcocks allow introduction of a stream of air to sweep the mercury vapor into a gas flow cell, or for analysis in an atomic absorption spectrophotometer.

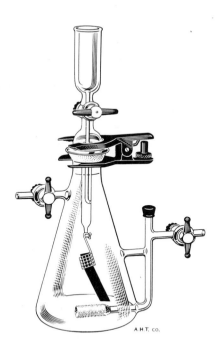

FIG. 12.8. Combustion flask for trace mercury analysis. (From Ref. 131, courtesy A. H. Thomas Co.)

Fujita and co-workers [132] determine organomercury fungicides in rice, hair, rat tissues, and vegetable oils by the oxygen-bomb combustion method (see Sec. 12.2A). The combustion products are absorbed in $\underline{M}$ HNO_3 and the mercury is determined spectrophotometrically with dithizone reagent at 490 nm.

Meyer [133] performs trace analysis of organomercury compounds in pharmaceutical preparations by mineralization with H_2SO_4-bromine water mixture. The mercury is determined with di-2-naphthylthiocarbazone. Lindstedt [134] determines mercury in urine by digestion with $H_2SO_4/$ $KMnO_4$ at room temperature, reduction of mercury(II) to elemental mercury by tin(II), evaporation of mercury vapor from the solution by an air current, and measurement at 253.7 nm. Petersen [135] has described a procedure for the determination of mercury in organic materials by neutron activation analysis.

Trace Analysis of Lead

Hwang and co-workers [136] determine lead in blood by direct flameless atomic absorption spectrometry; the results are comparable to those obtained by flame atomic absorption and by ultraviolet absorption spectrometry. Pagenkopf and co-workers [137] have described a new method to determine lead in fish by furnace atomic absorption, using the nonflame technique [138]. The fish muscle tissue is freeze-dried before digestion with HNO_3/H_2SO_4 mixture.

Mansell and Hiller [139] use the Parr acid digestion bomb [140] to decompose tetraethyl lead in gasoline. The apparatus consists of a cup and fitted cover made of Teflon plastic, contained in a stainless-steel bomb which is capable of withstanding up to 1200 lb/in.^{2} pressure and 150°C. This method eliminates possible loss of volatile lead compounds in wet digestion. The lead is then determined by atomic absorption spectrometry. Aue and Hill [141] determine volatile compounds of lead, iron, and tin as gas chromatographic effluents by a simple, dual-channel detector based on flame conductivity and emission.

12.4 SURVEY OF RECENT LITERATURE

A. Summary of Publications

A very large number of papers have been published on trace analysis during the past decade, probably because of public and governmental interest in this subject. Unlike the regular organic analysis, publications on trace analysis [228] are scattered in many different journals and institutional reports. It is beyond the scope of this chapter to cover all recent literature.

We shall present a brief survey only to indicate the trends in the field. Selected papers which have appeared during the past few years are summarized in the tables. In Table 12.1 are tabulated the publications that are concerned with the determination of traces of nonmetallic elements including carbon and hydrogen. In Table 12.2 are listed some metalloids which are of general interest owing to their relation to health and pollution of the environment. Table 12.3 presents papers dealing with trace metals.

TABLE 12.1. Recent Publications on Trace Analysis of Nonmetals. *

Element	Bulk material	Method	Reference
C	Aqueous, rock, biological	Combustion tube	41-43, 142-148
	Aqueous	Wet oxidation	148a
	Volatile, inert matrix	Gas chromatography	44, 149
	Biological	Flask combustion	9
H	Volatile	Gas chromatography	44
	Biological	Flask combustion	9
O	Petroleum	Combustion tube	47, 48
	Volatile	Gas chromatography	49
	Fluoro, polyethylene	Neutron activation	51, 150
N	Petroleum, feeding stuff, lipids	Kjeldahl	47, 53, 151-154
	Petroleum	Combustion tube	47, 155
	Petroleum	Oxyhydrogen flame	156, 157
	Petroleum	Hydrogenation	158
	Soil	Spectrophotometry	158a
	Nonnitrogenous	Gas chromatography	159
	Aqueous	Automated	160
	Plants	Mass spectrometry	55-57

*For publications during 1974-1976 see also Table 12.4.

TABLE 12.1. (Continued)

Element	Bulk material	Method	Reference
F	Plants	CaO/NaOH ashing	64-66
	Foodstuffs	Mg acetate ashing	161
	Biological	Diffusion	162-166a
	Petroleum	Na biphenyl	167
	Petroleum	Lamp flame	168
	Organic compounds	Oxygen bomb	169
	Aromatic compounds, biological	Neutron activation	170-171
C, Br, I	Liquids, polymers	Combustion tube	59, 172-174
	Biological	K_2CO_3 ashing	175, 176
	Polymers, biological	$NaNO_3/KNO_3$ fusion	177
	Biological	Catalytic	177a
	Aqueous, hydrocarbon	Oxygen bomb	178, 179
	Petroleum	Flame jet	180
	Pesticide residue	He plasma	181
	Plants, polymers, blood	X-ray fluorescence	77, 182-183
	Blood, herbicides	Mass spectrometry	69, 70
	Plants, hydrocarbons, polymers	Neutron activation	119, 184-185
S	Polymers, petroleum	Combustion tube	172-172b, 186
	Polymers, foodstuffs	Flask combustion	73, 187, 188
	Polymers, coal	Wet oxidation	73, 189
	Organic compounds	$KNO_3/NaNO_3$ fusion	177
	Fats, hydrocarbons, polymers	Hydrogenation	72, 190-195
	Petroleum, aqueous	Flame	168, 196, 197
	Drugs	He plasma	181

TABLE 12.2. Recent Publications on Trace Analysis of Metalloids. *

Element	Bulk material	Method	Reference
As	Plants	Wet oxidation	90
	Coal	Atomic absorption spectrometry	91
	Soil	Spectrophotometry	91a
	Plants, aqueous, soil	X-ray fluorescence	92
	Biological	Neutron activation	62, 93-93b
B	Foodstuffs	Atomic absorption spectrometry	198
	Plants	Spectrophotometry	199
P	Coal, biological	Wet oxidation	52, 95, 97
	Coke	Flask combustion	8
	Biological	Dry ashing	98
	Plants, polymers	Oxygen bomb	178
	Petroleum	Flame emission	83, 101
	Biological	Atomic absorption spectrometry	102
	Plants, fabrics	X-ray spectrometry	77, 103
	Biological	Mass spectrometry	69
	Polymers, plants, fuel	Neutron activation	63, 119, 200
Se	Plants, soil	Wet oxidation, spectrophotometry	201-204a
	Organic compounds	Flask combustion	205
	Biological, aqueous	Fluorimetry	85-88a
Si	Aqueous	Flame emission spectrometry	206
	Paper	Atomic absorption spectrometry	207
	Plants	Neutron activation	119

*For publications during 1974-1976 see also Table 12.4.

TABLE 12.3. Recent Publications on Trace Analysis of Metals.*

Element	Bulk material	Method	Reference
Hg	Pharmaceutical, biological	Wet oxidation	133, 134
	Biological	Combustion tube	128
	Biological	Flask combustion	130
	Fungicide residue, biological	Oxygen bomb	132, 132a
	Biological	Ultraviolet photometry	127
	Organic compounds	Neutron activation	135
Pb	Biological	Wet oxidation	137–137c
	Biological	Ashing	137d–f
	Petroleum	Plastic bomb	139
	Volatile	Gas chromatography	141
	Blood, biological	Atomic absorption spectrometry	136, 137
Other metals	Coal, biological	Wet oxidation	52, 208–210
	Biological, fabric	Ashing	211–212
	Polymer, biological	$KNO_3/NaNO_3$ fusion	177
	Petroleum, plants, biological	Oxygen bomb	178
	Wines, juices	Direct colorimetric	178a
	Petroleum, plants, blood	Extraction	213–218
	Biological	Flame	124
	Petroleum, plants	Emission spectrometry	219, 220
	Biological	Atomic absorption spectrometry	221–224a
	Biological, sugar	Polarography	211, 225, 225a
	Biological	Mass spectrometry	226
	Biological	X-ray emission	226a
	Biological, plants	Neutron activation	119–121, 227
	Biological	Gamma rays	227a

*For publications during 1974–1976 see also Table 12.4.

It will be noted in Tables 12.1-12.3 that the bulk materials subjected to trace analysis cover a wide range. Biological samples (blood, urine, tissues), natural products (plants), and fuel (coal, petroleum) are the common analytical specimens. It should be mentioned that many reports on trace element determination of industrial materials and commercial products are not available in the literature. For understandable reasons, some industrial laboratories do not wish to divulge their methods of analysis and their experimental data.

Inspection of Table 12.3 reveals several techniques for trace metal analysis that are not employed in the determination of nonmetals or metalloids. For instance, for some metals in certain bulk materials, simple extraction may be a satisfactory process of isolating the metallic element to be determined. Thus, in the analysis of gasoline, Prohaska [214] has determined lead by extraction with $\underline{M}$ iodine monochloride and direct titration with 5 x 10^{-3} $\underline{M}$ EDTA, and Kyriakopoulos [218] has extracted manganese by means of potassium chlorate-nitric acid mixture. Giraudet [217] has described a method to determine iron in serum by just incubating the sample with the color reagent (nitroso-R in acetate buffer) for 5 min and then measuring the absorbance at 720 nm. Trace amounts of organotin compounds have been extracted [215] from aqueous solutions by carbon tetrachloride after complexation with dithizone, and the tin is then determined by absorptiometric measurement at 570 nm.

With regard to the modes of finish, examples of the sophisticated physical methods are given in the tables. It should be recognized, however, that colorimetric and spectrophotometric techniques are frequently utilized. The papers are too numerous to be cited. Polarography and chelatometry are two techniques particularly suited for the determination of trace metals.

B. Evaluation of Methods

Evaluation of methods for trace element analysis is considerably more difficult and involved than that for the analysis of pure organic compounds. In general, it is practically impossible to know the exact value of the trace element, even though the analytical sample may be a synthetic mixture. There are always losses and contamination during the experiment. Furthermore, owing to the nature of the specimens, a statistical study is usually required to evaluate the analytical data.

As a rule, the workers who propose a new method for trace analysis also analyze their test samples by an existing method. This helps to prove the validity of the proposed method as well as to establish the advantage and limitation of the new procedure. Acceptance of the method, however, depends on whether or not there are pitfalls which subsequently reveal themselves. Collaborative studies, therefore, are valuable in evaluating these methods. For example, the Association of Official Analytical Chemists has published a collaborative study of the determination of arsenic in poultry [229], of phosphorus in meat products [229a], gelatin, dessert preparations, and mixes [230], and of lead in latex paint [230a]. The Institute of Brewing [231] has conducted a collaborative study of the Gutzeit and silver diethyldithiocarbamate methods for trace analysis of arsenic in beer and malt, resulting in the recommendation of the Gutzeit method. Kalembasa and Jenkinson [231a] have reported a comparative study of ti-trimetric and gravimetric methods for determining organic carbon in soil.

The British Society for Analytical Chemistry [232] has published recommended methods for the determination of trace elements, with spe-cial reference to fertilizers and feeding stuff. This organization also has recommended four methods of decomposition [106] that are suitable for most materials, including dyestuffs, medicinal compounds, rubber chemi-cals, synthetic polymers, biological material, and foodstuffs.

Barnett and Kahn [233] have reported on the comparison of atomic fluorescence with atomic absorption for the determination of a number of metallic elements. In the trace metal analysis of coal and coal ash [114], the comparison of dry ashing and wet digestion techniques for atomic ab-sorption spectrometry and the comparison of atomic absorption with spark source mass spectrometry have shown the discrepancies of these methods. Such comparative studies are useful references in selecting a method of analysis. For instance, Mossholder and co-workers [234] have investi-gated the analytical applications of premixed oxyhydrogen flames. Obser-vations on 14 elements indicate that, in spite of its high maximum temper-ature (2950°K), the premixed flame offers few advantages as an atomiza-tion cell for flame atomic absorption spectrometry. Even when solute vaporization is complete, elements like Ba, Fe, Na, and Ti which form stable monoxides cannot be appreciably atomized in the premixed flame.

Kaldor and Smith [234a] have compared the flame photometric and atomic absorption spectrometric methods for determining sodium and po-tassium in serum. For the determination of zinc in plants, emission spectrometry has been found to be as precise as polarography [234b]. Uthe and Armstrong [235] have compared the methods for determining mercury and organomercury compounds in environmental material, e.g., biological tissues. Florence et al. [236] have determined the beryllium content of standard reference orchard leaves by several methods involving various techniques of wet and dry ashing.

Anderson et al. [237] have described the preparation of gelatin multi-component trace-element reference material. Solutions of 25 elements are added to a gelatin solution and are thoroughly dispersed before the solution is allowed to gel. This material has been used to study the losses of arsenic, boron, mercury, and selenium during wet or dry ashing of proteinaceous substances.

C. Present Status of Trace Elemental Analysis

Presently, trace elemental analysis is in a state of flux. Intensive studies are being carried out. Understandably, much effort has been devoted to the toxic elements such as arsenic, cadmium, lead, and mercury. Table 12.4 shows a listing of reports that appeared during 1974-1976. Most reports deal with minor variations of the existing methods discussed in this book, or the application of a known procedure to some special materials. The table may serve as a guide to the reader who wants to find the latest information on the determination of a particular element by certain analytical procedures. On the other hand, it should be noted that relying on a single method may not be sufficient. For instance, while atomic absorption spectrometry is the technique in vogue, Cernik [238] has emphasized the importance of keeping the accuracy of atomic absorption determinations of lead in blood under surveillance by concurrently analyzing some of the samples at random by a reliable, independent method, e.g., cathode-ray polarography or anodic-stripping voltammetry.

TABLE 12.4. Partial List of Reports on Trace Elemental Analysis Published During 1974-1976.

Element determined	Material analyzed	Mode of finish	Reference
Nonmetals			
Carbon	Organics in water	Coulometry	239
Nitrogen	Biological	Color	240-242
	Biological	Gasometric	243
	Plants	Emission spectroscopy	244
Oxygen	Coal	Infrared spectroscopy	245
	Oil additives	Neutron activation	246
Sulfur	Petroleum	Titrimetric	247,248
	Plants, oils	Radioactivity	249-251
Fluorine	Petroleum	Titrimetric	252
	Animal feeds	Gas chromatography	253
Chlorine	Petroleum	Color	254,255
	Polymers	Turbidimetric	256
Bromine	Wheat flour	X-ray fluorescence	257

TABLE 12.4.   (Continued)

Element determined	Material analyzed	Mode of finish	Reference
Iodine	Milk	Color	258
	Biological	Kinetic	259
Phos-phorus	Oil additives	Color	260
	Grain pollen	Neutron activation	261
	Aqueous solution	Indirect atomic absorption	262
Arsenic	Plants, wines, vinegar	Color	263, 264
	Biological	Atomic absorption	265, 266
Boron	Biological	Emission spectroscopy	267
Selenium	Plants, papers, foods	Fluorimetric	268–271
	Biological	Neutron activation	272
	Petroleum products	Atomic absorption	273
Metals Aluminum	Plants	Color	274
Americium	Biological	Radioactivity	275
Antimony	Biological	Atomic absorption	276
Beryllium	Biological	Gas chromatography	277
Calcium	Blood, wines	Color	278, 279
	Biological, wines	Atomic absorption	280, 281
Cadmium	Petroleum, foods, biological	Atomic absorption	282–288
	Foods	Polarography	289
	Biological	Neutron activation	290
Chromium	Blood	Atomic absorption	291, 292
	Biological	Chemiluminescence	293
Cobalt	Pharmaceutical	Color	294
	Plants, animal foods	Atomic absorption	295, 296

TABLE 12.4. (Continued)

Element determined	Material analyzed	Mode of finish	Reference
Copper	Biological, foods	Color	297-301
	Plants, edible oils	Atomic absorption	302-304
	Water	Polarography	305
Gold	Blood	Atomic absorption	306
	Pharmaceutical	Polarography	307
Iron	Blood	Color	308-311
	Blood	Atomic fluorescence	312
	Blood	Coulometry	313
Lead	Foods, blood, urine, petroleum	Atomic absorption	314-327
	Blood	Atomic fluorescence	328
	Blood, urine	Polarography	329
	Meat	X-ray fluorescence	330
Lithium	Blood	Atomic absorption	331,332
Magnesium	Wines, herbage	Atomic absorption	333,334
Manganese	Petroleum, biological	Atomic absorption	335-337
	Plants	Neutron activation	338
Mercury	Biological	Coulometry	339
	Foods, biological	Atomic absorption	340-349
	Plants	Mass spectroscopy	350
	Biological	Neutron activation	351
Molybdenum	Plants, biological	Color	352-354
	Plants	Atomic absorption	355
	Biological	Neutron activation	356
Plutonium	Urine	Radioactivity	357

TABLE 12.4. (Continued)

Element determined	Material analyzed	Mode of finish	Reference
Thallium	Biological	Color	358
	Blood	Atomic absorption	359
	Urine	Polarography	360
Tin	Foods	Color	361
	Foods	Polarography	362
	Biological	Neutron activation	363, 364
Zinc	Petroleum, biological, wines	Atomic absorption	365–369
	Biological	Atomic fluorescence	370

A new development in trace elemental analysis involves the use of suitable instruments and techniques whereby several, or scores of, elements can be determined simultaneously. Among the techniques employed are atomic absorption or emission spectrometry, mass spectrometry, polarography, neutron or proton activation, and X-ray fluorescence. The reader is referred to the latest biennial review [238a] on organic elemental analysis published in 1978.

12.5 EXPERIMENTAL: DETERMINATION OF FLUORINE AT THE PARTS–PER–MILLION LEVEL USING THE DIFFUSION TECHNIQUE

A. Principle

After the fluorine present in the organic substance has been converted to the fluoride form, the sample is placed in a polystyrene petri dish. Perchloric acid is added and the dish is immediately covered with a polystyrene lid, the underside of which is coated with a layer of sodium hydroxide. The covered dish is kept at 65°C overnight whereupon hydrogen fluoride diffuses from the sample and all fluorine is retained on the lid as sodium fluoride.

The latter is then rinsed from the lid and fluoride ions are subsequently determined either colorimetrically [371] with Eriochrome cyanine R, or potentiometrically by use of the fluoride specific-ion electrode.

B. Equipment

Polystyrene petri dishes (100 mm x 10 mm) and lids (Fisher Scientific Co., Pittsburgh, Pa.).
Drying oven, controlled to ±2°C at 65°C.
Infrared lamp, or infrared drying apparatus [34].

For ashing:

Nickel crucibles, 50-ml capacity.
Muffle furnace.

For colorimetric determination:

Spectrophotometer, visual range (e.g., Beckman DU, or Bausch and Lomb Spectronic 20).

For titrimetric determination:

Orion Fluoride Electrode, Model 94-09.
Calomel reference electrode (sleeve type), filled with saturated KCl.
pH meter.
Plastic beakers, 150-ml capacity.
Magnetic stirrer.

C. Reagents

Dilute perchloric acid: Mix two volumes of 70% perchloric acid with one volume of distilled water and cool to room temperature.
Ethanolic sodium hydroxide solution, 0.5 $\underline{M}$: Dissolve 2.0 g of sodium hydroxide pellets in 20 ml of distilled water, cool, and dilute to 100 ml with USP ethyl alcohol.
Sodium fluoride standard solution: (a) 2 μg F/ml: Dissolve 4.421 mg of pure sodium fluoride in exactly 1 liter of distilled water. (b) 50 μg F/ml: Dissolve 110.525 mg of pure sodium fluoride in exactly 1 liter of distilled water.

For ashing:

Calcium oxide, fluorine-free.
Sodium hydroxide, fluorine-free.

For colorimetric determination:

> Eriochrome cyanine R solution: Dissolve 900 mg of Eriochrome cyanine R in 500 ml of distilled water.
>
> Zirconium solution: Dissolve 132.5 mg of $ZrOCl_2 \cdot 8H_2O$ in 40 ml of distilled water, add 350 ml of concentrated hydrochloric acid, and dilute to 500 ml with water.
>
> Zirconium-Eriochrome cyanine R reagent: Mix equal volumes of the above two solutions. Prepare this reagent daily.

For titrimetric determination:

> TISAB (buffer solution): Add 57 ml of glacial acetic acid, 58 g of sodium chloride, and 0.3 g of sodium citrate to 500 ml of distilled water. Mix thoroughly. Adjust the pH of the solution to 5.0 by means of 5.0 $\underline{M}$ NaOH. Cool to room temperature and then dilute to 1 liter.

D. Procedure

Ashing

(See Sec. 12.5F, Note 1.) Accurately weigh 0.5-1.0 g of the pulverized organic material (e.g., tea leaves [64], grass, seeds) into the nickel crucible. Add 0.1 g of calcium oxide as fixative and char the mixture under the infrared lamp. Then mineralize in the muffle furnace at 600°C. Fuse the resulting ash in the same crucible with about a fivefold amount of sodium hydroxide, thus releasing completely the organically bound fluorine in the form of ionic fluoride.

Preparation of the Sodium Hydroxide Coated Polystyrene Lid

Pipet 3 ml of 0.5 $\underline{M}$ ethanolic NaOH onto the inside of the polystyrene lid and rotate the lid in order to get a uniform coating of sodium hydroxide on the surface. Place the lid under the infrared lamp or on the infrared drying apparatus [34], in a position to keep the lid at 65°C until its surface is completely dry.

Diffusion

Transfer the material after ashing (or accurately weigh 0.5-1 g sample containing 10-500 μg F) into the polystyrene petri dish. Spread the sample evenly over the bottom of the dish and add 25 ml of dilute perchloric acid to wet the sample (Note 2). Immediately cover the dish with the sodium hydroxide coated lid. Place the dish in the drying oven set at 65-66°C overnight to permit the hydrogen fluoride to diffuse into the sodium hydroxide. After 20 h, carefully lift the lid. Rinse the lid thoroughly with distilled water into a 100-ml volumetric flask, keeping the volume below 60 ml.

Determination of Fluoride by the Colorimetric Method

Add exactly 20 ml of the zirconium-Eriochrome cyanine R reagent to the 100-ml volumetric flask containing the fluoride. Dilute to the mark with distilled water and mix. Then measure the absorbance of the solution at 356 nm. Prepare a calibration graph using the same conditions with 0, 10, 30, 50, 70, 80, and 100 μg F, respectively, from the sodium fluoride standard solutions. When the Beckman DU spectrophotometer is employed, set the slit at 0.04, and read the absorbance of the samples by balancing the instrument on the 0 μg F sample with the absorbance set at exactly 0.500. Because fluoride ions bleach the zirconium-Eriochrome cyanine R reagent, a decrease in absorbance of the sample solution indicates the presence of fluoride.

Determination by Potentiometry

Neutralize the fluoride sample solution in the 100-ml volumetric flask with 6 $\underline{M}$ hydrochloric acid. Add 10 ml of TISAB buffer and make up to volume with distilled water. Determine the fluoride concentration utilizing the fluoride specific-ion electrode as described in Sec. 5.6.

E. Calculation

$$\text{ppm F} = \frac{\mu\text{g F found}}{\text{g sample}}$$

F. Notes

Note 1: Some organic materials containing ppm amounts of F can be analyzed directly without ashing. For routine analysis of a large number of samples of similar nature, it is recommended to compare the results of ashing and nonashing procedures.

Note 2: If the sample contains large amounts of alkali, place a litmus paper in the dish to test the adequacy of the perchloric acid.

G. Comments

The colorimetric method is more suitable for low-range fluoride, while the potentiometric method gives better results for high-range samples.

12.6 EXPERIMENTAL: DETERMINATION OF TOTAL MERCURY IN FISH
       BY PYROLYSIS AND ULTRAVIOLET PHOTOMETRY

A. Principle

Trace amounts (0.05-3.0 ppm) of mercury in fish can be pyrolytically converted to the elemental form and quantitated photometrically. The homo-
genized fish tissue is combusted at 900°C in a flowing air stream and the
vapors are passed over copper oxide at 850°C to ensure complete oxidation.
Possible interfering gases which would absorb in the ultraviolet region of
253.7 nm, such as oxides of sulfur and nitrogen, are removed by silver
wire heated at 450°C, a caustic scrubber, and an Ascarite filter. Halides
which could recombine with the mercury formed are also removed by the
silver wire. The flowing air stream containing mercury vapor is then
conducted through an ultraviolet photometer. Electronic integration of the
detector signal is employed to improve analysis statistics. The procedure
given below has been published by Thomas et al. [128], Central Analytical
Department, Olin Corporation, New Haven, Connecticut.

B. Apparatus

Figure 12.9 shows diagrammatically the assembly of the apparatus. The
combustion tube (36 in. long, 8-mm i.d.) is made of quartz. The combustion boat (2 in. long, 3/16 in. wide, 5/32 in. deep) is made of nickel.
The single electric furnace is a multiple unit with three sections which are
respectively 9, 12, and 15 in. long.
    A Laboratory Data Control Mercury Monitor is used as the ultraviolet
photometer. This instrument has an absorption cell length of 30 cm and
an internal volume of 14 ml. Since this unit has a full-scale response of
0.005 μg Hg, an attenuated range giving a linear response from 0.01 to
0.10 μg Hg is used. The output of the photometer is connected to a Perkin-
Elmer Model 165 recorder. For the digital readout system, either an
Infotronics Model CRS-11 HSB or an Autolab 6300 Integrator can be used.
    A 100-ml round-bottom flask is used for the caustic bubbler. A four-
way stopcock connected to the magnesium perchlorate-Ascarite tube enables
a constant flow to be maintained through the combustion tube without passing through the optical cell when samples are not being pyrolyzed, thus
extending the life of the absorbers. With the geometry described, the solid
absorbers should be changed after 4 h of operation.
    The amount of sample to be pyrolyzed must be balanced with the air
flow rate and the length of effective oxidant. For the geometry described,
50-150 mg of fish tissue can be completely pyrolyzed at 900°C in a 9 liters/
min air stream using a 10-in. length of copper oxide heated at 850°C. A

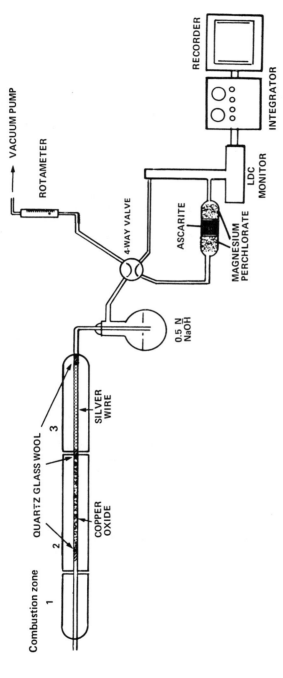

FIG. 12.9. Combustion apparatus with ultraviolet mercury detector. 1, first furnace, 900°C combustion zone; 2, second furnace, 850°C packed with 10-in. copper oxide; 3, third furnace, 450°C packed with 7-in. silver wire. (From Ref. 128, courtesy Anal. Chem.)

7-in. length of fine silver wire heated at 450°C will effectively remove halides and oxides of sulfur. The dimensions of the remaining part of the train are not critical.

C. Reagents

Copper oxide, AR, wire form.
Silver wool, 0.005-in.
Magnesium perchlorate, anhydrous.
Ascarite.
Sodium hydroxide solution, 0.5 $\underline{M}$.
Mercuric acetate standard solution: Dissolve 1.595 g of mercuric acetate, ACS reagent grade, in distilled water containing 5 ml of concentrated nitric acid and diluted to 100 ml. Dilute 1:100 with distilled water for 100 μg Hg/ml solution, and then 5:100 for a 5 μg Hg/ml working solution.

D. Procedure

Conditioning the Combustion Train

After the apparatus has been allowed to warm up, pull air through the combustion tube at a rate of about 9 liters/min. Keep the air flow rate constant during the experiment.

Preparation of the Calibration Graph

Prepare a calibration graph with 5-20 μl of the 5 μg Hg/ml standard solution. Combust each standard and record the area measurement from the integrator. (The recorder is used to make a permanent trace of the values obtained.) The graph (integrating area vs. μg Hg) should show a straight line from 0.01 to 0.10 μg of mercury.

Analysis of Fish Samples

Accurately weigh 50-150 mg of frozen fish homogenates into a nickel combustion boat. Push the boat and contents into the hot combustion zone and keep them there until the readout system indicates that pyrolysis is complete.

E. Calculation

$$\text{ppm Hg} = \frac{\mu\text{g Hg found x 1000}}{\text{mg fish}}$$

12.7 EXPERIMENTAL: DETERMINATION OF TRACE MERCURY IN COAL
      BY COLD VAPOR ATOMIC ABSORPTION SPECTROMETRY

A. Principle

Coal samples containing 0.1 ppm of mercury are burned in air in a combustion tube at 600°C. The vapors are further combusted at 800°C in a
stream of oxygen and then passed through acid permanganate solution where
mercury is retained. Subsequently, mercury in the absorbent solution is
determined by cold vapor atomic absorption spectrometry. The procedure
given below has been reported by Lo and Bush [135a], Division of Laboratories and Research, New York State Department of Health, Albany, New
York. The experimental details were verified by B. Bush in October 1974.

B. Equipment

    Combustion train: The apparatus for combustion of coal is shown in
       Fig. 12.10. The combustion tube is constricted and bent at the exit,
       and the bent tube is connected to a piece of Teflon tubing (14 cm
       long, 1 cm i.d.). Eight 8 x 2-mm slots are made at the end of the
       bent tube to reduce back pressure so that gas can be released as
       small bubbles, thus reducing the possibility of splashing.
    Atomic absorption spectrometer, Varian Techtron Model AA5.

C. Reagents

    Air cylinder.
    Oxygen cylinder.
    Acid potassium permanganate solution.

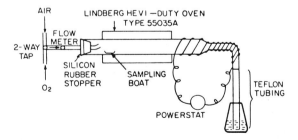

FIG. 12.10. Apparatus for combustion of coal, after Lo and Bush.
(From Ref. 135a, courtesy J. Ass. Offic. Anal. Chem.)

D. Procedure

Combustion

Accurately weigh about 1 g of coal into the combustion boat which consists of a split piece of Vycor tubing, 7 cm long, 1.5 cm i.d. Heat the
furnace to 450°C and adjust the Powerstat so that the bent tube is heated to
a red glow (800°C). Regulate the flow rates of both air and oxygen to 1.0-
1.2 ft^3/h. Pipet 50 ml of acid permanganate solution into the 125-ml
Erlenmeyer flask. Immerse the tip of the Teflon tubing into the absorbing
solution until it almost touches the bottom of the flask. Now introduce the
combustion boat and contents, and push them to a position about 0.5 cm
inside the heated part of the combustion tube (see Fig. 12.10). Immediately
pass air at the specified rate by replacing the silicon rubber stopper. Turn
the furnace control to high; let the temperature rise quickly to 600°C and
then keep it constant. Open the furnace lid occasionally to observe the
burning of the coal. When there is a sustained red glow, switch to oxygen.
Continue burning until the coal has been in the furnace for a total of 15 min.

Determination

After combustion, remove the Erlenmeyer flask, and rinse the Teflon
tubing with deionized water. Heat the absorbent solution on an electric
plate just to boiling; then let it simmer gently for 5 min to remove any
uncombusted organic compounds. Analyze for mercury by the cold vapor
atomic absorption technique, using the manual method [372], with the
Varian spectrometer.

E. Calculation

$$\text{ppm Hg} = \frac{\mu\text{g Hg found}}{\text{g coal}}$$

12.8 EXPERIMENTAL: DETERMINATION OF TRACE CARBON IN INORGANIC MATRIX

A. Principle

The inorganic material containing trace amounts of carbon or organic
matter is combusted over WO_3 in a stream of nitrogen. The combustion
products (CO_2, H_2O, oxides of sulfur, etc.) are conducted through sections
of Ag, MnO_2, and Anhydrone, respectively, whereupon every product is

removed but the CO_2. The pure CO_2 is then converted to CH_4 by reduction over a nickel catalyst in an atmosphere of hydrogen, and the CH_4 is measured by a flame ionization detector which is connected to a recorder. The area of the peak is related to a previously prepared calibration graph obtained by combusting known amounts of carbon. The procedure described below was developed with the collaboration of Albert C. Mayer, Jr., Central Analytical Department, Olin Corporation. It has been applied to the determination of trace carbon in sulfuric acid, phosphoric acid, inorganic solids, and waste effluents.

B. Equipment

Combustion furnace, 88 cm long, provided with three independent heating units (see Sec. 2.6) (Hevi-Duty, Milwaukee, Wisconsin).
Reduction furnace, 33 cm long (Hevi-Duty, Milwaukee, Wisconsin).
Combustion tube, quartz, 11-mm o.d. x 9-mm i.d. x 120-cm ℓ (P. W. Blackburn, Inc., Dobbs Ferry, New York): It has male ball joints at both ends, and a nitrogen inlet tube 4.5 cm from the left end connected to a quartz projection (7-mm o.d. x 5-mm i.d. x 70-mm ℓ) by means of Kovar seal and Swage-Lok tube fitting. Another quartz projection (7-mm o.d. x 5-mm i.d. x 30-mm ℓ) from the combustion tube, located 20 cm from the nitrogen inlet, serves as the sample inlet. This is fitted with an overlapping rubber septum. The diagrammatic sketch of the combustion train is shown in Fig. 12.11.

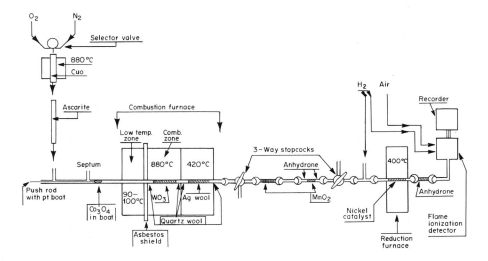

FIG. 12.11. Combustion train for trace carbon determination.

Stainless steel push rod, 2-mm diameter x 84-cm ℓ, connected to the combustion tube by means of a Kovar seal, Swage-Lok tube fitting, and female ball joint.

Reduction tube, quartz, 11-mm o.d. x 9-mm i.d. x 45-cm ℓ, with male ball joints at both ends.

Scrubber tubes (for MnO_2 or Anhydrone), 11-mm o.d. x 9-mm i.d. x 12.7-cm ℓ, with Pyrex female ball joints.

Three-way stopcocks, 11-mm o.d. x 9-mm i.d. x 10.5-cm ℓ, with Pyrex male and female ball joints.

Hydrogen inlet tube, Pyrex, 11-mm o.d. x 9-mm i.d. x 7.4-cm ℓ, connected to the hydrogen source by means of a Kovar seal and Swage-Lok tube fitting.

Kovar seals (device for connecting metal to glass) (Ace Glass Co., Vineland, New Jersey).

Swage-Lok tube fittings (used in conjunction with Kovar seals for connecting metal to glass and vice versa) (Norwalk Valve and Fitting Co., Fairfield, New Jersey).

Selector valve (valve which allows either nitrogen or oxygen to flow into combustion tube), Model S (George W. Dahl Co., Bristol, Rhode Island).

Platinum boats, No. 701 (Engelhard Industries, Carteret, New Jersey). Boat is fused to end of stainless-steel rod.

Syringe, 50-μl capacity, with Teflon Luer Lock; platinum needle with Teflon hub, 2-in. ℓ, point style No. 1 (Hamilton, Inc., Reno, Nevada): The syringe and needle are resistant to attack by concentrated mineral acids (H_2SO_4, H_3PO_4, HCl).

Glass capillaries with rods (used to introduce solid samples), size B (from A. H. Thomas Co., Philadelphia).

Molecular sieve tube, 3-mm o.d. x 2-mm i.d. x 12.7-cm length, with female ball joints (see Fig. 12.12): This tube is used if the sample releases CO_2 slowly, resulting in a broad peak, as discussed in Sec. 12.8F, part (4).

Flame ionization detector: Total hydrocarbon Analyzer, Model 23-500, (Gow-Mac Instrument Co., Madison, New Jersey).

Recorder: Flat bed, dual-range (0-1 mV; 0-5 mV) (Leeds & Northrup Co., North Wales, Pennsylvania).

C. Reagents

MgO.
Tungstic oxide (95% WO_3 on 5% alumina).
Ag wool.
Silver tungstate on MgO, used when Ag wool is not available.

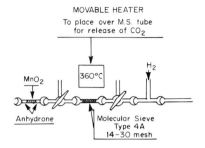

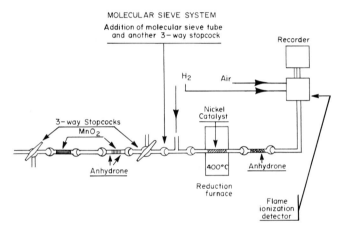

FIG. 12.12. The molecular sieve system for trace carbon determination.

Quartz wool.
Cobaltic oxide (Co_3O_4).
MnO_2.
Anhydrone ($Mg(ClO_4)_2$).
Ascarite.
CuO.
Nickel catalyst (Envirotech Corp., Mountain View, California).
Prepurified nitrogen: The nitrogen supply is passed through CuO at
    880°C and then Ascarite at room temperature before being intro-
    duced into the combustion train.
Hydrogen, for reduction of CO_2 and for flame ionization detector.
Air, for flame ionization detector.

Oxygen, for regeneration of WO_3.

Potassium acid phthalate standard solution: Prepare a standard solution containing 200 ppm C by dissolving 42.51 mg of potassium acid phthalate in distilled water and making up to volume in a 100-ml volumetric flask.

D. Procedure

Preparation of the Combustion Train

Fill the combustion, scrubber, and reduction tubes as specified in Fig. 12.11. Add cobaltic oxide to the platinum boat until it is half full. Assemble the train by clamping all male and female joints as shown in Fig. 12.11. Set the combustion furnace temperature to 880°C, the low-temperature furnace to 90-100°C, the Ag wool area to 420°C, and the reduction furnace to 400°C. Adjust N_2, H_2, and air pressures to 20, 7, and 20 lb/in.^{2}, respectively. Press the igniter button to ignite the H_2 + air mixture in the flame ionization detector. Before introducing the sample, position valve (1) so that prepurified N_2 is flowing through the system, stopcock (2) is opened to the atmosphere, and stopcock (3) is opened to the nickel catalyst. Place the platinum boat directly underneath the rubber septum by sliding the stainless steel push-rod further into the combustion tube.

Introduction of the Sample

For liquids: remove the septum and introduce 5-30 μl of liquid sample into the platinum boat by means of an acid-resistant syringe. For solids: carefully introduce the sample by means of a glass capillary fitted with a solid glass push-rod; obtain the weight of the sample by weighing the capillary before and after the sample introduction.

Course of Combustion

After the sample has been added, close stopcock (2) to the atmosphere and open it to the detector. Turn the recorder on. When the recorder indicates a stable baseline, move the boat into the low-temperature zone, and record the volatile carbon if desired. When the recorder again returns to the baseline, move the boat into the high-temperature zone and record the nonvolatile carbon. The sum of volatile and nonvolatile carbon values gives the total carbon present in the sample.

Preparation of the Calibration Graph

Prepare a calibration graph by combusting known amounts of carbon (addition of varying amounts of standard solution of potassium acid phthalate or other suitable compound) in the 0–1000 ppm range. Plot the carbon concentration on the absicissa vs. the peak area on the ordinate. (The peak area is obtained by multiplying the peak height by the peak width at one-half of the peak height.)

Regeneration of WO_3

At the end of the day, open stopcock (2) to the atmosphere and switch valve (1) to oxygen. Oxygen then flows through the combustion tube during the night, regenerating the WO_3 that may have been depleted.

E. Calculation

For Liquid Samples

Obtain the ppm C present in the sample by comparing the peak area of the sample to the peak area vs. carbon concentration (in ppm) calibration graph (equal volumes of standard solution and sample are used).

For Solid Samples

Assume 1 μl of standard solution = 1 mg. Obtain the ppm C present in the sample from the calibration graph. Then correct for the difference in weight between the standard and the sample. Example: If the calibration graph was prepared from 10-μl standards and a 5.0-mg sample gave a peak area corresponding to 100 ppm C, then

$$\text{Corrected carbon concentration (ppm C)} = \frac{100 \times 10}{5.0} = 200$$

F. Comments

(1) No appreciable blank values were found when the septum was removed for the introduction of samples. Since injection of liquid samples through the rubber septum sometimes led to erratically high results (possibly due to minute pieces of rubber breaking off) and since for the introduction of solids the septum had to be removed anyhow, it was decided to remove the septum for the introduction of all samples.

(2) While this method was originally designed for the determination of trace carbon in H_2SO_4 and H_3PO_4, we have found that it is equally effective in the determination of carbon in concentrated HCl, waste effluents, brines, and inorganic solids.

(3) The time of analysis is approximately 5 min. The precision and accuracy is $\pm 10\%$.

(4) Occasionally a broad peak was observed in the determination of some samples. Since the area measurement of a sample of this type might be inaccurate, it was felt that if all the CO_2 could be stopped and then released when the combustion was complete, better accuracy could be attained. This is accomplished by retaining the CO_2 on molecular sieve type 4A (14-30 mesh) at room temperature. To release the CO_2, the molecular sieve is heated rapidly at 360°C. The apparatus is illustrated in Fig. 12.12.

(5) Cobaltic oxide (Co_3O_4) serves not only as a combustion catalyst but also absorbs liquid samples preventing them from creeping out of the platinum boat.

(6) The reader is referred to the articles published by Carter and Goodell [373] on the determination of organic carbon in sulfuric acid, by Dobbs and co-workers [374] on the measurement of carbon in water, and by Porter and Volman [375] on the conversion of CO to CH_4.

REFERENCES

1. W. W. Meinke and J. K. Taylor (eds.), Analytical Chemistry: Key to Progress on National Problems, National Bureau of Standards Special Publication 351, U.S. Government Printing Office, Washington, D.C., 1972.
2. "Organic Pollutant Analysis," 26th Summer Symposium on Analytical Chemistry, Oxford, Ohio, June 1973.
2a. "Accuracy in Trace Analysis," Symposium at the National Bureau of Standards, Gathersburg, Md., October 1974.
2b. J. H. Runnels, R. Merryfield, and H. B. Fisher, Anal. Chem. 47:1258 (1975).
2c. H. E. Knauer and G. E. Milliman, Anal. Chem. 47:1263 (1975).
2d. W. K. Robbins and H. H. Walker, Anal. Chem. 47:1269 (1975).
2e. R. A. Hofstader, O. I. Milner, and J. H. Runnels (eds.), Analysis of Petroleum for Trace Metals, American Chemical Society, Washington, D.C., 1976.
3. J. H. Yoe and H. J. Koch (eds.), Trace Analysis, Wiley, New York, 1957.
4. E. J. Underwood, Trace Elements in Human and Animal Nitrition, 3rd Ed., Academic, New York, 1971.

5. O. I. Milner, Analysis of Petroleum for Trace Elements, Pergamon, New York, 1963.

5a. D. J. D. Nicholas and A. R. Egan (eds.), Trace Elements in Soil-Plant-Animal Systems, Academic, New York, 1975.

5b. Chemical Determination of Aluminium, Boron, Calcium and Magnesium, Copper and Zinc, Iron and Manganese, and of Sulphur in Soil and Plant Material (annotated bibliographies), Commonwealth Agricultural Bureau, Slough, England, 1974.

6. J. W. Mitchell, Anal. Chem. 45:492A (1973).

7. E. Sawicki, Photometric Organic Analysis, Wiley, New York, 1970.

8. B. P. Kirk and H. C. Wilkinson, Talanta 19:80 (1972).

9. R. E. Ober, A. R. Hansen, D. Mourer, J. Baukema, and G. W. Gwynn, J. Appl. Radiat. Isotopes 20:703 (1969).

10. B. L. Horeker, T. S. Ma, and I. Haas, J. Biol. Chem. 136:775 (1940).

11. J. Majewska, A. Milosz, and H. Krzystek, Chem. Anal. (Warsaw) 17:41 (1972).

11a. V. F. Gaylor, reported at the American Chemical Society National Meeting, Chicago, August 1973.

12. A. M. Vogel and J. J. Quatrone, Jr., Anal. Chem. 32:1754 (1960).

13. A. L. Conrad, Mikrochim. Acta 38:514 (1951).

14. American Society for Testing and Materials, Specification D-808-63 (1966).

15. S. Fujiwara and H. Narasaki, Anal. Chem. 40:2032 (1968).

15a. A. S. Baker and R. L. Smith, J. Agr. Food Chem. 22:103 (1974).

15b. Metallic Impurities in Organic Matter Subcommittee (report), Analyst 101:62 (1976).

15c. R. H. Thompson and W. J. Blanchflower, Lab. Pract. 23:362 (1974).

15d. C. Feldman, Anal. Chem. 46:1606 (1974).

15e. S. P. Bessman, Anal. Biochem. 59:524 (1974).

15f. J. G. van Raaphorst, A. W. van Weers, and H. M. Haremaker, Analyst 99:523 (1974).

15g. H. Windermann and U. Müller, Mitt. Geb. Lebensmittelunters. Hyg. 66:64 (1975).

15h. T. Chow, C. C. Patterson, and D. Settle, Nature 251:159 (1974).

15i. Yanaco Plasma Ashing Apparatus, Yanagimoto Mfg. Co., Kyoto, Japan (1975).

15j. K. Hozumi (ed.), Low-Temperature Plasma Chemistry (in Japanese), Nankodo, Tokyo, 1976; private communication, August 1976.

16. R. Belcher, Talanta 15:357 (1968).

17. E. Emich, Mikrochemie 13:285 (1933).

18. W. Schöniger, Microchem. J. 11:469 (1966).

19. L. Blom and L. Edelhausen, Anal. Chim. Acta 13:120 (1955).

20. H. Weisz and U. Fritsche, Mikrochim. Acta 1973:361.
21. R. Belcher and R. Goulden, Mikrochim. Acta 1953:290.
22. W. I. Awad, S. S. M. Hassan, and M. B. Elsayes, Mikrochim. Acta 1969:688.
23. S. S. M. Hassan and S. A. I. Thoria, Mikrochim. Acta 1974:561.
24. M. H. Hashmi, Mikrochim. Acta 1970:359.
25. J. L. Lambert, in McGraw-Hill Yearbook of Science, McGraw-Hill, New York, 1973, p. 361.
26. H. V. Malmstadt and T. P. Hadjiioannou, Anal. Chem. 35:2157 (1963).
27. I. Iwasaki, S. Utsumi, and T. Ozawa, Bull. Chem. Soc. Japan 26:108 (1953).
28. E. B. Sandell and I. M. Kolthoff, Mikrochim. Acta 1:9 (1937).
29. V. J. Pileggi and G. Kessler, Clin. Chem. 14:339 (1968).
30. T. S. Ma and W. Nazimowitz, Mikrochim. Acta 1969:345.
30a. A. A. Aleksiev, P. R. Bonchev, and I. Todorov, Arch. Oral Biol. 18:1461 (1973).
31. P. J. Ke, R. J. Thibert, R. J. Walton, and D. K. Soules, Mikrochim. Acta 1973:569.
31a. T. P. Hadjiioannou, private communication, May 1976.
31b. R. A. Sheikh and A. Townshend, Talanta 21:401 (1974).
32. G. H. Morrison (ed.), Trace Analysis: Physical Methods, Wiley, New York, 1965.
32a. R. Jenkins, J. Inst. Petrol. 48:246 (1962).
32b. W. Wildanger, Z. Lebensmitt. 155:321 (1974).
32c. Y. K. Chau and K. Lum-Shue-Chan, Water Res. 8:383 (1974).
32d. J. D. Winefordner, Chemtech 1975(2):123.
33. J. Polesuk, J. M. Amadeo, and T. S. Ma, Mikrochim. Acta 1973:512.
33a. F. L. Fricke, O. Rose, Jr., and J. A. Caruso, Talanta 23:317 (1976).
33b. L. R. Layman and G. M. Hieftje, Anal. Chem. 47:194 (1975).
33c. R. Belcher, S. Karpek, T. A. K. Nasser, K. P. Ronjitker, M. Shahidullah, A. Townshend, S. L. Bogdanski, A. Calokerinos, I. Z. Al-Zamil, and M. Q. Al-Abachi, reported at the Federation of Analytical Chemistry and Spectroscopy Societies 3rd Annual Meeting, Philadelphia, November 1976.
33d. J. Donau, Monatsheft 34:949 (1913).
33e. J. C. Chambers and B. E. McClellan, Anal. Chem. 48:2061 (1976).
34. T. S. Ma and V. Horak, Microscale Manipulations in Chemistry, Wiley, New York, 1976.
34a. W. Wolf, W. Mertz, and R. Masironi, Agr. Food Chem. 22:1037 (1974).

34b. L. E. Kopito, M. A. Davis, and H. Shwachman, Clin. Chem. 20:205 (1974).

34c. M. Lachica and J. Robles, An. Edafol. Agrobiol. 33:91 (1974).

34d. R. T. Lofberg and E. A. Levri, Anal. Lett. 7:775 (1974).

34e. V. D. Anand, J. A. Lott, G. F. Grannis, and J. E. Mercier, Amer. J. Clin. Pathol. 59:717 (1973).

34f. F. M. F. G. Olthuis, K. Kruisinga, and J. B. J. Soons, Clin. Chem. Acta 49:123 (1973).

34g. P. W. J. M. Boumans and F. J. M. J. Maessen, Appl. Spectrosc. 24:241 (1970).

35. J. Y. Hwang, Anal. Chem. 44 December:21A (1972).

36. J. W. Mitchell, Anal. Chem. 45:492A (1973).

37. Y. Takahashi, R. T. Moore, and R. J. Joyce, American Laboratory (1972), July 31.

38. American Public Health Association, Standard Methods for the Examination of Water and Wastewater, American Public Health Association, Inc., New York, 1967, pp. 405, 505.

39. American Society for Testing and Materials, 1971 Annual Book of ASTM Standards, American Society for Testing and Materials, Philadelphia, 1971, Part 23, p. 211.

40. R. M. Emery, E. B. Welch, and R. F. Christman, J. Water Pollut. Contr. Fed. 43:1834 (1971).

41. C. E. van Hall, J. Safranko, and V. Stenger, Anal. Chem. 35:315 (1963).

41a. C. E. Carr, J. Sci. Food Agr. 24:1091 (1973).

42. I. Lysyj, K. H. Nelson, and P. R. Newton, Water Res. 1:233 (1968).

42a. S. J. Kalembasa and D. S. Jenkinson, J. Sci. Food Agr. 24:1085 (1973).

43. N. Oda, S. Ono, and H. Matsumori, Japan Analyst 18:854 (1969).

44. S. N. Pennington and H. D. Brown, J. Gas Chromatogr. 6:505 (1968).

45. J. Theer, Chem. Tech. (Berlin) 20:484 (1968).

45a. W. Merz, International Laboratory, 1977, January, p. 49.

45b. E. R. Russell, Anal. Chim. Acta 88:171 (1977).

45c. H. Swift, Analyst 102:217 (1977).

45d. Carbon Dioxide Coulometer, Coulometrics Inc., Wheat Ridge, Color.

46. W. H. Jones, Anal. Chem. 25:1449 (1953).

47. A. J. Smith, G. Meyers, Jr., and W. C. Shaner, Jr., Mikrochim. Acta 1972:217.

48. P. Gouverneur and A. C. Bruijn, Talanta 16:827 (1969).

49. I. Klesment, Mikrochim. Acta 1969:1237.

50. S. A. Liebman, D. H. Ahlstrom, C. D. Nauman, R. Averitt, J. L. Walker, and E. J. Levy, Anal. Chem. 45:1360 (1973).

460 Trace Analysis

51. J. J. Lauff, E. R. Champlin, and E. P. Przybylowicz, Anal. Chem. 45:52 (1973).
52. N. K. Prasad, Chem. Ind. (London) 1972(2):81.
53. A. R. Law and R. L. Norton, J. Ass. Offic. Anal. Chem. 54:764 (1971).
54. C. V. Lundeen, A. S. Viscomi, and F. H. Field, Anal. Chem. 45:1288 (1973).
55. G. Proksch, Plant Soil 31:380 (1969); International Atomic Energy Agency Report IAEA-SM-151/11 (1972).
56. R. Fielder and G. Proksch, Anal. Chim. Acta 60:277 (1972).
57. R. Fielder, G. Proksch, and A. Koepf, Anal. Chim. Acta 63:435 (1973).
58. E. Kunkel, "Experience With the Wickbold Combustion Apparatus for Organic Elemental Analysis," in Colloquium of Heraeus-Schott Quarzschmelze GmbH, Hanau, Germany, 1966.
59. W. Lädrach, F. van de Croats, and P. Gouverneur, Anal. Chim. Acta 50:219 (1970).
60. W. Krijgsman, W. P. van Bennekom, and B. Griepink, Mikrochim. Acta 1972:42.
61. W. Krijgsman, G. de Groot, W. P. van Bennekom, and B. Griepink, Mikrochim. Acta 1972:364.
62. G. Lunde, J. Amer. Oil Chem. Soc. 49:44 (1972).
63. F. T. Bunus and S. Murgalescu, Talanta 19:372 (1972).
64. T. S. Ma and J. J. Gwirtsman, Int. J. Environ. Anal. Chem. 2:133 (1972).
65. "Tentative Method of Analysis for Fluorine Content of the Atmosphere and Plant Tissues," Health Lab. Sci. 6:64 (1969).
66. W. P. Schmidt and R. A. Paulson, National Bureau of Standards Technical Note, No. 585, U.S. Government Printing Office, Washington, D.C., 1972, p. 48.
67. J. A. Hargreaves, G. S. Ingram, and D. L. Cox, Analyst 95:177 (1970).
68. S. Ono, M. Suzuki, K. Sasajima, and S. Iwata, Analyst 95:260 (1970).
69. N. L. Gregory, Anal. Chem. 44:231 (1972).
70. S. C. Tong, W. H. Gutenman, L. E. St. John, Jr., and D. J. Lisk, Anal. Chem. 44:1069 (1972).
71. J. Slanina, E. Buysman, J. Agterdenbos, and B. Griepink, Mikrochim. Acta 1971:657.
72. J. Slanina, P. Vermeer, J. Agterdenbos, and B. Griepink, Mikrochim. Acta 1973:607.
73. W. I. Stephen, Anal. Chim. Acta 50:413 (1970).
74. A. E. Mendes-Bezerra and P. E. Uden, Analyst 94:308 (1969).

75. J. Dokladalova and M. Nekovarova, Chem. Tech. (Berlin) 21:490 (1969).

76. S. Mlinko and E. Dobis, Acta Chim. Acad. Sci. Hung. 61:133 (1969).

77. M. Fieldes and R. J. Furkert, N.Z. J. Sci. 14:280 (1971).

78. R. N. Heistand and C. T. Blake, Mikrochim. Acta 1972:212.

79. F. C. A. Killer and K. E. Underhill, Analyst 95:505 (1970).

80. R. Kellner and E. Pell, reported at the 6th International Symposium on Microtechniques, Graz, 1970.

81. Leco Sulfur Titrator, Laboratory Equipment Corp., St. Joseph, Mich.

82. Y. A. Gawargious and A. B. Farag, Talanta 19:64 (1972).

83. C. Veillon and J. Y. Park, Anal. Chim. Acta 60:293 (1972).

84. G. F. Kirkbright, M. Marshall, and T. S. West, Anal. Chem. 44:2379 (1972).

85. H. H. Taussky, A. Washington, E. Zubillaga, and A. T. Milhorat, Microchem. J. 10:470 (1966).

86. M. Levesque and E. D. Vendette, Can. J. Soil Sci. 51:85 (1971).

87. J. H. Wiersma and G. F. Lee, Environ. Sci. Technol. 5:1203 (1971).

88. J. A. Raihle, Environ. Sci. Technol. 6:621 (1972).

88a. D. Arthur, Can. Inst. Food Sci. Technol. J. 5:165 (1972).

89. R. D. Beaty, Anal. Chem. 45:234 (1973).

90. R. M. Sachs, J. L. Michael, F. B. Anastasia, and W. A. Wells, Weed Sci. 19:412 (1971).

91. G. I. Spielholtz, G. C. Toralballa, and R. J. Steinberg, Mikrochim. Acta 1971:618.

91a. G. F. Collier, J. Sci. Food Agr. 24:1115 (1973).

92. T. M. Reymont and R. J. Dubois, Anal. Chim. Acta 56:1 (1971).

93. A. R. Byrne, Anal. Chim. Acta 59:91 (1972).

93a. G. Lunde, J. Sci. Food Agr. 24:1021 (1973).

93b. A. R. Byrne and T. Vakeselj, Rep. Nucl. Inst. Jozef Stefan, IJS-R-617 (1973).

94. D. J. Myers and J. Osteryoung, Anal. Chem. 45:267 (1973).

95. D. S. Kirkpatrick and S. H. Bishop, Anal. Chem. 43:1707 (1971).

96. D. Townsand, B. Livermore, and H. Jenkin, Microchem. J. 16:456 (1971).

97. M. C. Shaw, Anal. Biochem. 44:288 (1971).

98. J. Tusl, Analyst 97:111 (1972).

99. T. S. Ma and J. D. McKinley, Jr., Mikrochim. Acta 1953:1.

100. G. Ghimicescu and V. Dorneanu, Mikrochim. Acta 1972:68.

101. W. N. Elliot, C. Heathcote, and R. A. Mostyn, Talanta 19:359 (1972).

102. G. Linden, S. Turk, and B. Tarodo de la Fuente, Chim. Anal. 53:244 (1971).

103. I. N. Smirnov and A. G. Khislavskii, Zavod. Lab. 37:924 (1971).
104. S. P. Cram, "Amino acid Complexation of Trace Metals," reported at the Symposium on Organic Pollutant Analysis, Oxford, Ohio, 1973.
105. Y. K. Chau, "Complexing Capacity of Natural Water--Its Significance and Measurement," reported at the Symposium on Organic Pollutant Analysis, Oxford, Ohio, 1973.
106. S. C. Jolly (ed.), Official Standard and Recommended Methods of Analysis, Heffer, Cambridge, England, 1963, p. 3.
107. A. D. Thomas and L. E. Smythe, Talanta 20:469 (1973).
108. M. Pinta, Detection and Determination of Trace Elements, Ann Arbor Science Publishers, Ann Arbor, Mich., 1973.
108a. D. J. Lisk, Science 184:1137 (1974).
109. A. M. Hartstein, R. W. Freedman, and D. W. Platter, Anal. Chem. 45:611 (1973).
109a. Y. E. Araktingi, C. L. Chakrabarti, and I. S. Maines, Spectrosc. Lett. 7:97 (1974).
110. M. T. Glenn, J. Savory, S. A. Fein, R. D. Reeves, C. J. Molnar, and J. D. Winefordner, Anal. Chem. 45:203 (1973).
111. M. Puyo and R. Thiel, Bull. Cent. Rech. Pau 1:453 (1967).
111a. G. P. Morie and P. E. Morrisett, Beitr. Tabakforsch. 7:302 (1974).
111b. E. L. Henn, Anal. Chim. Acta 73:273 (1974).
112. F. T. N. Varghese, A. Lipton, and G. J. Huxham, Lab. Pract. 18:419 (1969).
112a. K. H. Falchuk, M. A. Evenson, and B. L. Vallee, Anal. Biochem. 62:255 (1974).
112b. R. S. Tse, S. C. Wong, and S. S. L. Wong, Anal. Chem. 48:234 (1976).
113. A. J. Ahearn (ed.), Trace Analysis by Mass Spectroscopy, Academic, New York, 1973.
114. R. J. Guidoboni, Anal. Chem. 45:1275 (1973).
115. J. A. McHugh and J. F. Stevens, Anal. Chem. 44:2187 (1972).
116. L. J. Moore and L. A. Machlan, Anal. Chem. 44:2291 (1972).
117. D. de Soete, R. Gijbels, and J. Hoste, Neutron Activation Analysis, Wiley, New York, 1972.
117a. E. Orvini, T. E. Gills, and P. D. LaFleur, Anal. Chem. 46:1294 (1974).
118. G. J. Lutz, R. J. Boreni, R. S. Maddock, and J. Wing (eds.), Activation Analysis--A Bibliography Through 1971, National Bureau of Standards Technical Note, No. 467, U.S. Government Printing Office, Washington, D.C., 1972.
119. J. P. Garrec, Rapp. CEA, R-3636 (1968).

120. W. A. Haller, L. A. Rancitelli, and J. A. Cooper, J. Agr. Food Chem. 16:1036 (1968).

121. K. Fritze and R. Robertson, J. Radioanal. Chem. 1:463 (1968).

122. R. A. Zweidinger, L. Barnett, and C. G. Pitt, Anal. Chem. 45:1563 (1973).

123. G. Kaiser, E. Grallath, P. Tschöpel, and G. Tölg, Z. Anal. Chem. 259:257 (1972).

124. F. S. Chuang, J. R. Sarbeck, P. A. St. John, Jr., and J. D. Winefordner, Mikrochim. Acta 1973:523.

125. W. R. Seitz and D. M. Hercules, Anal. Chem. 44:2143 (1972).

126. D. L. Levit, Anal. Chem. 45:1291 (1973).

127. J. F. Uthe, A. I. Armstron, and M. P. Stainton, J. Fish Res. Bd. Can. 27:805 (1970).

128. R. J. Thomas, R. A. Hagstrom, and E. J. Kuchar, Anal. Chem. 44:512 (1972).

129. E. G. Pappus and L. A. Rosenberg, J. Ass. Offic. Anal. Chem. 49:782 (1966).

130. W. R. Hatch and W. L. Ott, Anal. Chem. 40:2085 (1968).

131. Combustion Flask for Mercury Analysis, A. H. Thomas Co., Philadelphia.

132. M. Fujita, Y. Takeda, T. Terao, O. Hoshimo, and T. Ukita, Anal. Chem. 40:2042 (1968).

132a. E. W. Bretthauer, A. A. Moghissi, S. S. Snyder, and N. W. Mathews, Anal. Chem. 46:445 (1974).

133. R. Meyer, Ann. Pharm. Fr. 28:271 (1970).

134. G. Lindstedt, Analyst 95:264 (1970).

135. B. R. Petersen, Dansk. Kemi 49:171 (1968).

135a. F. C. Lo and B. Bush, J. Ass. Offic. Anal. Chem. 56:1509 (1973).

136. J. Y. Hwang, P. A. Ullucci, and C. J. Mokeler, Anal. Chem. 45:795 (1973).

137. G. K. Pagenkopf, D. R. Neuman, and R. Woodriff, Anal. Chem. 44:2248 (1972).

137a. S. Horiguchi, K. Teramoto, K. Shinagawa, and G. Endo, J. Osaka City Med. Cent. 22:299 (1973).

137b. M. Kiboku and M. Aihara, Japan Analyst 22:1581 (1973).

137c. Y. Marumo, T. Oikawa, and T. Niwaguchi, Japan Analyst 22:1024 (1973).

137d. G. Reusmann and J. Westphalen, Staub. Reinhalt. Luft 33:435 (1973).

137e. W. Oelschlaeger and E. Frenkel, Landw. Forsch. 26:281 (1973).

137f. D. J. Snodin, J. Ass. Publ. Analysts 11:112 (1973).

138. G. F. Kirkbright, Analyst 96:609 (1971).

139. R. E. Mansell and T. A. Hiller, Anal. Chem. 45:975 (1973).

140. Parr Instrument Co., Moline, Ill.
141. W. A. Aue and H. H. Hill, Jr., Anal. Chem. 45:729 (1973).
142. T. F. Egan, Microchem. J. 13:646 (1968).
142a. A. Colombo and R. Vivian, Microchem. J. 18:589 (1973).
143. J. Blazejczak and B. M. van der Weide, Bull. Cent. Rech. Pau
      2:163 (1968).
143a. H. Schneider and E. Nold, Z. Anal. Chem. 269:113 (1974).
144. F. R. Cropper, D. M. Heinekey, and A. Westwell, Analyst 92:436
      (1967).
144a. F. Ehrenberger, Z. Anal. Chem. 267:17 (1973).
145. O. Holm-Hansen, J. Coombs, B. E. Volcani, and P. M. Williams,
      Anal. Biochem. 19:561 (1967).
146. H. C. E. van Leuven, Anal. Chim. Acta 49:364 (1970).
147. R. K. Patterson, Anal. Chem. 45:605 (1973).
148. A. Anusiem and P. A. Hersch, Anal. Chem. 45:592 (1973).
148a. J. M. Baldwin and R. E. McAtee, Microchem. J. 19:179 (1974).
149. D. L. Fanter and C. J. Wolf, Anal. Chem. 45:565 (1973).
150. I. P. Lisovskii and L. A. Smakhtin, Zh. Anal. Khim. 24:749 (1969).
151. A. J. Smith, F. F. Cooper, Jr., J. O. Rice, and W. C. Shaner,
      Jr., Anal. Chim. Acta 40:341 (1968).
152. G. R. N. Jones, Lab. Pract. 16:1486 (1967).
153. G. H. Sloane-Stanley, J. Biochem. 104:293 (1967).
154. A. Lewandowski, M. Madrowa, and J. Skirbiszewski, Chem. Anal.
      (Warsaw) 12:1043 (1967).
155. J. P. Wineburg, Anal. Chem. 40:1744 (1968).
156. P. Gouverneur, O. I. Snoek, and M. Heeringa-Kommer, Anal.
      Chim. Acta 39:413 (1967).
157. P. Gouveneur and F. van de Croats, Analyst 93:782 (1968).
158. I. J. Oita, Anal. Chem. 40:1753 (1968).
158a. M. W. Brown, J. Sci. Food Agr. 24:1119 (1973).
159. J. Franc, B. Trtik, and K. Placek, J. Chromatogr. 36:1 (1968).
160. D. G. Kramme, R. H. Griffen, C. G. Hartford, and J. A. Corrado,
      Anal. Chem. 45:405 (1973).
161. S. Henry, Rev. Ferment, Ind. Aliment. 23:80 (1968).
162. D. R. Taves, Anal. Chem. 40:204 (1968).
163. P. J. Ke, L. W. Regier, and H. E. Power, Anal. Chem. 41:1081
      (1969).
164. D. R. Taves, Talanta 15:1015 (1968).
165. L. Torma and B. E. Ginther, J. Ass. Offic. Anal. Chem. 51:1181
      (1968).
166. J. Tusl, Collect. Czech. Chem. Commun. 35:1001 (1970).
166a. D. M. Paez and B. A. Gil, Rev. Ass. Bioquim. (Argentina) 38:54
      (1973).

167. M. Miller and D. A. Keyworth, Talanta 14:1287 (1967).
168. V. G. Plyusmin, A. S. Filatova, and Z. M. Titova, Tr. Inst. Khim. (Sverdlovsk) 1968:84.
169. W. E. Dahl, Anal. Chem. 40:416 (1968).
170. S. P. Cram and J. L. Brownlee, Jr., J. Gas Chromatogr. 6:313 (1968).
170a. S. Ohno, M. Suzuki, M. Kadota, and M. Yatazawa, Mikrochim. Acta 1973:61.
171. P. K. Wilkniss and V. J. Linnenbom, Limnol. Oceanogr. 13:530 (1968).
172. V. R. Negina, E. P. Krasheninnikova, A. T. Mikhailina, K. S. Ratnikova, and M. T. Dokuchaeva, Zh. Anal. Khim. 22:1552 (1967).
172a. J. S. Hetman, Bull. Cent. Rech. Pau 7:83 (1973).
172b. K. Hoshino, Japan Analyst 22:866 (1973).
173. A. Dirscherl, Mikrochim. Acta 1968:316.
174. P. Maltese, L. Clementini, and A. Mori, Chim. Ind. (Milano) 49:1070 (1967).
175. I. Yanagisawa and H. Toshikawa, Clin. Chim. Acta 21:217 (1968).
176. B. Paletta and K. Panzenbeck, Clin. Chem. Acta 26:11 (1969).
177. H. J. M. Bowen, Anal. Chem. 40:969 (1968).
177a. P. J. Ke, R. J. Thibert, R. J. Walton, and D. K. Soules, Mikrochim. Acta 1973:569.
178. S. Fujiwara and H. Narasaki, Anal. Chem. 40:2031 (1968).
178a. O. Mitoseru and I. Herinean, Ind. Aliment. (Bucharest) 24:143 (1973).
179. A. Turuta and T. Crisan, Rev. Chim. (Bucharest) 19:610 (1968).
180. G. Kainz and E. Wachberger, Mikrochim. Acta 1968:596.
181. C. A. Bache and D. J. Lisk, Anal. Chem. 39:786 (1967).
182. F. Leuteritz and G. Brunner, Plaste Kautsch. 14:887 (1967).
182a. A. Visapa, Paperi Puu 55:385 (1973).
183. G. Hauk, Arch. Toxikol. 23:273 (1968).
184. E. T. Bramlitt, Anal. Chem. 38:1669 (1966).
184a. M. H. Friedman, T. M. Farber, and J. T. Tanner, Anal. Chim. Acta 67:277 (1973).
185. R. Malvano and S. Kivievinski, J. Radioanal. Chem. 3:257 (1969).
186. W. W. Marsh, Jr., Anal. Lett. 3:341 (1970).
187. J. Majewska, Chem. Anal. (Warsaw) 13:29 (1968).
188. J. Belisle, C. B. Green, and L. D. Winter, Anal. Chem. 40:1006 (1968).
189. M. M. Podorozhanskii, E. M. Zeidlits, and I. I. Eru, Zavod. Lab. 33:697 (1967).
190. J. Baltes, Fette Seifen Anstrichm. 69:512 (1967).
190a. C. L. Kimbell, British Patent 1,357,453 (1972).

191. A. Turuta, T. Crisan, and M. Pal, Rev. Chim. (Bucharest) 18:306 (1967).

192. M. Jawoski and E. Chromniak, Chem. Anal. (Warsaw) 11:705 (1966).

193. L. L. Farley and R. A. Winkler, Anal. Chem. 40:962 (1968).

194. J. Soucek and I. Slovicek, Chem. Prum. 19:472 (1969).

195. J. Slanina, J. Agterdenbos, P. Vermeer, and B. Griepink, Mikrochim. Acta 1970:1225; 1973:607.

196. British Standards Institution, BS 4350 (1968).

197. K. M. Aldous, R. M. Dagnall, and T. S. West, Analyst 95:417 (1970).

198. W. Holak, J. Ass. Offic. Anal. Chem. 54:1138 (1971).

198a. G. I. Speilholtz, G. C. Toralballa, and J. T. Wilson, Mikrochim. Acta 1974:649.

199. Z. E. Deikova and Z. F. Andreeva, Dokl. Mosk. Sel. Akad. K.A. Timiryazeva 1971:281.

200. M. H. Rison, W. H. Barber, and P. Wilkniss, Anal. Chem. 39:1028 (1967).

201. A. Smoczkiewiczowa, J. Augustyniak, and W. Meissner, Chem. Anal. (Warsaw) 12:629 (1967).

202. R. C. Ewan, C. A. Baumann, and A. L. Pope, J. Agr. Food Chem. 16:212 (1968).

203. D. I. Ryabchikov, I. I. Nazarenko, and L. I. Anikina, Zh. Anal. Khim. 23:1242 (1968).

204. M. Karvanek and B. Mankovska-Vorlova, Sb. Vys. Sk. Chem. Technol. Praze E20:5 (1968).

204a. O. E. Olson, J. Ass. Offic. Anal. Chem. 56:1073 (1973).

205. R. Engler and G. Tölg, Z. Anal. Chem. 235:151 (1968).

206. R. W. Morrow and J. A. Dean, Anal. Lett. 2:133 (1969).

207. C. M. Paralusz, Appl. Spectrosc. 22:520 (1968).

208. L. Truffert, M. Favert, and Y. Le Gall, Ann. Falsif. Expert. Chim. 60:275 (1967).

208a. H. Khalifa, M. T. Foad, Y. L. Awad, and M. E. Georgy, Microchem. J. 18:617 (1973).

208b. M. T. Tsao and R. P. Beliles, Amer. J. Clin. Pathol. 59:160 (1973).

208c. G. Boenig and H. Heigener, Landw. Forsch. 26:81 (1973).

209. R. J. Thibert and M. Sawar, Anal. Biochem. 25:440 (1968).

210. J. Agterdenos, L. van Broekhoven, B. A. H. G. Juette, and J. Schuring, Talanta 19:341 (1972).

211. V. D. Melekhim and E. M. Roizenblat, Lab. Delo 1969:107.

211a. D. I. Levit, Anal. Chem. 45:1291 (1973).

212. A. I. Chernova, Zavod. Lab. 34:1072 (1968).

213. C. E. Lambdin and W. V. Taylor, Anal. Chem. 40:2196 (1968).

214. G. Prohaska, Nafta (Zagreb) 19:461 (1968).

215. L. Chromy and K. Uhacz, J. Oil Colour Chem. Ass. 51:494 (1968).

216. M. Karvanek, Sb. Vys. Sk. Chem. Technol. Praze E21:5 (1969).

217. P. Giraudet, J. Pre, and P. Cornillot, Clin. Chim. Acta 22:429 (1968).

218. G. B. Kyriakopoulos, J. Inst. Petrol. 54:369 (1968).

219. N. A. Azizov and E. A. Pometun, Dokl. Akad. Nauk Tadzh. SSR 11:25 (1968).

220. J. H. Rossouw, C. F. J. van der Welt, and E. P. Marais, S. Afric. J. Agr. Sci. 11:363 (1968).

221. W. W. Harrison, J. P. Yurachek, and C. A. Benson, Clin. Chim. Acta 23:83 (1969).

221a. M. S. Vigler and V. F. Gaylor, Appl. Spectrosc. 28:342 (1974).

222. H. L. Kohn and J. S. Sebestyen, At. Absorption Newslett. 9:33 (1970).

222a. F. Amore, Anal. Chem. 46:1597 (1974).

223. H. Matsumoto, K. Tsunematsu, and T. Shiraishi, Japan Analyst 17:703 (1968).

224. A. E. Woods, R. D. Crowder, J. T. Coates, and J. J. Wittrig, At. Absorption Newslett. 7:85 (1968).

224a. J. P. Matousek and K. G. Brodie, Anal. Chim. 45:1606 (1973).

225. W. Mauch, K. Gierschner, and U. Feier, Ind. Obst-Gemuesever-wert. 54:1 (1969).

225a. S. W. Bishara, Y. A. Gawargious, and B. N. Faltaoos, Anal. Chem. 46:1103 (1974).

226. W. R. Wolf, M. L. Taylor, B. M. Hughes, T. O. Tiernan, and R. E. Sievers, Anal. Chem. 44:616 (1972).

226a. M. Berti, G. Buso, P. Colautti, G. Moschini, B. M. Stievano, and C. Tregnaghi, Anal. Chem. 49:1313 (1977).

227. A. Morgan and A. Holmes, Radiochem. Radioanal. Lett. 9:329 (1972).

227a. R. B. Boulton and G. T. Ewan, Anal. Chem. 49:1297 (1977).

228. T. S. Ma and M. Gutterson, Anal. Chem. 42:111R (1970); 44:453R (1972); 46:437R (1974).

229. W. H. Buttrill, J. Ass. Offic. Anal. Chem. 56:1144 (1973).

229a. W. M. Gantenbein, J. Ass. Offic. Anal. Chem. 55:123 (1972).

230. B. Royer, J. Ass. Offic. Anal. Chem. 55:581 (1972).

230a. W. K. Porter, J. Ass. Offic. Anal. Chem. 57:614 (1974).

231. Institute of Brewing Analysis Committee, J. Inst. Brew. 77:365 (1971).

231a. S. J. Kalembasa and D. S. Jenkinson, J. Sci. Food Agr. 24:1085 (1973).

232. Society for Analytical Chemistry, Analytical Methods Committee, "Determination of Trace Elements," Heffer, Cambridge, England, 1963.

233. W. B. Barnett and H. L. Kahn, Anal. Chem. 44:935 (1972).

234. N. V. Mossholder, V. A. Fassel, and R. N. Kniseley, Anal. Chem. 45:1614 (1973).

234a. J. Kaldor and B. Smith, Aust. J. Med. Technol. 4:51 (1973).

234b. T. F. Borovik-Romanova and E. A. Belova, Zh. Anal. Khim. 28:1828 (1973).

235. J. F. Uthe and F. A. J. Armstrong, Toxicol. Environ. Chem. Rev. 2:45 (1974).

236. T. M. Florence, Y. J. Farrar, L. S. Dale, and G. E. Batley, Anal. Chem. 46:1874 (1974).

237. D. H. Anderson, J. J. Murphy, and W. W. White, Anal. Chem. 48:116 (1976).

238. A. A. Cernik, Chem. Brit. 10(2):58 (1974).

238a. T. S. Ma and M. Gutterson, Anal. Chem. 50:86R (1978).

239. Total Carbon Apparatus (pamphlet), Coulometrics Inc., Wheat Ridge, Colo., 1976.

240. L. Janicke, Anal. Biochem. 61:623 (1974).

241. K. H. Nicholls, Anal. Chim. Acta 76:208 (1975).

242. H. Schafer and N. F. Olson, Anal. Chem. 47:505 (1975).

243. K. Bunnig, Glas.-Instrum.-Tech. Fachz. Lab. 18:1236 (1974).

244. C. P. Lloyd-Jones, G. A. Hudd, and D. G. Hill-Cottingham, Analyst 99:580 (1974).

245. W. Thuerauf, Erdöl. Kohle Erdgas Petrochem. 27:135 (1974).

246. T. Vaudlik, V. Kliment, and V. Scasnar, Radioisotopy 14:537 (1973).

247. J. A. Krueger, Anal. Chem. 46:1338 (1974).

248. Z. Sliepcevic and Z. Stefanac, Z. Anal. Chem. 269:31 (1974).

249. D. R. Peirson, Can. J. Bot. 52:177 (1974).

250. A. Trost, Erdöl Kohle Erdgas Petrochem. 27:431 (1974).

251. A. R. Pouraghabagher and A. E. Profis, Anal. Chem. 46:1223 (1974).

252. J. N. Wilson and C. Z. Marczewski, Erdöl Kohle Erdgas Petrochem. 26:647 (1973).

253. K. Ranfft, Z. Anal. Chem. 269:18 (1974).

254. T. Fernandez, J. M. Rocha, N. Rufino, and A. Garcia-Luis, An. Quim. 70:722 (1974).

255. H. Narasaki and K. Takahashi, Japan Analyst 23:172 (1974).

256. J. Z. Falcon, J. L. Love, J. L. Gaeta, and A. G. Altenou, Anal. Chem. 47:171 (1975).

257. R. A. Martin, G. G. Seaman, and A. Ward, Cereal Chem. 52:138 (1975).

258. M. M. Joerin, Analyst 100:7 (1975).

259. K. Lauber, Anal. Chem. 47:769 (1975).

260. Y. A. Gawargious, R. M. Habib, and S. A. El-Mergawy, Mikrochim. Acta 1975 (II):493.

261. P. Fawcett, D. Green, and G. Shaw, Radiochem. Radioanal. Lett. 17:121 (1974).

262. Y. Kidani, H. Takemura, and H. Koike, Japan Analyst 23:212 (1974).

263. P. F. Reay, Anal. Chim. Acta 72:145 (1974).

264. M. D. Garrids, M. L. Gil, and C. Llaguno, An. Bromatol. 26:167 (1974).

265. R. M. Ortheim and H. H. Bovee, Anal. Chem. 46:921 (1974).

266. H. R. Griffin, M. D. Hocking, and D. G. Lowery, Anal. Chem. 47:229 (1975).

267. E. H. Daughtry, Jr. and W. W. Harrison, Anal. Chim. Acta 72:225 (1974).

268. O. E. Olson, I. S. Palmer, and E. E. Carey, J. Ass. Offic. Anal. Chem. 58:117 (1975).

269. P. R. Haddad and L. E. Smythe, Talanta 21:859 (1974).

270. R. J. Ferretti and O. A. Levander, J. Agr. Food Chem. 24:54 (1976).

271. C. C. Y. Chan, Anal. Chim. Acta 82:213 (1976).

272. A. R. Byrne and L. Kosta, Talanta 21:1083 (1974).

273. H. H. Walker, J. H. Runnels, and R. Merryfield, Anal. Chem. 48:2056 (1976).

274. L. A. Lancaster and R. Balasubramaniam, J. Sci. Food Agr. 25:381 (1974).

275. F. E. H. Crawley, Int. J. Appl. Radiat. Isotopes 26:137 (1975).

276. J. A. Goleb and C. R. Midkiff, Jr., Appl. Spectrosc. 29:44 (1975).

277. G. M. Frame, R. E. Ford, W. G. Scribner, and T. Cturtnicek, Anal. Chem. 46:534 (1974).

278. L. G. Morin, Amer. J. Clin. Pathol. 61:114 (1974).

279. M. L. Rojkin and M. C. Olguin de Mariani, Bioquim. Atlant. 1974:1708.

280. H. Kuntziger, A. Antonetti, S. Couette, C. Coureau, and C. Amiel, Anal. Biochem. 60:449 (1974).

281. M. Peres, Anais Inst. Vinho Porto 1974:157.

282. H. Woidich and W. Pfannhauser, Z. Lebensmitt. 155:72 (1974).

283. W. Olschlager and L. Bestenlehner, Landw. Forsch. 27:62 (1974).

284. E. Schumacher and F. Umland, Z. Anal. Chem. 270:285 (1974).

285. F. D. Posma, J. Balke, R. F. M. Herber, and E. J. Stuik, Anal. Chem. 47:834 (1974).

286. R. T. Ross and J. G. Gonzalez, Anal. Chim. Acta 70:443 (1974).

287. F. J. Langmyhr, A. Sundli, and J. Jonsen, Anal. Chim. Acta 73:81 (1974).

288. D. R. Boline and W. G. Schrenk, Appl. Spectrosc. 30:607 (1976).

289. S. A. K. Hsieh, G. J. K. Wong, and T. S. Ma, Mikrochim. Acta 1976(II):253.

290. R. E. Jervis, B. Tiefenbach, and A. Chattopadhyay, Can. J. Chem. 52:3008 (1974).

291. R. S. Pekarek, E. C. Hauer, R. W. Wannemacher, Jr., and W. R. Beisel, Anal. Biochem. 59:283 (1974).

292. B. Grafflage, G. Buttgereit, W. Kubler, and H. M. Mertens, Z. Klin. Chem. Klin. Biochem. 12:287 (1974).

293. R. T. Li and D. M. Hercules, Anal. Chem. 46:916 (1974).

294. B. Janik and T. Gancarczyk, Acta Pol. Pharm. 31:61 (1974).

295. N. K. Sorensen, Tidsskr. Plant. 78:156 (1974).

296. S. A. Popova, L. Bezur, and E. Pungor, Z. Anal. Chem. 271:269 (1974).

297. M. D. Barnett and B. Brozovic, Clin. Chim. Acta 58:295 (1974).

298. M. E. M. S. de Silva, Analyst 99:408 (1974).

299. M. Nabrzyski, Anal. Chem. 47:552 (1975).

300. R. Fried and J. Hoeflmayr, Münch. Med. Wschr. 116:113 (1974).

301. K. T. Lee, Y. F. Chang, and S. F. Tan, Mikrochim. Acta 1976(II):505.

302. W. J. Simmons and J. F. Loneragan, Anal. Chem. 47:566 (1975).

303. M. K. Kundu and A. Prevot, Anal. Chem. 46:1591 (1974).

304. R. A. Jacob and L. M. Klevay, Anal. Chem. 47:741 (1975).

305. Y. K. Chau, R. Gachter, and K. Lum-Shue-Chan, J. Fish Res. Bd. Can. 31:1515 (1974).

306. F. J. M. J. Maessen, F. D. Posma, and J. Balke, Anal. Chem. 46:1445 (1974).

307. G. M. Schmid and G. W. Bolger, Clin. Chem. 19:1002 (1973).

308. P. R. Bonchev, D. Raikova, A. A. Aleksiev, and D. Yankova, Clin. Chim. Acta 57:37 (1974).

309. E. W. Rice and H. E. Fenner, Clin. Chim. Acta 53:391 (1974).

310. H. Y. Yee and J. F. Goodwin, Clin. Chem. 20:188 (1974).

311. A. Aleksiev, P. Bonchev, and D. Raikova, Mikrochim. Acta 1974:751.

312. W. E. Rippetoe, V. I. Muscat, and T. J. Vickers, Anal. Chem. 46:796 (1974).

313. S. W. McClean and W. C. Purdy, Anal. Chim. Acta 69:425 (1974).

314. A. Food, B. Young, and C. Meloan, J. Agr. Food Chem. 22:1034 (1974).

315. H. L. Huffman, Jr. and J. A. Caruss, J. Agr. Food Chem. 22:824 (1974).

316. J. F. Reith, J. Engelsma, and M. van Ditmarsch, Z. Lebensmitt. 156:271 (1974).
317. D. Mack, Deut. Lebensmitt.-Rundsch. 71:71 (1975).
318. A. A. Cernik, Brit. J. Ind. Med. 31:239 (1974).
319. J. Ebert and H. Jungmann, Z. Anal. Chem. 272:287 (1974).
320. A. N. Clark and D. J. Wilson, Arch. Environ. Health 28:292 (1974).
321. A. Sapek, Chem. Anal. (Warsaw) 19:687 (1974).
322. J. K. Kapur and T. S. West, Anal. Chim. Acta 73:180 (1974).
323. G. Velghe, M. Verloo, and A. Cottenie, Z. Lebensmitt. 156:77 (1974).
324. M. A. Evenson and D. D. Pendergast, Clin. Chem. 20:163 (1974).
325. N. P. Kubasik and M. T. Volosin, Clin. Chem. 20:300 (1974).
326. S. Yamazoe, N. Ikeda, and S. Oshima, Anal. Lett. 7:53 (1974).
327. B. C. Bowen and H. Foote, Rep. Inst. Petrol., IP 74-010 (1974).
328. H. G. C. Human and E. Norval, Anal. Chim. Acta 73:73 (1974).
329. W. Kisser, Mikrochim. Acta 1974:545.
330. E. Forochner, W. Wildanger, L. Beitz, I. Hasse, and L. Muller, Fleischwirtschaft 54:529 (1974).
331. J. B. Lopez and J. E. Buttery, Lab. Pract. 23:557 (1974).
332. J. K. Grime and T. J. Vickers, Anal. Chem. 47:432 (1975).
333. M. R. Peres and J. Pereira, Anais. Inst. Vinho Porto 1974:143.
334. N. T. Faithfull, Lab. Pract. 23:177 (1974).
335. W. K. Robbins, Anal. Chem. 46:2177 (1974).
336. M. Suzuki and W. E. C. Wacker, Anal. Biochem. 57:605 (1974).
337. G. L. Everett, T. S. West, and R. W. Williams, Anal. Chim. Acta 70:204 (1974).
338. S. C. Yang and C. H. Chang, Rep. Taiwan Sugar Expt. Sta. 58:17 (1972).
339. T. J. Rohm and W. C. Purdy, Anal. Chim. Acta 72:177 (1974).
340. E. T. Hall, J. Ass. Offic. Anal. Chem. 57:1068 (1974).
341. S. Nishi, Y. Horimoto, and N. Nakano, Japan Analyst 23:386 (1974).
342. K. Bok, Z. Lebensmitt. 155:209 (1974).
343. B. Wiadrowska and T. Syrowatka, Rocz. Panstw. Zakl. Hig. 25:701 (1974).
344. H. Woidich and W. Pfannhauser, Z. Lebensmitt. 155:271 (1974).
345. M. Berode, W. U. Neumeier, and A. Mirimanoff, Mitt. Geb. Lebensmittelunbers. Hyg. 65:427 (1974).
346. D. Siemer and R. Woodriff, Anal. Chem. 46:597 (1974).
347. H. Nakamachi, K. Okamoto, and I. Kusumi, Japan Analyst 23:10 (1974).
348. B. Krinitz and W. Holak, J. Ass. Offic. Anal. Chem. 57:568 (1974).
349. T. Giovanoli-Jakubczak, M. R. Greenwood, J. C. Smith, and T. W. Clarkson, Clin. Chem. 20:222 (1974).

350. R. Alvarez, Anal. Chim. Acta 73:33 (1974).
351. M. H. Friedman, E. Miller, and J. T. Tanner, Anal. Chem. 46:236 (1974).
352. E. G. Bradfield and J. F. Stickland, Analyst 100:1 (1975).
353. B. F. Quin and R. R. Brooks, Anal. Chim. Acta 74:75 (1975).
354. J. Cardenas and L. E. Mortenson, Anal. Biochem. 60:372 (1974).
355. J. Stupar, F. Dolinsek, M. Spenko, and J. Furlan, Landw. Forsch. 27:51 (1974).
356. C. A. Weers, H. A. van der Sloot, and H. A. Das, J. Radioanal. Chem. 20:529 (1974).
357. G. N. Stradling, D. S. Popplewell, and G. J. Ham, Int. J. Appl. Radiat. Isotopes 25:217 (1974).
358. S. N. Tiwari, S. P. Harpalani, and S. S. Triparthi, Mikrochim. Acta 1975(I):13.
359. F. Amore, Anal. Chem. 46:1597 (1974).
360. A. R. Curtis, J. Ass. Offic. Anal. Chem. 57:1366 (1974).
361. M. Glathe, Chem. Mikrobiol. Technol. Lebensm. 3:125 (1974).
362. W. E. Eipeson and K. Paulus, Lebensm.-Wiss. Technol. 7:47 (1974).
363. A. R. Byrne, J. Radioanal. Chem. 20:627 (1974).
364. P. C. Ooms, R. van Woerkom, and H. A. Das, J. Radioanal. Chem. 23:33 (1974).
265. H. U. Hopp, Erdöl Kohle, Erdgas Petrochem. 27:435 (1974).
366. G. O. Thorneberry, Anal. Biochem. 60:358 (1974).
367. W. Oelschlaeger and H. J. Lantzsch, Landw. Forsch. 27:31 (1974).
368. K. T. Lee and E. Jacob, Mikrochim. Acta 1974(I):65.
369. Institute of Brewing Analysis Committee, J. Inst. Brew. 80:486 (1974).
370. D. Kolihova and V. Sychra, Chem. Listy 68:1091 (1974).
371. R. J. Rowley and G. H. Farrah, J. Amer. Ind. Hyg. Ass. 23:314 (1962).
372. W. R. Hatch and W. L. Ott, Anal. Chem. 40:2085 (1968).
373. M. K. Carter and C. Goodell, American Laboratory 1974; July, p. 43.
374. R. A. Dobbs, R. H. Wise, and R. B. Dean, Anal. Chem. 39:1255 (1967).
375. K. Porter and D. H. Volman, Anal. Chem. 34:748 (1962).

Appendix
Nitrogen Determinations
in the U. S. Pharmacopeia

NITROGEN DETERMINATION

Some alkaloids and other nitrogen-containing organic compounds fail to yield all of their nitrogen upon digestion with sulfuric acid; therefore these methods cannot be used for the determination of nitrogen in all organic compounds.

Method I

Nitrates and Nitrites Absent — Place about 1 g of the substance, accurately weighed, in a 500-ml Kjeldahl flask of hard glass. The material to be tested, if solid or semi-solid, may be wrapped in a sheet of nitrogen-free filter paper for convenience in transferring it to the flask. Add 10 g of powdered potassium sulfate or anhydrous sodium sulfate, 500 mg of powdered cupric sulfate, and 20 ml of sulfuric acid. Incline the flask at an angle of about 45°, and gently heat the mixture, keeping the temperature below the boiling point until frothing has ceased. Increase the heat until the acid boils briskly, and continue the heating until the solution has been clear green in color or almost colorless for 30 minutes. Allow to cool,

*These procedures are reproduced from the United States Pharmacopeia XIX (official from July 1, 1975). The use of portions of the text of USP XIX is by permission of the USP Convention. The Convention is not responsible for any inaccuracy of quotation or for any false or misleading implication that may arise from separation of excerpts from the original context or by obsolescence resulting from publication of a supplement.

add 150 ml of water, mix the contents of the flask, and again cool. Add cautiously 100 ml of sodium hydroxide solution (2 in 5), in such manner as to cause the solution to flow down the inner side of the flask to form a layer under the acid solution. Add a few pieces of granulated zinc, and connect the flask, by means of a Kjeldahl connecting bulb, with a condenser, the delivery tube from which dips beneath the surface of 50 ml of boric acid solution (1 in 25) contained in a conical flask or a wide-mouth bottle of about 500-ml capacity. Mix the contents of the Kjeldahl flask by gentle rotation, and distil until about two-thirds of the contents of the flask has distilled over. Add about 3 drops of methyl red-methylene blue TS to the contents of the receiving vessel, and titrate with 0.5 $\underline{N}$ sulfuric acid. Perform a blank determination, and make necessary corrections. Each ml of 0.5 $\underline{N}$ acid consumed is equivalent to 7.003 mg of nitrogen.

When the nitrogen content of the substance is known to be low, the 0.5 $\underline{N}$ sulfuric acid may be replaced by 0.1 $\underline{N}$ sulfuric acid. One ml of 0.1 $\underline{N}$ sulfuric acid is equivalent to 1.401 mg of nitrogen.

Nitrates and Nitrites Present — Place a quantity of the substance, accurately weighed, corresponding to about 150 mg of nitrogen, in a 500-ml Kjeldahl flask of hard glass, and add 25 ml of sulfuric acid in which 1 g of salicylic acid previously has been dissolved. Mix the contents of the flask, and allow the mixture to stand for 30 minutes with frequent shaking. To the mixture add 5 g of powdered sodium thiosulfate, again mix, then add 500 mg of powdered cupric sulfate, and proceed as directed under Nitrates and Nitrites Absent, beginning with "Incline the flask at an angle of about 45°."

When the nitrogen content of the substance is known to exceed 10%, 500 mg to 1 g of benzoic acid may be added, prior to digestion, to facilitate the decomposition of the substance.

Method II

Apparatus — Select a unit of the general type known as a semi-micro Kjeldahl apparatus, by which the nitrogen is first liberated by acid digestion and then transferred quantitatively to the titration vessel by steam distillation.

Procedure — Place an accurately weighed or measured quantity of the material, equivalent to 2 to 3 mg of nitrogen, in the digestion flask of the apparatus. Add 1 g of a powdered mixture of 10 parts of potassium sulfate and 1 part of cupric sulfate, and wash down any adhering material from the neck of the flask with a fine jet of water. Add 7 ml of sulfuric acid, allowing it to rinse down the wall of the flask, then, while swirling the flask, add 1 ml of 30 percent hydrogen peroxide cautiously down the side of the flask. (Do not add hydrogen peroxide during the digestion.)

Heat the flask over a free flame or an electric heater until the solution has a clear blue color and the sides of the flask are free from carbonaceous material. Cautiously add to the digestion mixture 20 ml of water, cool the solution, and arrange for steam distillation. Add through a funnel 30 ml of sodium hydroxide solution (2 in 5), rinse the funnel with 10 ml of water, tightly close the apparatus, and begin the distillation with steam at once. Receive the distillate in 15 ml of boric acid solution (1 in 25), to which has been added 3 drops of methyl red-methylene blue TS and sufficient water to cover the end of the condensing tube. Continue the distillation until the distillate measured 80 to 100 ml. Remove the absorption flask, rinse the end of the condensing tube with a small quantity of water, and titrate the distillate with 0.01 $\underline{N}$ sulfuric acid. Perform a blank determination, and make any necessary correction. Each ml of 0.01 $\underline{N}$ acid is equivalent to 0.140 mg of nitrogen.

When a quantity of material containing more than 2 to 3 mg of nitrogen is taken, 0.02 $\underline{N}$ or 0.1 $\underline{N}$ sulfuric acid may be employed, provided that at least 15 ml is required for the titration. If the total dry weight of material taken is greater than 100 mg, increase proportionately the quantities of sulfuric acid and sodium hydroxide.

Author Index

Numbers refer to pages on which complete references are found.

Subject Index